Pesticide Residues in Food and Drinking Water

Human Exposure and Risks

Wiley Series in Agrochemicals and Plant Protection

Series Editors:

Terry Roberts, *Consultant, Anglesey, UK.*
Junshi Miyamoto (deceased), *Formerly of Sumitomo Chemical Ltd, Japan*

Previous Titles in the Wiley Series in Agrochemicals and Plant Protection:

The Methyl Bromide Issue (1996), ISBN 0 471 95521 3.
  Edited by C. H. Bell, N. Price *and* B. Chakrabarti
Pesticide Remediation in Soils and Water (1998), ISBN 0 471 96805 6.
  Edited by P. Kearney *and* T. R. Roberts
Chirality in Agrochemicals (1998), ISBN 0 471 98121 4.
  Edited by N. Kurihara *and* J. Miyamoto
Fungicidal Activity (1998), ISBN 0 471 96806 4.
  Edited by D. Hutson *and* J. Miyamoto
Metabolism of Agrochemicals in Plants (2000), ISBN 0 471 80150 X.
  Edited by Terry R. Roberts
Optimising Pesiticide Use *Edited by* Michael F. Wilson

Pesticide Residues in Food and Drinking Water

Human Exposure and Risks

Edited by
DENIS HAMILTON
Department of Primary Industries, Brisbane, Australia

and

STEPHEN CROSSLEY
Food Standards Australia New Zealand, Canberra, Australia

John Wiley & Sons, Ltd

Copyright © 2004   John Wiley & Sons Ltd, The Atrium, Southern Gate, Chichester,
West Sussex PO19 8SQ, England

Telephone (+44) 1243 779777

Email (for orders and customer service enquiries): cs-books@wiley.co.uk
Visit our Home Page on www.wileyeurope.com or www.wiley.com

Other Wiley Editorial Offices

John Wiley & Sons Inc., 111 River Street, Hoboken, NJ 07030, USA

Jossey-Bass, 989 Market Street, San Francisco, CA 94103-1741, USA

Wiley-VCH Verlag GmbH, Boschstr. 12, D-69469 Weinheim, Germany

John Wiley & Sons Australia Ltd, 33 Park Road, Milton, Queensland 4064, Australia

John Wiley & Sons (Asia) Pte Ltd, 2 Clementi Loop #02-01, Jin Xing Distripark, Singapore 129809

John Wiley & Sons Canada Ltd, 22 Worcester Road, Etobicoke, Ontario, Canada M9W 1L1

Wiley also publishes its books in a variety of electronic formats. Some content that appears
in print may not be available in electronic books.

Library of Congress Cataloging-in-Publication Data

Pesticide residues in food & drinking water : human exposure and risks / edited by Denis
  Hamilton and Stephen Crossley.
     p. cm. – (Agrochemicals and plant protection)
   Includes bibliographical references and index.
   ISBN 0-471-48991-3 (pbk. : alk. paper)
    1. Pesticides – Toxicology. 2. Pesticide residues in food. 3. Drinking water – Contamination.
   I. Hamilton, Denis. II. Crossley, Stephen. III. Series.

RA1270.P4P4823 2003
615.9′51 – dc21

                                                                    2003053814

British Library Cataloguing in Publication Data

A catalogue record for this book is available from the British Library

ISBN 0-471-48991-3

Typeset in 10/12pt Times by Laserwords Private Limited, Chennai, India
Printed and bound in Great Britain by Antony Rowe Ltd, Chippenham, Wiltshire
This book is printed on acid-free paper responsibly manufactured from sustainable forestry
in which at least two trees are planted for each one used for paper production.

Dedication

We dedicate this book to the safety of food and drinking water and the security of supplies for future generations. Hope for the future comes to mind in the image of Julia, grand-daughter of Denis.

Denis Hamilton and Stephen Crossley

Contents

Contributors

Árpád Ambrus
Training and Reference Centre for Food and Pesticide Control, Joint FAO/IAEA Division of Nuclear Techniques in Food and Agriculture, FAO/IAEA Agriculture and Biotechnology Laboratory, Food Contaminants and Pesticide Analysis Unit, A-2444 Seibersdorf, Vienna, Austria

Sir Colin Berry
Department of Morbid Anatomy and Histopathology, The Royal London Hospital, London E1 1BB

Stephen Crossley
Food Standards Australia New Zealand, PO Box 7186, Canberra, MC ACT 2610, Australia

Les Davies
Chemical Review and International Harmonisation Section, Therapeutic Goods Administration, Australian Department of Health and Ageing, PO Box 9848, Canberra, MC ACT 2601, Australia

John A. Edgar
Livestock Industries, CSIRO, Private Bag 24, Geelong, Victoria 3320, Australia

Denis Hamilton
Animal and Plant Health Service, Department of Primary Industries, 80 Ann Street, GPO Box 46, Brisbane, Queensland 4001, Australia

Jack Holland
Risk Assessment and Policy Section, Department of Environment and Heritage, Canberra, ACT 2604, Australia

Sheila Logan[1]
Chemical Review and International Harmonisation Section, Therapeutic Goods Administration, Australian Department of Health and Ageing, PO Box 9848, Canberra, MC ACT 2601, Australia

Utz Mueller
Chemical Review and International Harmonisation Section, Therapeutic Goods Administration, Australian Department of Health and Ageing, PO Box 9848, Canberra, MC ACT 2601, Australia

---

[1]Currently (2003) United Nations Environment Programme (UNEP), Geneva, Switzerland.

Michael O'Connor
*Chemical Review and International Harmonisation Section, Therapeutic Goods
Administration, Australian Department of Health and Ageing, PO Box 9848,
Canberra, MC ACT 2601, Australia*

Matthew O'Mullane
*Chemical Review and International Harmonisation Section, Therapeutic Goods
Administration, Australian Department of Health and Ageing, PO Box 9848,
Canberra, MC ACT 2601, Australia*

Barbara J. Petersen
*Exponent, Inc., NW Suite 1100, 1730 Rhode Island Avenue, Washington, DC
20036, USA*

Phil Sinclair
*Risk Assessment and Policy Section, Department of Environment and Heritage,
Canberra, ACT 2604, Australia*

Michael W. Skidmore
Syngenta AG, Jealotts Hill, Bracknell, Berkshire, RG42 6ET, UK

Gabriele Timme
*Bayer CropScience AG, Landwirtschaftszentrum Monheim, Geschäftsbereich
Pflanzenschutz, PF-E-Registrierung, Alfred Nobel Strasse 50, D-51368
Leverkusen, Germany*

J. Robert Tomerlin
*USEPA – Registration Division, US Environmental Protection Agency, Ariel
Rios Building, 1200 Pennsylvania Avenue, N.W. Washington, DC 20460, USA*

Kim Travis
Syngenta AG, Jealotts Hill, Bracknell, Berkshire RG42 6ET, UK

Wim H. van Eck[2]
Ministry of Health, Welfare and Sport, 2500 EJ The Hague, The Netherlands

Birgitt Walz-Tylla
*Bayer CropScience AG, Landwirtschaftszentrum Monheim, Geschäftsbereich
Pflanzenschutz, PF-E-Registrierung, Alfred Nobel Strasse 50, D-51368
Leverkusen, Germany*

Michael Watson
*Department of Toxicology, Ricerca Inc., 7528 Auburn Road, PO Box 1000,
Painesville, OH 440770-1000, USA*

---

[2]Currently (2003) World Health Organization, Geneva, Switzerland.

Series Preface

There have been tremendous advances in many areas of research directed towards improving the quantity and quality of food and fibre by chemical and other means. This has been at a time of increasing concern for the protection of the environment, and our understanding of the environmental impact of agrochemicals has also increased and become more sophisticated thanks to multi-disciplinary approaches.

Wiley has recognized the opportunity for the introduction of a series of books within the theme 'Agrochemicals and Plant Protection' with a wide scope that will include chemistry, biology and biotechnology in the broadest sense. This series is effectively a replacement for the successful 'Progress in Pesticide Biochemistry and Toxicology' edited by Hutson and Roberts that has run to nine volumes. In addition, it complements the international journals *Pesticide Science* and *Journal of the Science of Food and Agriculture* published by Wiley on.behalf of the Society of Chemical Industry.

Volumes already published in the series cover a wide range of topics including environmental behaviour, plant metabolism, chirality, and a volume devoted to fungicidal activity. In addition, subsequent topics for 2003–2004 include optimization of pesticide use, operator exposure and dietary risk assessment. These together cover a wide scope and form a highly collectable series of books within the constantly evolving science of plant protection.

As I write this preface, I am deeply saddened by the recent death of Dr Junshi Miyamoto, who contributed so much to this series as my Co-Editor-in-Chief. More significantly, Junshi will be remembered for his lifetime achievements in agrochemical biochemistry, toxicology and metabolism – and not least for the energy he displayed in international activities aimed at harmonizing knowledge within the field of agrochemicals. He leaves us with a wealth of scientific publications.

Terry Roberts
Anglesey
July 2003

THE SERIES EDITORS

Dr Terry R Roberts is an independent consultant, based in Anglesey, North Wales. He was Director of Scientific Affairs at JSC International based in Harrogate, UK from 1996 to **2002**, where he provided scientific and regulatory consulting services to the agrochemical, biocides and related industries with an emphasis on EU registrations.

From 1990 to 1996 Dr Roberts was Director of Agrochemical and Environmental Services with Corning Hazleton (now Covance) and was with Shell Research Ltd for the previous 20 years.

He has been active in international scientific organizations, notably OECD, IUPAC and ECPA, over the past 30 years. He has published extensively and is now Editor-in-Chief of the *Wiley Series in Agrochemicals and Plant Protection.*

Dr Junshi Miyamoto (deceased) was Corporate Advisor to the Sumitomo Chemical Company, for 45 years, since graduating from the Department of Chemistry, Faculty of Science, Kyoto University. After a lifetime of working in the chemical industry, Dr Miyamoto acquired a wealth of knowledge in all aspects of mode of action, metabolism and toxicology of agrochemicals and industrial chemicals. He was a Director General of Takarazuka Research Centre of the Company covering the areas of agrochemicals, biotechnology as well as environmental health sciences. He was latterly Present of Division of Chemistry and the Environment, IUPAC, and in 1985 received the Burdick Jackson International Award in Pesticide Chemistry from the American Chemical Society and in 1995, the Award of the Distinguished Contribution of Science from the Japanese Government. Dr Miyamoto published over 190 original papers and 50 books in pesticide science, and was on the editorial board of several international journals including *Pesticide Science.*

Preface

We were delighted to be given the opportunity to edit a volume about pesticide residues and safety in food and drinking water. Everyone is interested in the age-old questions, 'Is the food safe to eat?' and 'Can I drink the water?'.

Our experience with pesticide assessment at national and international (Codex Alimentarius) levels has taught us about the processes. The interesting thing is that new questions continue to arise as refinements are needed or as new approaches are tried, thus resulting in public health risk assessment methods that have made substantial progress in recent years. Editing this volume has helped clarify for us the role of each scientific discipline in the larger process.

The health risk for a particular substance (synthetic or natural) depends on the dose, and so the idea is to find a value for its dietary intake that is almost certain not to result in human harm of any kind and, furthermore, to keep its actual dietary intake as low as reasonably achievable. Results from the many scientific studies are brought together in the decision-making process. The aim is to ensure that approved pesticide uses do not produce pesticide residues in food and drinking water that are unsafe. Where there is scientific uncertainty, decisions are always based on the cautious side. Finally, explaining the risk is a very difficult area and we must remember that communication is a 'two-way street' – the specialist must listen as well as inform.

We give heartfelt thanks to the authors, who have contributed their expertise in chemistry, toxicology, environmental science, metabolism, food processing, diets, risk analysis, medicine, public administration and communication. We must especially thank those authors who prepared their manuscripts first, for their patience while they waited for us to tidy up the remainder. We were impressed by the energy and work expended by all authors and we hope there is some reward in seeing the final product. Thank you Terry Roberts (Joint Editor-in-Chief) for your helpful suggestions in the final stages of editing.

The reader will notice differences in writing styles between chapters. Manuscripts were edited for clarity and a degree of consistency, but the authors' individual styles in sentence construction and expression have generally been maintained.

We learned of the untimely death of Dr Junshi Miyamoto (Joint Editor-in-Chief) recently. He originally invited us to prepare this volume and he provided encouragement throughout. He was an inspiration in his International Union of Pure and Applied Chemistry (IUPAC) work and helped everyone with his clear sense of strategic direction. Dr Miyamoto, we will miss you.

We trust that you, the reader, will find the material interesting and informative. We hope that it answers questions you might have, and may stimulate further examination of this fascinating topic.

DENIS HAMILTON AND STEPHEN CROSSLEY

June 2003

1 Introduction

[1]DENIS HAMILTON and [2]STEPHEN CROSSLEY
[1] Department of Primary Industries, Brisbane, Australia
[2] Food Standards Australia New Zealand, Canberra, Australia

Pesticide Residues in Food and Drinking Water: Human Exposure and Risks. Edited by Denis Hamilton and
Stephen Crossley
© 2004 John Wiley & Sons, Ltd   ISBN: 0-471-48991-3

INTRODUCTORY REMARKS

As consumers, we do not want pesticide residues in our food because they have no nutritional value and can potentially pose a risk to health. However, we need pesticides to ensure that a consistent supply of economical and high quality food is available and sometimes residues will remain in the food supply. As a compromise, we require that the amounts of these residues in our food and drinking water will not be harmful to our health and should be no more than absolutely necessary. Risk assessment, which uses scientific processes to meet these requirements, has progressed considerably in recent years.

This book aims to describe the issues surrounding pesticide residues in food and drinking water and, in particular, the issues associated with human exposure and consumer risk assessment. In broad terms, consumer risk assessment encompasses three areas of scientific disciplines – human toxicology, pesticide residue chemistry and dietary consumption – which are explored in further detail within this book.

This chapter will briefly introduce the contents of the book and will discuss some of the commonly asked questions associated with pesticide residues.

WHAT ARE PESTICIDES?

The term 'pesticide' covers a wide range of substances, including insecticides, acaricides, fungicides, molluscicides, nematocides, rodenticides, and herbicides. Pesticides[1] are not necessarily single chemicals of natural or synthetic origin but may be micro-organisms (e.g. fungi or bacteria) or components thereof (e.g. endotoxins from *Bacillus thuringiensis*), or even so-called 'macro-organisms', e.g. predatory wasps such as *Trichogramma evanescens*, specifically bred in large numbers to control caterpillars, aphids and other sucking insects. Pesticides are used widely in agriculture since significant economic damage can occur when insects, nematodes, fungi and other micro- and macro-organisms affect food and commodity crops. The quantity and types of pesticides required to ensure high crop yield and unblemished produce acceptable to the consumer vary, depending on climatic conditions, pest species and pest burdens.

Many pesticides of natural origin have been used throughout the history of agriculture. The pesticidal or repellent action of some plants forms the basis of an age-old practice of companion planting, where the proximity of one plant is

---

[1]The Food and Agriculture Organization of the United Nations (FAO) has defined a pesticide as a substance or mixture of substances intended for preventing, destroying or controlling any pest, including vectors of human or animal disease, unwanted species of plants or animals causing harm or otherwise interfering with the production, processing, storage, transport, or marketing of food, agricultural commodities, wood and wood products or animal feedstuffs, or substances which may be administered to animals for the control of insects, arachnids or other pests in or on their bodies. Also included in the FAO definition are substances intended for use as plant growth regulators, defoliants, desiccants, or agents for thinning fruit or preventing the premature fall of fruit, and substances applied to crops either before or after harvest to protect the commodity from deterioration during storage or transport (FAO, 2003).

used to increase the yield of another plant which may be subject to attack by pests. Alternatively, pesticidal extracts from a particular plant type can be applied on or around another to control pests; examples include pyrethrum extracts (from a variety of daisies) or extracts from neem trees (*Azadirachta indica*). Other naturally occurring inorganic (e.g. arsenic or sulfur) or organic compounds (e.g. nicotine or strychnine) have been used for their pesticidal actions; many of these are extremely hazardous (i.e. poisonous) and pose a significant risk to users and to consumers of the produce, as well as a risk of accidental poisoning.

HISTORY OF PESTICIDE USE AND REGULATION

Large-scale use of pesticides began after World War II with the widespread use of organochlorine and organophosphorus compounds. Other chemical groups were subsequently developed and are used in agriculture today (e.g. triazine herbicides, carbamate insecticides and synthetic pyrethroids). However, pesticides are not a new development and have been used for centuries. For example, sulfur was used in classical Roman times for pest control in agriculture (Smith and Secoy, 1976). In the 19th century, highly toxic, mainly inorganic, compounds of copper, arsenic, lead and sulfur were used for the control of fungal diseases and insects.

HAZARD AND RISK

The World Health Organization (WHO) in 1995 provided specialist definitions for hazard and risk and associated terms such as risk assessment (WHO, 1995). These specialist meanings are used in assessing and explaining the risks of biological and chemical contaminants of food, including pesticide residues. They should not be confused with the normal dictionary meanings of risk and hazard, where the words 'risk' and 'hazard' are often synonymous.

Under the WHO definitions, risk assessment can be split into four different parts. First, in hazard identification, the possible adverse health effects of the chemical are identified from toxicological studies. Secondly, in hazard assessment, the toxic effects and characterization of the biological response in terms of the dose, i.e. the dose–response relationship, are considered and acceptable levels of dietary intake are derived. Thirdly, in exposure assessment, referred to as the 'dietary intake estimate' in this book, the dietary exposure of residues resulting from the consumption of food and drinking water containing residues is estimated. Finally, in risk characterization, the estimated dietary intake is compared with the acceptable levels of dietary intake or dose that were derived as part of the hazard assessment. In simple terms, if the dietary intake is less than this dose, then the risk is acceptable.

SCOPE OF THE BOOK

This section gives an overview and briefly introduces each chapter in the book: environmental fate, metabolism, food processing, toxicology, dietary consumption, chronic and acute dietary intakes, natural compounds, international standards and explaining the risks.

OVERVIEW

Studies of the environmental fate, metabolism and food processing provide basic information for studying residue levels in food. Whereas toxicology describes the hazard, the dietary consumption, in combination with residue levels, provides the dietary intake. Chronic and acute consumer intake estimates compare dietary exposure with acceptable intakes derived from the toxicology. Natural compounds, for proprietary reasons, have not usually been studied as thoroughly as synthetic compounds and therefore the safety of these compounds is frequently less well known. The risk assessment of residues in food must be acceptable at the international level to protect the consumer and to prevent disruption of the international trade in food. The final chapter deals with the very important topic of risk communication.

Most pesticide residues occur in food as a result of the direct application of a pesticide to a crop or farm animal or the post-harvest treatments of food commodities such as grains to prevent pest attack. Residues also occur in meat, milk and eggs from the consumption by farm animals of feed from treated crops. However, residues can also occur in foods from environmental contamination and spray drift. In addition, transport of residues and sediment, e.g. in storm water run-off or leaching through the soil to ground water, may also contaminate drinking water sources.

Since the publication of Rachel Carson's book *Silent Spring* in the 1960s (Carson, 1965), there has been increased public concern about the impact of pesticides on the environment. Much of this concern was associated with the organochlorine pesticides such as dichlorodiphenyltrichloroethane (DDT) and dieldrin. These compounds have both high environmental persistence and high fat solubility which commonly lead to residues occurring in meat, milk and eggs. Most countries have now withdrawn the registration of these persistent organochlorine pesticides. However, residues are occasionally detected in food because of the environmental contamination that remains from historical usage of the chemical. For example, animals grazing on contaminated land readily consume residues, which can be detected in the fat. Grazing cattle may consume 1 kg of soil per head per day and so will ingest the residue directly from the soil as well as residue in the pasture or forage itself. Of the crops grown in soil contaminated with organochlorines, root crops are the most likely to take up residues.

It is possible to estimate dietary intake from the environmental fate, metabolism and food processing experimental data that are commonly submitted by the agrochemical companies. However, these estimates are usually large overestimates of dietary intake as a result of the 'worst-case' assumptions that are included. The most realistic estimate of dietary intake can be obtained by conducting a *Total*

Diet Study. These studies are conducted by a number of countries (WHO, 1999) and many still look at the levels of organochlorine residues in our overall diets. In general, some organochlorine pesticides are no longer detected and the dietary intake of others is slowly declining.

Another potential route by which residues can result in food is through spray drift at the time of pesticide application. Spray drift results in very little residue in our diet since the rate of application is usually far lower than on the directly treated crop. Nevertheless, the contamination can be devastating for an individual farmer whose crops become unsaleable as a result.

Environmental Fate

Studies of environmental fate aim to determine what happens to the pesticide once it has been applied by investigating the behaviour of the compound in soil and water systems. Of particular importance to the overall dietary intake is the potential for the compound to leave residues in water. The environmental properties of pesticides likely to result in contamination of surface water and ground water are persistence, mobility and water solubility. A widely used herbicide such as atrazine has these properties and is frequently detected in surface and ground waters. In contrast to food where most residues result from direct treatment, residues in drinking water usually result from this indirect environmental contamination. Dejonckheere *et al*. (1996) showed that, even though atrazine was often detected in drinking water in Belgium, its estimated dietary intake constituted only 0.3 % of the acceptable level, known as the *Acceptable Daily Intake* (ADI).

Pesticides are transformed in soil, water and air into metabolites and other degradation products. The transformations may be microbiological (metabolism), hydrolysis (reaction with water) or photolysis (broken down by sunlight). Transformation usually proceeds through small changes to the parent pesticide molecule through to complete mineralization to carbon dioxide, water, chloride, phosphate and so on. For some pesticides, the initial transformation products may also be residues of concern in food or drinking water and should be included in the risk assessment process. Some transformation products are more persistent than the parent pesticide, e.g. dichlorodipenylethylene (DDE) is more persistent than DDT.

Pesticide Metabolism

The metabolism of a pesticide compound is studied by administering a radio-labelled compound to the test animal or the test crop and then, after a suitable interval, examining the distribution of the radio-label. Tissues, milk and eggs are examined in farm animal studies, whereas in plants, the plant foliage, fruit, seeds or roots are examined. The next stage is to investigate the nature of the residue – how much is still unchanged parent pesticide and what are the identities and amounts of metabolites and transformation products. Toxicological decisions are required on which metabolites need to be included with the parent pesticide in the risk assessment and which metabolites can be ignored because their amounts and toxicity are insignificant.

Plant and animal metabolic systems may conjugate the pesticide or a transformation product, i.e. chemically bond it to a natural compound such as a sugar. The conjugate will have different physical properties, e.g. a sugar conjugate is likely to be more water soluble, thus facilitating its elimination by an animal in the urine.

The results of metabolism studies are absolutely crucial before residue and food processing trials can begin. The metabolism studies tell us which compounds must be included in the residue tests of the processed samples. In some cases, the metabolite of one pesticide is another pesticide in its own right, hence suggesting that the risk assessment of the two should be combined.

Food Processing

The level and nature of residues in food can also be affected by commercial or domestic processing and preparation of the food. For example, food preparation will remove surface residues from some foods, e.g. mangoes or citrus, where surface residues are discarded with the peel. Specific studies are commonly conducted to investigate if the nature of the residue changes during processing and how much of the residue remains in the processed products. These food processing studies are a very important aspect of dietary intake estimates, particularly for those commodities that are consumed only after processing, e.g. cereal grains, or substantially after processing, e.g. grapes consumed as wine.

Changes to the nature of the residue during processing and the identification of transformation products, are commonly determined by studying the hydrolysis of the pesticide (reaction with water) at typical cooking temperatures. Hydrolysis experiments tell us which compounds must be included in the residue tests of the food processing studies.

The food processing studies themselves should simulate commercial processing practices as far as practicable. Thorough cleaning is often the first step in commercial processes and, depending on the nature of the residue, has the potential to remove a good part of surface residues, e.g. tomatoes and apples are vigorously washed before juicing, and wheat is cleaned to remove traces of grit and stones before milling. Experience tells us that residue levels in wheat bran are usually higher than in the original grain, while residues in flour are lower than in the grain – results which are hardly surprising since most residues are found on the grain surface. Fat-soluble residues tend to partition into the crude oil when oilseeds are processed. Water-insoluble residues tend to be depleted in clear fruit juices while attaching themselves to the pomace when apples or grapes are processed. Similarly, water-soluble residues in grapes have a greater chance of reaching wine than water-insoluble residues.

Toxicological Assessment

Toxicity studies aim to characterize the nature and extent of toxic effects caused by the pesticide and to find doses that cause no adverse effects in the test animals (*No Observed Adverse Effect Level* (NOAEL)). A wide range of studies from acute (i.e. short-term) to chronic (i.e. long-term) on laboratory animals is necessary, with dosing regimes and animal examination designed to investigate all kinds of effects such as tumour initiation and production, changed bodyweight gain, increased liver weight, changed blood properties, enzyme inhibition and foetal abnormalities.

The acceptable level of long-term dietary exposure, referred to as the *Acceptable Daily Intake* (ADI) for humans may be calculated by using a safety factor (usually 100) from the NOAEL for the most sensitive animal species (ADI = NOAEL/100). The ADI is used in the chronic risk assessment and is expressed as an amount of chemical per kilogram of bodyweight. Similarly, the acceptable level of short-term dietary exposure, referred to as the *acute reference dose* (acute RfD), for humans is calculated, where appropriate, with a safety factor applied to the NOAEL for the most sensitive animal species in the short-term toxicity tests. The acute RfD is used in the acute risk assessment and is also expressed as an amount of chemical per kilogram of bodyweight.

Diets and Food Consumption

Dietary intake of pesticide residues is calculated from residue levels in each food and the food consumption per person per day (i.e. the diet). Various methods have been used to assess diets for the human populations and for population sub-groups, e.g. children and infants.

At the international level, food balance sheets are used as a first estimate of *per capita* food consumption (WHO, 1997). The food balance sheets are based on a country's food production, imports and exports. Several countries' food balance data have been aggregated to produce regional diets, e.g. the European diet. Because waste at the household level is not considered, food balance sheets are usually overestimates of long-term average food consumption. In addition, dietary data for processed foods are sometimes missing, which prevents the use of processing studies for refining intake estimates beyond the raw commodity stage. Food balance sheet data do not, however, take into account differences in the diet within a population, the different consumption patterns of particular population sub-groups, e.g. infants and seasonal differences in consumption; nor do they allow for high consumption of a specific food by some individuals.

Some countries have surveyed thousands of households (e.g. household food consumption budget method), chosen to represent the population, in order to get a more accurate measure of food consumption over 24 hours. Other surveys have been based on detailed records of individual consumer's consumption of food over a 24 hour to 7 day period (e.g. diary record method). The subsequent analysis of these survey data provides not only information on average consumption of many foods over the whole population, but also provides dietary data for various sub-populations such as infants, toddlers, men, women and ethnic groups. The detailed surveys provide data on the diets of those people who consume much more of a food than average (high-percentile consumers), which is particularly useful for acute dietary intake estimates. The detailed survey data are also used by some countries in their chronic intake estimates.

Chronic Dietary Intake

Chronic intake or exposure assessment (intake and exposure mean the same thing for residues in food and drinking water) provides us with the estimated amount of residue consumed daily with our food and drinking water in the long term. In theory, this is for a lifetime of dietary intake and in practice it is for at least several years of continuous dietary intake. It is concluded that the dietary intake of residues is safe if it is less than the ADI derived from the toxicology studies.

Accurate chronic intake estimates are difficult because crucial information may be missing and then 'conservative or worst-case assumptions' are substituted for data. For example, often only a small portion (no more than 1–5 %) of a crop is treated with a specific pesticide on a national basis, but in the absence of solid information we assume conservatively that it is all treated. As previously explained, because the information is not available, we assume that all of the crop is treated at the maximum rate permitted on the label and harvested at the minimum time interval permitted. Dietary intake estimates with these assumptions will produce values much higher than intakes in reality, but the estimates can be still useful for deciding if the intake is acceptable or needs more detailed investigation.

Total diet studies measure residue levels in food purchased at retail level and prepared for consumption. They provide the most realistic estimates of chronic residue intake and usually give much lower values than those calculated with the conservative assumptions.

Acute Dietary Intake

The focus of dietary risk assessment for pesticide residues has generally been on the risks arising from chronic dietary intake. However, recent attention has focused on the potential for acute dietary intake from pesticide residues. Two developments have led to this recent attention.

First, as chronic dietary intake methodology has improved, there has been a move away from 'worst-case' estimates of chronic intake. Whereas in the past there were always large conservative assumptions to account for lack of data, now with more data available the chronic intakes are more realistic and this has directed more attention to a greater need for an explicit consideration of acute dietary intake. Secondly, recent research, especially in the UK, has shown that residue levels in individual carrots, apples or other fruits and vegetables are quite variable and that, for example, an individual carrot may have residue levels which are two to five times as high as the average residue level in its fellow carrots from the same field (PSD, 1997). In these circumstances, it is a legitimate question to ask about the effects of short-term residue intake that may be much higher, in a single meal or on a daily basis, than the chronic dietary intake. The methodology of acute dietary intake estimates aims to answer this question.

For an acutely toxic pesticide we need to take into account the person who eats a large portion of a specific food at one meal, or over a short time such as 24 hours and the highest possible residue that may occur in that food. Acute dietary intake estimate methodology takes all of these factors into account to calculate an estimated short-term dietary intake for each food. We conclude that this short-term intake is safe if it is less than the acute RfD derived from the toxicology studies.

Natural Toxicants

Plants, fungi and bacteria produce low-molecular-weight secondary chemicals (natural toxicants) thought to be aimed primarily at protecting the producing organisms from predators and competitors, e.g. aflatoxins produced by certain fungi. Such chemicals may be considered as natural pesticides. When such pesticides are present in food as intrinsic components or as contaminants, they raise food safety issues parallel to and, in many cases, of greater concern to public health than those posed by residues of manufactured pesticides. Unlike manufactured pesticides, natural pesticides have evolved for maximum deterrence without regard to their poisonous effects on mammals. Consequently, many natural pesticides are extremely poisonous to mammals, e.g. cyanogenic glycosides (cyanide-producing), present in the cassava plant and glycoalkaloids found in potato tubers under certain stress conditions (Johnston, 1991). This is illustrated by one incident in 1979 in which 78 boys in Lewisham, England became ill after eating a school meal which included potatoes with high glycoalkaloid levels. Seventeen of the children required hospital treatment (Consumers Association, 1994).

When organisms producing natural pesticides infect or contaminate food or drinking water, human exposure and risk assessment studies, as probing and as rigorous as those to which synthetic pesticides are subjected, are justified but are not usually available. Where data are available then they are often found to be far from benign. For example, Professor Bruce Ames of the University of Berkeley has cited 27 natural pesticides known to cause cancer in rodents, that are found in concentrations exceeding 10 mg/kg in several foodstuffs (Johnston, 1991).

International Standards

International agreements on pesticide residues in food rely on the work of the Codex Alimentarius Commission, established by the FAO and the WHO in 1962 to set standards for food in trade. The purpose of the Codex Food Standards Programme is the protection of the health of consumers and ensuring fair practices in the international trade in food. The main reason given by national governments for non-acceptance of Codex pesticide residue standards has been 'concern with dietary intake of residues'. Consequently, the Codex Committee on Pesticide Residues has devoted time and energy to improving the risk assessment process for residues in food. The current methodology for chronic risk assessment (WHO, 1997) is now generally accepted at the international level and attention has turned to acute dietary intake.

Codex maximum residue limits (MRLs) are recognized by the World Trade Organization (WTO) as the standards applying to food commodities in international trade and are assumed in the event of a trade dispute to represent the international consensus. National governments may be tempted to seek a trade advantage for their local industries by imposing unjustified standards on food to 'protect the health of their consumers'. The fine line between genuine health standards and standards imposed as a non-tariff trade barrier is not always clear, particularly where the details and methods are somewhat obscure. Codex procedures and detailed evaluations for each pesticide are published and have become increasingly transparent. Indeed, it is possible to trace the data and the reasoning supporting each standard for pesticide residues in food. Furthermore, the detailed calculations of the dietary intake are also now published.

Explaining the Risks

It is difficult for the public to understand the level of the risks associated with pesticide residues in their food and drinking water or for the regulatory or agrifood industry to effectively communicate the relative risks and benefits. Indeed, the risks associated with chemical residues is a complex matter and the technical complexity probably adds to consumer concern. In these authors' opinion, some people perceive the risks from pesticide residues to be much higher than justified from a detailed study of the evidence, while others are totally indifferent. We have therefore tried to explain the situation as openly, transparently and sincerely as possible and hope that people wanting to understand can make good use of the information presented. The following questions and discussion may help in this respect.

QUESTIONS ABOUT PESTICIDE RESIDUES IN FOOD AND DRINKING WATER

In this section we present discussion and answers to some common concerns about pesticide residues in food and water. Although the questions are simple and straightforward, the answers are not simple, because the subject is complex.

WHERE CAN I OBTAIN RELIABLE INFORMATION ON PESTICIDE RESIDUES?

Reliable information on pesticide residue issues is publicly available. However, when assessing any such information, it is worth examining the interests of organizations or groups making the information available, in order to see if those interests might influence the views expressed. Individuals in each stakeholder group (e.g. consumers, regulators and agrochemical companies) might have a very wide range of views, but the emphasis of the information made available is likely to be coloured by the interests of that particular stakeholder group. A plausibility check on such public statements therefore needs to take into account the interests involved. The following text helps to explain the interests of some of the stakeholder groups to help in this process:

- The agrochemical industry has made huge investments in generating scientific data to meet government regulatory requirements and has a commercial interest in presenting their pesticides as safe and effective.
- Consumer groups and activists need regular exposés of unsafe residues in food to maintain their profiles. Safety concerns raised by activists are frequently based on evidence that is taken out of perspective.
- Research scientists seeking research grants may try to influence research funding bodies by correctly timed and purpose-designed press releases or may overemphasize a safety concern in order to secure funding.
- The media are interested in selling newspapers or television time, which means priority for colourful and sensational stories. It is not generally in their interests to provide a completely objective balance to such stories.

One of the best sources of information on pesticide residues is from national regulatory authorities, many of whom make summaries of the evaluation of pesticides registration data available at a nominal cost, e.g. the United Kingdom's pesticide disclosure documents, available from the UK Pesticide Safety Directorate. These are very useful sources of detailed information on individual pesticides. They commonly include a summary not only of pesticide residue related data, but also other information, such as the exposure of operators or users and the effects on wildlife and the wider ecosystem.

A further authoritative source, which is also free from any national emphasis, is that of the Food and Agricultural Organization (FAO) and the World Health Organization (WHO). These international organizations jointly publish excellent material on pesticide residues and toxicology written by independent reviewers. The FAO and WHO systems rely on an expert panel of scientists chosen to be representative of a geographical spread of countries around the world, but principally chosen for their expertise. The scientists systematically review proprietary and published data, prepare summaries and explain reasoning and conclusions in a transparent manner. The FAO and WHO publications are an excellent starting point for information about toxicology and residues in food of particular pesti-cides. However, only a limited number of pesticide compounds have been dealt with by the FAO and the WHO although these are generally those compounds which have the greatest propensity for leaving residues in food. Recent FAO and WHO reports are available directly from their respective websites.

Information on pesticide residues is also available on a number of official websites via the Internet. For example, the United States Food and Drug Admin-istration (FDA) conducts a large-scale pesticide residue monitoring programme which is published in both paper and electronic form.

WHAT IS THE DIFFERENCE BETWEEN A RISK AND A HAZARD?

As outlined earlier in this chapter, in scientific terms, the words 'risk' and 'hazard' have specific and different meanings, as has been elaborated by the WHO (WHO, 1995).

To explain the difference in the two terms, let us consider a simple example, that of a high mountain such as Mount Everest. Mount Everest clearly poses a significant **hazard** given the number of lives that have been lost in attempting to conquer its peak. However, Mount Everest does not pose any **risk** unless you try to climb it, i.e. the risk is a function not only of the intrinsic hazard but also of the level of exposure. If you do not attempt to climb the mountain or you just stay in base camp, the level of exposure is zero or small and the level of risk will also be zero or small, respectively.

A pesticide chemical can be considered in the same way. Although it may be very toxic and therefore an extreme hazard, the level of risk to the consumer associated with the chemical will be dependent on the level of exposure, referred to as the dietary intake. If the chemical leaves no residues in the food, then there is no risk to the consumer. If on the other hand, the use of the chemical leads to high residues in food, then this will result in a risk. A risk assessment is then required in order to decide if the risk is low and acceptable in scientific terms.

In conclusion, the hazard that a chemical poses can be considered as being dependent on its intrinsic properties. On the other hand, the risk that a chemical poses also depends on the level of exposure, e.g. dietary intake, and can be thought of as the probability of an adverse outcome.

IF MY FOOD IS SAFE, DOES IT FOLLOW THAT THERE IS NO RISK?

No, the food may still pose a low level of risk despite being perfectly safe to eat. Indeed everything that we do in life has a risk associated with it and it is impossible to eliminate all the risks associated with eating food. Each type of food contains different risks, e.g. the risk of heart disease associated with saturated fats contained in most dairy and other farm animal products to the risk associated with the toxicity of the natural components of food.

Food is considered 'safe' when the level of risk is sufficiently low as to be considered minimal or negligible. This is analogous to a driver of a car who considers it 'safe' to drive along a quiet road in a well-maintained car; this, though, would not be risk free. In a similar way for pesticide residues, it is generally accepted by the scientific community that this 'safe' level of minimal or negligible risk is achieved when the dietary intake is within the ADI, and, when applicable, within the acute RfD.

WHY ARE PESTICIDE RESIDUES COMMONLY PERCEIVED TO POSE A SIGNIFICANT RISK TO CONSUMER SAFETY?

Total diet studies indicate that the level of pesticide residues as consumed are very low and are generally well within acceptable exposures, commonly a very low percentage of the ADI. However, when surveyed, the general public frequently perceive the risk associated with pesticide residues to be similar to that of smoking or driving a car. To understand why this is the case and whether this perception is justified, one needs to understand the factors that commonly influence the perception of risk by consumers. This perception is, perhaps, influenced by three main factors:

- the level of understanding of the nature of the risk by the consumer
- the amount of control that the consumer has over the risk
- the degree to which the consumer benefits from the risk.

To illustrate these three factors, let us consider again the example of a consumer driving a car to the grocers. The consumer has a relatively good understanding of the level of risk associated with the driving of a car. However, crucially, the consumer has control over the car and the associated risks and is also the beneficiary of the trip to the grocers.

In contrast, if we consider the case of pesticide residues in food, a consumer may have little understanding of how the risk is assessed and what it means. In addition, the consumer has only very limited control (perhaps some home-grown vegetables) and may believe that the only beneficiaries from the use of the pesticide are the farmers and agrochemical companies.

The above example helps to explain the apparent significant difference between an evidence-based evaluation of the risks posed by pesticide residues with the common public perception. However, this does not mean that regulators can be complacent about the risks since pesticide residues, and therefore dietary intake, can be high if pesticides are not properly controlled and significant misuse occurs. An example of this was in June 1992, when the illegal and gross misuse of the compound 'aldicarb' on cucumbers in Ireland led to at least 29 people being poisoned, with some requiring hospital treatment. A similar case was also reported in California involving watermelons contaminated with aldicarb. Luckily, incidents of this kind are rare; however, they do illustrate how pesticide residues can pose unacceptable risks to human health when they are used in a way that differs significantly from the product label recommendations or statutory conditions of use, i.e. illegal misuse.

WHY IS CAUTION NEEDED WHEN INTERPRETING 'WORST-CASE' SCENARIOS USED IN THE EVALUATION OF PESTICIDES?

We should distinguish decisions relating to what is typically or actually happening from those that are based on 'worst-case' scenarios. What is 'worst-case'? In practice, the range of circumstances and possibilities is very wide; the worst-case scenario is the circumstance which will lead to the most extreme result but still has a theoretically possible chance of occurring in practice.

In making decisions about pesticides, regulators commonly use worst-case assumptions, particularly when more realistic evidence is not available. However, when we run a series of worst-case possibilities layered one on the other the estimated end result can be quite remote from reality, and yet the perception can be that such an end result is typical. For example, for pesticide residues we commonly see dietary intake estimates based on assumptions that a person consumes throughout a lifetime food always containing pesticide residues at the maximum allowable concentration. The purpose of such a calculation is to show that if safety is achieved under this worst-case then it will be safe under other circumstances. It is, of course, totally impossible to produce residues consistently at the maximum allowed, and only in a minority of cases are more than a few percent of crops treated with a specific pesticide. Furthermore, it is quite impossible that someone consumes every day a range of foods that have all been treated and that all of these have been harvested to contain residues at the maximum residue limit (MRL).

It is recognized that these 'worst-case calculations' can act as useful tools for regulatory agencies, who may decide that if the 'worst-case' is acceptable, then the risk is minimal and no further scientific studies are needed. However, difficulties arise when people misunderstand or misinterpret the worst-case scenarios and present them as a typical case and representative of the real situation.

HOW ARE SAFETY FACTORS DECIDED?

Safety factors, sometimes known as 'uncertainty factors', are used to convert the *no-observed-effect levels* (NOELs) or the *no-observed-adverse-effect levels* (NOAELs) from the animal toxicology studies to an ADI for humans. Safety factors are also incorporated into the derivation of acute RfDs.

The USA FDA (Food and Drug Administration, 1955) explained the basis for the safety factor then adopted and which is still largely in force today. The FDA, in predicting the quantity of a poisonous compound that may be consumed over a long period without hazard to man, deemed it reasonable and advisable to assume the following:

- that man is ten times more prone to injury from the compound than other warm-blooded animals;
- that the most sensitive humans are ten times more susceptible to injury from the compound than the average human.

Therefore, in dealing with new compounds to which humans have not been exposed extensively, it is proper to apply a combination of these two factors and use a combined safety factor of 100. A safety factor of less than 100 may be used if data on physiological or other effects on humans are available. A safety factor greater than 100 may be desirable if unusually alarming reactions have occurred from exposure of humans, or other animals, to the compound.

A WHO publication (WHO, 1990) reiterated the interpretation of the 100-fold safety factor as two 10-fold factors, i.e. one for inter- and one for intra-species variability, and explained the factors that might influence a choice of other safety factors. For example, when relevant human data are available, the 10-fold factor for inter-species variability may not be necessary. The quality of a study or difficulties of interpretation may suggest the choice of a higher safety factor.

ARE MIXTURES OF RESIDUES MORE TOXIC THAN THE INDIVIDUAL COMPONENTS?

Risk assessment for pesticide residues normally deals with one pesticide at a time or, at most, with a small group of related pesticides perhaps with the same or closely related residues. Questions have been posed about the toxicity of mixtures, such as, 'is the toxicity of a mixture higher than the added toxicities of the individual compounds?'.

This question is not easy to answer and, because of the multitude of possibilities, there can never be enough empirical data to cover each different combination of residues. Mumtaz et al. (1993) posed the question as to whether from a public health perspective the risk from mixtures is overestimated, underestimated or is realistic, and looked at possible mechanisms.

For example, if compound A reduces the liver function so that the liver detoxifies compound B much more slowly, we would expect compound B to be more toxic in the presence of compound A. However, if compound B is metabolized by the liver to a more toxic compound, then compound A would reduce the toxicity of B. In practice, the timing of administration, the doses, absorption, transport within the body and numerous complex mechanisms will all influence the process and make the simple explanation conceptually useful but unlikely to be more than part of the story.

The Joint FAO/WHO Meeting on Pesticide Residues replied to a question about the possible combined effects of pesticides (JMPR, 1996). The JMPR noted that interactions between pesticide residues, other dietary constituents and environmental contaminants could occur and the outcome, which cannot be predicted reliably, may be enhanced, mitigated or additive toxicity. The JMPR report concluded that the safety factors that are used for establishing ADIs should provide a sufficient margin of safety to account for potential synergism (i.e. effects that exceed the sum of their combined effects).

WHY DO WE USE THE TERM 'ORGANIC FOOD'?

In the late 18th century, natural substances were classified according to the three 'kingdoms of nature', namely animal, vegetable and mineral (von Meyer, 1898), although a number of substances were found to be common to animals and plants and were classified as organic compounds, i.e. produced by organisms. In 1828, Wohler produced urea, an organic substance, from ammonium cyanate, an inorganic substance, demonstrating at least in this case and subsequently for others that production of an 'organic substance' did not necessarily require an organism. The terms 'organic chemistry' and 'organic compound' are, however, still retained for carbon compounds, the main components of plants and animals.

In the early 19th century a 'vital principle' was invoked to explain the ability of organisms to produce complex organic substances. Liebig (1842) expressed the opinion that the processes in plants and animals could best be explained in chemical terms and that 'vital principle' was of equal value with the terms 'specific' and 'dynamic' in medicine, i.e. 'vital principle' is just a learned name, not an explanation:

> .... everything is specific which we cannot explain, and dynamic is the explanation of all which we do not understand; the terms having been invented merely for the purpose of concealing ignorance by the application of learned epithets.

The terms 'organic farming' and 'organic food' appear to be a revival of the idea of drawing a distinction between substances produced in nature and those produced artificially or synthetically. There may be an intuitive belief that humans have been extensively exposed to natural compounds over the ages and that our metabolism and biological system are adjusted to them and render them safe. The belief may extend to synthetic compounds that, by the same logic, will be new to human metabolic systems and therefore cannot be detoxified and will be hazardous.

Biological systems are very complex and adaptable. A simplistic approach, such as an association of 'natural' with 'good' and 'synthetic' with 'bad' is useful in advertising but is difficult to justify when we begin looking at details of individual cases. This issue is discussed further later in this chapter.

Gardner (1957) described the organic farming movement in the USA, which maintained that food loses its health value if it is grown in soil that has been devitalized by chemical fertilizers and that artificial fertilizers and sprays had caused almost all of the nation's health disorders, including cancer.

'Organic food' and 'organically produced' are now useful marketing concepts. The market will supply the wants of those consumers especially concerned about the safety of pesticide residues in their food and who are willing to pay a premium for reassurance from vendors of the produce.

ARE PESTICIDE USES ON FOOD CROPS ADEQUATELY TESTED FOR CONSUMER SAFETY?

Before a pesticide is registered for use, the government pesticide regulatory authority requires the submission of a wide range of test data. These data are evaluated and an independent scientific assessment is conducted to ensure that the use of the pesticide is safe to the consumer, the user and the environment (including wildlife). Consumer safety is of crucial importance and pesticides are not registered if the scientific assessment indicates that residues in food pose an unacceptable risk.

Pesticide uses and the resulting residues in food and drinking water are highly regulated, particularly in the developed world, thus reflecting the high level of political and public interest. However, as previously discussed, the public tends to perceive the risks as higher than the scientifically assessed risks based on a detailed evaluation of the data by government authorities.

A further important consideration regarding the regulation of pesticide residues is that trade is involved. Governments and export industries may find that extensive data on residue levels and their safety are required by importing countries to gain trade access.

Political and trade interests combine to ensure that pesticides are extensively tested and studied before registrations are granted and that extensive regulatory requirements are developed. As a comparison, the use of veterinary drugs on food-producing animals is generally not of such high political and public interest (growth promotants are an exception) and the data requirements for veterinary drug uses are commonly less than those for pesticide uses.

WHAT IS 'ALARA'?

Exposure to chemicals in food and drinking water and to chemicals in the workplace are regulated by the use of two general principles. First, exposure should not exceed a pre-determined daily dose derived from a no-effect-level in animal experiments with the application of a safety factor. Secondly, exposure should be no higher than necessary when good practices are followed, i.e. 'as low as reasonably achievable' (ALARA). Permitted legal limits for residues in food and permitted legal exposure to chemicals by workers mostly derive from the 'as low as reasonably achievable' principle.

WHAT IS THE RELATION BETWEEN MRLs AND FOOD SAFETY?

The maximum residue limit (MRL) or tolerance for a pesticide residue is the maximum concentration of a pesticide residue legally permitted in or on a food commodity. MRLs are based on the highest residues expected in or on a food commodity when the pesticide is used according to registered label instructions.

Label instructions originate from the application rate, interval between treatment and harvest, method of application, etc. found necessary for effective pest control under practical conditions but leaving a residue which is the smallest amount practicable, i.e. *as low as reasonably achievable*.

Foods derived from commodities that comply with the respective MRLs are intended to be toxicologically acceptable. Before an MRL is established, it must pass the hurdles of risk assessment (Figure 1.1).

It follows from the procedure used for establishing MRLs that they are based on the registered uses of a pesticide and have no direct calculated relationship to the ADI (acceptable daily intake) of the pesticide. The acceptability from a food safety point of view of the recommended limits for a particular pesticide is assessed from the long-term dietary intake of that pesticide, which is compared to the permissible intake of residue calculated from the ADI for a consumer, while the short-term intake is compared with the acute RfD for a consumer.

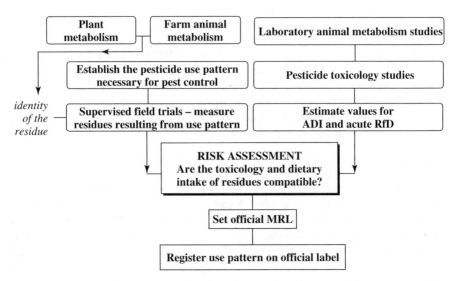

Figure 1.1 Risk assessment process before registration for pesticide residues in food: ADI, acceptable daily intake; acute RfD, acute reference dose; MRL, maximum residue limit or tolerance. Reprinted from Hamilton, D. J., Food contamination with pesticide residues, in *Encyclopedia of Pest Management*, 2002, Figure 1, p. 287, by courtesy of Marcel Dekker, Inc

HOW ARE 'NO DETECTED RESIDUES' INCLUDED IN THE DIETARY INTAKE ESTIMATE CALCULATIONS?

Analytical methods are used to measure the concentrations of pesticide residues in foods. Major progress has been made in the development of analytical methods for pesticide residues since the early days of pesticide residue regulation in the 1950s and 1960s. Colorimetric methods were the best methods available at that time. These methods had high limits of detection (LOD) by modern standards, being commonly around 1 mg/kg and even higher. If the residue levels were higher than the LOD, the analyst would report the values, but for lower concentrations in the food the analyst could only report 'not detected'. Pesticide regulatory officials often interpreted 'not detected' as 'nil,' but the real value could have been anywhere from zero up to the limit of detection.

Modern analytical methods mostly using gas chromatography (GC) or high performance liquid chromatography (HPLC) with very sensitive detectors routinely measure residue concentrations a hundred- or a thousand-fold lower than previously, i.e. at $1-10$ µg/kg in food commodities. In principle, the same problem still exists, i.e. the method cannot 'see' residue levels below the lower limit. However, in many cases for dietary intake estimates, levels below the LOD are now sufficiently low as to be of little or no concern.

What values can we use in dietary intake calculations when the analyst reports 'not detected' or more likely now, 'less than limit of quantification' (LOQ)?

Some regulators use a conservative assumption that the actual residue is just below the LOQ and so justify use of the LOQ in the calculation. This is a plausible assumption when many of the values exceed the LOQ with some at 'less than the LOQ'. It is not plausible when all values are 'less than the LOQ' because the natural spread of values in a residue data population will ensure that if the highest value is just below the LOQ, the average or typical value will be much lower.

Some regulators use other assumptions such as '$\frac{1}{2}$ of the LOQ' or 'zero'. The '$\frac{1}{2}$ of the LOQ' has no scientific justification, but is a recognition that the LOQ is an unrealistic estimate of typical residue levels in the circumstances. Assumptions of 'zero residues' can be justified when there is supporting evidence apart from the analyses themselves. For example, if a pesticide is destroyed by processing (e.g. cooking), the assumption of 'zero residues' is reasonable for these processed foods.

In assessing residues below the LOQ in supervised trials, the FAO Panel of the Joint Meeting on Pesticide Residues (JMPR) uses the LOQ unless there is scientific evidence that residues are 'essentially zero' (FAO, 2002). The supporting evidence would include residues below the LOQ from trials at exaggerated treatment rates (i.e. above the maximum application rate) or relevant information from the metabolism studies.

In total diet studies, the pesticide treatment history of the samples is commonly not known although the reason samples have no detectable residues is probably because the pesticide had not been used. In these circumstances, two estimates of dietary intake are sometimes made, one with residue results at 'less than the LOQ' set at the LOQ and one with these residues assumed to be at zero. If the two estimates arrive at two conclusions (acceptable and unacceptable intake), then more research is required on the analytical method to achieve a lower LOQ.

LOQs for pesticide residues in water are typically 100-fold or more lower than for the same residues in food, but the daily dietary consumption of drinking water is normally taken as two litres for an adult, which is higher than for any individual food. Experience shows that the LOQs for residues in drinking water do not normally lead to the sort of 'no detected residue' problem described above, in dietary intake estimates.

HOW DOES DIETARY INTAKE FROM RESIDUES IN DRINKING WATER COMPARE WITH THAT FROM RESIDUES IN FOOD?

The residues found in drinking water are of those compounds with some water solubility and their presence is likely to be as a result of widespread use in the water catchment area. The type of pesticides most commonly found in drinking water are herbicides with many uses in agriculture and other situations such as on railway lines and roadways.

Some compounds with sufficient water solubility and weak binding to soil particles are mobile down through the soil profile to ground water. Aldicarb, an insecticide, and atrazine, a herbicide, are two examples that have been found in ground water and in many places where ground water is used for drinking water.

Levels of residues found in drinking water are usually much lower than those found in food commodities and even when combined with the relatively high consumption of water, the estimated dietary intakes are usually very low.

Pesticide residues occur in drinking water mainly from environmental contamination, which is in contrast to residues in food where most residues occur from direct uses on crops producing food or animal feed. National authorities use various methods to set regulatory limits for pesticide residues (Hamilton et al., 2003).

First, a drinking water residue limit may be calculated directly from the ADI (acceptable daily intake) by assuming a person of stated body weight (say 70 kg) consumes two litres of water per day and the intake is a percentage of the ADI (say 10 %).

Secondly, if the authority decides that residues should not occur in drinking water the limit may be set at the LOQ (limit of quantification) of the analytical method. An LOQ limit for a particular pesticide will usually be lower than a limit calculated from the ADI.

Thirdly, the authority may decide to set the limits by legislation.

Fourthly, where the pesticide has a direct use in drinking water, e.g. for mosquito control, the limit may be set at the level required for the pesticide to be effective for its intended use. It must also pass the risk assessment test for consumer safety.

ARE NATURAL CHEMICALS BENIGN AND SYNTHETIC CHEMICALS HARMFUL?

The common perception of the public is of nature as being benign, whereas man-made things are perceived as having destroyed our harmonious relationship with nature. In the area of chemicals, this idea is extended to suggest that natural chemicals are either benign or have low toxicity and that man-made synthetic chemicals are harmful. In truth, this belief does not live up to scrutiny with some of the most toxic chemicals known to man being produced naturally by plants and animals as part of their defence mechanisms (Ames, 1992). Indeed, it has been reported that the botulinus toxin produced naturally by *Clostridium botulinus* is approximately 30 000 times more toxic than 2,3,7,8-tetrachlorodibenzodioxin (TCDD) which is thought to be one of the most toxic man-made poisons; TCDD is the most toxic of the dioxin group of chemicals (Faust, 1990).

Of those natural chemicals that are consumed in food on a regular basis, many are found to be carcinogenic (cancer-causing) in rodent toxicological studies that are commonly required by regulators for man-made pesticides. Examples of these include D-limonene in orange juice, 5-/8-methoxypsoralen in parsley and parsnips, and caffeic acid found in a large number of crops, including apples, carrots, grapes and potatoes (Johnston, 1991). It has been reported that there are probably at least half a million naturally occurring chemicals in the food that we eat, ranging from low-molecular-weight flavour compounds to macromolecular proteins and polysaccharides (Fenwick and Morgan, 1991).

In response to the rhetorical question 'Are natural chemicals benign and synthetic chemicals harmful?', the answer is clearly 'no' since the statement is a gross simplification. Indeed, each chemical needs to be treated on a case-by-case basis in the scientific risk assessment. Scientific risk assessments are justified even for natural chemicals commonly found in food, when the dietary intake by consumers may increase significantly.

REFERENCES

Ames, B. N. (1992). Pollution, pesticides and cancer, *J. AOAC Int.*, **75**, 1–5.
Carson, R. (1965). *Silent Spring*, Penguin Books, Harmondsworth, Middlesex, UK.
Consumers' Association (1994). Toxins in food, *Which? Magazine*, August 1994, 17–19 (Consumer's Association, Hertford, UK).
Dejonckheere, W., Steurbaut, W., Drieghe, S., Verstraeten, R. and Braeckman, H. (1996). Pesticide residue concentrations in the Belgian diet, 1991–1993, *J. AOAC Int.*, **79**, 520–528.
FAO (2002). Submission and Evaluation of Pesticide Residues Data for the Estimation of Maximum Residue Levels in Food and Feed, *FAO Plant Production and Protection Paper*, Vol. 170, p. 76.
FAO (2003). *International Code of Conduct on the Distribution and Use of Pesticides*, (Revised version), Food and Agriculture Organization of the United Nations, Rome, p. 6.

Faust, E. W. (1990). Staying alive in the 20th century, presentation given at the *World Environment Energy and Economic Conference*, Winnepeg, Manitoba, Canada, 17–20 October, 1990.

Fenwick, R. and Morgan, M. (1991). Natural toxicants in plant foods, *Chem. Br.*, **27**, 1027–1029.

Food and Drug Administration (1955). Tolerances and exemptions from tolerances for pesticide chemicals in or on raw agricultural commodities, *Fed. Reg.*, **20**, 1473–1508 (11 March, 1955).

Gardner, M. (1957). *Fads and Fallacies in the Name of Science*, Dover Publications Inc., New York, pp. 224–226.

Hamilton, D. J. (2002). Food contamination with pesticide residues, in *Encyclopedia of Pest Management* (ed Pimental), Marcel Dekker, Inc., New York, pp. 286–289.

Hamilton, D. J., Ambrus, A., Dieterle, R. M., Felsot, A. S., Harris, C. A., Holland, P. T., Katayama, A., Kurihara, N., Linders, J., Unsworth, J. and Wong, S.-S. (2003). Regulatory limits for pesticide residues in water, *Pure Appl. Chem.*, **75**, 1123–1155.

JMPR (1996). Interactions of pesticides, *Pesticide Residues in Food – 1996*, FAO Plant Production and Protection Paper 140, Food and Agriculture Organization of the United Nations, Rome, p. 13.

Johnston, J. (1991). Pesticides: public responsibilities, *Chem. Br.*, **27**, 111–112.

Liebig, J. (1842). *Chemistry in its Application to Agriculture and Physiology*, translated by Playfair, L., 2nd Edn, Taylor and Walton, London, pp. 56–58.

Mumtaz, M. M., Spies, I. G., Clewell, H. J. and Yang, R. S. H. (1993). Risk assessment of chemical mixtures: biologic and toxicologic issues, *Fundamental Appl. Toxicol.*, **21**, 258–269.

PSD (1997). *Organophosphorus Residues in Carrots: Monitoring of UK Crops in 1996/7 and Carrots Imported between November and May 1996*, Pesticides Safety Directorate, York, UK.

Smith, A. E. and Secoy, D. M. (1976). A compendium of inorganic substances used in European pest control before 1850, *J. Agric. Food Chem.*, **24**, 1180–1186.

von Meyer, E. (1898). *History of Chemistry*, translated by McGowan, G., 2nd Edn, Macmillan and Co. Ltd, New York, pp. 246–252.

WHO (1990). *Principles for the Toxicological Assessment of Pesticide Residues in Food*, Environmental Health Criteria 104, World Health Organization, Geneva, Switzerland, pp. 76–80.

WHO (1995). *Application of Risk Analysis to Food Standards Issues*, Report of the Joint FAO/WHO Expert Consultation, WHO/FNU/FOS/95.3, World Health Organization, Geneva, Switzerland.

WHO (1997). *Guidelines for Predicting Dietary Intake of Pesticide Residues* (revised), WHO/FSF/FOS/97.7, World Health Organization, Geneva, Switzerland.

WHO (1999). *Report of a Workshop on Total Diet Studies* (Kansas City, MO, USA, July 1999), World Health Organization, Geneva, Switzerland.

ACRONYMS APPEARING IN THE BOOK

3-PBA	3-phenoxybenzoic acid
ADI	acceptable daily intake
ADP	adenosine diphosphate

ALARA	as low as reasonably achievable
ANZFA	Australia New Zealand Food Authority
ARC	anticipated residue contribution
ARfD	acute reference dose
ATP	adenosine triphosphate
BCF	bioconcentration factor
Bt	*Bacillus thuringiensis*
CAC	Codex Alimentarius Commission
CCFAC	Codex Committee on Food Additives and Contaminants
CCGP	Codex Committee on General Principles
CCPR	Codex Committee on Pesticide Residues
CSFII	Continuing Survey of Food Intakes by Individuals (USA)
CXL	Codex Alimentarius Maximum Residue Limit
DEEM™	Dietary Exposure Evaluation Model
EBDC	ethylene bisdithiocarbamate
EC	emulsifiable concentrate
EC	European Community
EDI	estimated daily intake
EMDI	estimated maximum daily intake
EMRL	extraneous maximum residue limit
EPA	(US) Environmental Protection Agency
ETU	ethylenethiourea
FAO	Food and Agriculture Organization of the United Nations
FBS	(FAO) food balance sheet
FDA	(US) Food and Drug Administration
FFDCA	Federal Food, Drug and Cosmetic Act (USA)
FIFRA	Federal Insecticide, Fungicide and Rodenticide Act (USA)
FQPA	Food Quality Protection Act (USA)
GAP	Good Agricultural Practice
GATT	General Agreement on Tariffs and Trade
GC	gas chromatography
GEMS/Food	Global Environment Monitoring System – Food Contamination Monitoring and Assessment Programme (WHO)
GIT	gastrointestinal tract
GLP	Good Laboratory Practice
GSH	glutathione
GUS	Gustafson Ubiquity Score
HPLC	high performance liquid chromatography
IARC	International Agency for Research on Cancer (WHO)
IEDI	international estimated daily intake
IESTI	international estimated short-term intake
IGR	insect growth regulator

IPCS	International Programme for Chemical Safety (WHO)
IPPC	International Plant Protection Convention
IUPAC	International Union of Pure and Applied Chemistry
JECFA	Joint FAO/WHO Expert Committee on Food Additives
JMPR	Joint FAO/WHO Meeting on Pesticide Residues
LOD	limit of determination
LOD	limit of detection
LOEL	lowest-observed-effect level
LOQ	limit of quantification
MRL	maximum residue limit
NAFTA	North American Free Trade Association
NDNS	National Diet and Nutrition Survey (USA)
NEDI	national estimated daily intake
NESTI	national estimated short-term intake
NMR	nuclear magnetic resonance (spectroscopy)
NOAEL	no-observed-adverse-effect level
NOEL	no-observed-effect level
NRA	National Registration Authority for Agricultural and Veterinary Chemicals (Australia)
NTMDI	national theoretical maximum daily intake
OC	organochlorine (pesticide)
OECD	Organization for Economic Co-operation and Development
OIE	International Office of Epizootics
OP	organophosphorus (compound)
OPPTS	Office of Prevention, Pesticides and Toxic Substances (USA)
PAs	1,2-dehydropyrrolizidine alkaloids
PDP	Pesticide Data Program (USA)
PHI	pre-harvest interval
PMTDI	provisional maximum tolerable daily intake
PSD	Pesticide Safety Directorate (UK)
PTU	propylenethiourea
PTWI	provisional tolerable weekly intake
RAC	raw agricultural commodity
SC	suspension concentrate
SOP	standard operating procedure
SPS	sanitary and phytosanitary (measures)
STMR	supervised trials median residue
STMR–P	supervised trials median residue for processed foods
TBT	(Agreement on) Technical Barriers to Trade
TMDI	theoretical maximum daily intake
TMRC	theoretical mean residue concentration
TRR	total radioactive residue
UDMH	1,1-dimethylhydrazine

UF	uncertainty factor
UNECE	United Nations Economic Commission for Europe
USDA	United States Department of Agriculture
WHO	World Health Organization
WTO	World Trade Organization

2 Environmental Fate of Pesticides and the Consequences for Residues in Food and Drinking Water

JACK HOLLAND and PHIL SINCLAIR
Department of Environment and Heritage, Canberra, Australia

Pesticide Residues in Food and Drinking Water: Human Exposure and Risks.   Edited by Denis Hamilton and Stephen Crossley
© 2004 John Wiley & Sons, Ltd   ISBN: 0-471-48991-3

PESTICIDE PATHWAYS TO FOOD AND DRINKING WATER

This chapter examines the pathways by which pesticides may move into food and drinking water following their deliberate release into the environment, i.e. their fate, transport and methods of transformation. The latter processes are important because they limit the lifetime of pesticides in the environment. Mobility is important because it determines where pesticides may move in that lifetime. Not only the original pesticide, but also its degradation products need consideration, as degradation products may still retain potential to cause harmful effects, may persist after the parent has gone and may move where the parent has not.

Pesticides may reach food and drinking water in a variety of ways, with the most obvious being through direct contamination of produce as the result of deliberate application to control pests on the growing crop. This route also includes direct application to animals, for example, through processes such as treatment of salmon for lice in aquaculture or livestock for ectoparasites such as treatment of sheep for lice. It is not proposed to deal with this most obvious route in this chapter, as it will be well covered in those following.

Rather, the focus here will be on the mechanisms by which pesticides may contaminate the wider environment, such as adjacent crops and particularly water through spray drift and surface run-off, or leaching to ground water. Spray drift will be discussed below, but the latter two will be dealt with later in this chapter. Volatility may also result in contamination of the environment, but is less important as a local contamination factor. It will be dealt with as one of the wider dispersion mechanisms to areas distant from where the chemical is used.

Contamination of food may also occur through uptake by the roots of plants of pesticide residues in the soil within the field in which the crop is growing. For persistent chemicals, this may occur in crops planted long after the treated crop has been harvested. Consumption by stock of plants containing residues may lead to contamination of meat or milk. Pesticide residues in various agricultural produce may also be transported over large distances through trade and subsequently be re-released to the environment or contaminate other food products, e.g. through manufacturing processes or use as stock feed. Residues in storage or transport facilities could also conceivably contaminate subsequently stored produce.

Gene transfer, either directly through the technique of genetic engineering, or through subsequent unplanned outcrossing, is a controversial new means by which pesticides may reach food. For example, there have been claims that a toxin derived from the bacterium *Bacillus thuringiensis* (*Bt*) has been found in corn products for human consumption (Preston, 2000). Protein toxins such as this may be incorporated into crop plants such as corn and cotton to provide resistance to insect pests. Various strains of *Bt* have already been used in agricultural sprays for crop protection, so transfer of toxin genes is in effect an alternative means of pesticide application.

Pesticide residues may be transported long distances from the target area in air and water, potentially globally in some cases. In air, in addition to spray drift, movement may occur through volatilization from spray droplets or treated areas and by transport of pesticide picked up in fog or on dust particles. In water, movement may occur through dissolution or through transport of adsorbed residues on soil particles or organic matter. Once released, pesticides may alter or degrade due to various biotic and abiotic processes, and these products may also be transported and potentially reach food and drinking water. Rather than being accessible for transport, residues of pesticides and their products may also bind strongly in soil and remain in place there indefinitely. These transport and degradation or transformation pathways are discussed further below.

MOBILITY, TRANSPORT AND DEGRADATION IN THE ATMOSPHERE

TRANSPORT IN SPRAY DRIFT

The best known and most studied of the mechanisms for off-target contamination of food and drinking water is spray drift. While there are a number of books and articles on the subject (e.g. Mathews, 1992), a definition of 'spray drift' is difficult to find.

The US EPA (1999) has recently defined pesticide drift, concluding that:

EPA defines spray drift as the physical movement of a pesticide through air at the time of pesticide application or soon thereafter, to any site other than that intended for application (often referred to as off-target). EPA does not include in its definition the movement of pesticides to off-target sites caused by erosion, migration, volatility or contaminated soil particles that are windblown after application, unless specifically addressed on a pesticide product label with respect to drift control requirements.

This definition clearly refers to spray drift only as what may be termed *primary* drift, which is the off-site movement of spray droplets before deposition. It does not cover *vapour* drift, or any other form of secondary drift that may occur after deposition, which is predominantly specific to the active ingredient, whereas spray drift is primarily a generic phenomenon.

The extent that a chemical may drift off-target will depend on a number of variables such as formulation, weather conditions, type of nozzle and droplet size, and in particular, the application method, which can usually be divided into aerial, orchard and ground boom spraying. Of these, the aerial method potentially leads to the greatest drift, particularly if small droplet sizes are used, as is popular in some countries such as Australia. However, it is important to recognize that all methods can lead to drift, particularly if the pesticide is applied poorly, or under adverse conditions.

The main factors leading to drift from aerial application have recently been summarized in the literature as a result of work done by the Spray Drift Task

Force in the US (Bird *et al.*, 1996; Spray Drift Task Force, 1997; Hewitt *et al.*, 2002). From an analysis of literature results and experimental work carried out under the Task Force, these papers conclude that droplet size is consistently the primary application variable controlling off-target drift during low-flight applications. Weather, especially wind speed, is also very important, as is spray release position (drift is worse with greater release height and longer spray booms). Bird *et al.* (1996) recommend that clear specification of nozzles and operating conditions is the best approach for effective management of off-site drift and deposition from aerial applications. A valuable outcome of this work has been the development of the AgDRIFT® aerial spray prediction model (Teske *et al.*, 2002).

The Spray Drift Task Force has also examined drift from orchard and ground spraying (Hewitt, 2000). Again, droplet size is one of the most important factors affecting spray drift. For an orchard airblast sprayer, the characteristics of the crop canopy (height and shape, foliage density and the amount of open space between trees) are also important since the spray is released from within, rather than above the canopy. As may be expected, droplet size, boom height, wind speed and direction are significant factors affecting spray drift from ground (boom) application.

Salyani and Cromwell (1992) have also attempted to quantify spray drift in Florida orchards comparing both fixed-wing aircraft and helicopters with high- and low-volume airblast ground sprayers. Averaged over all distances and replications, the highest and lowest drift fall-out was from the fixed-wing and low-volume ground sprayer, respectively. However, the highest and lowest airborne drift was from the low- and high-volume ground sprayers. While this might seem surprising, it needs to be noted that very fine particles may stay airborne indefinitely and not deposit.

Drift from ground boom spraying, and in particular how to minimize this phenomenon, has been discussed in two recent articles (Marrs and Frost, 1995; Mathews and Piggott, 1999). In general, the potential for drift is lower from this application technique, with most of the spray deposited within a few metres from the edge of the boom. Studies have also been carried out in Germany from which tables have been derived to allow estimation of the extent of spray drift at certain distances from a variety of ground application methods to orchard crops, cereals, etc. (Ganzelmeier and Rautmann, 2000).

A constant theme throughout the above papers is the need for a (downwind) buffer zone to minimize the impact of off-target drift. A windbreak such as a hedgerow to filter any airborne droplets is also an important tool to reduce drift (Longley *et al.*, 1997; Longley and Sotherton, 1997).

As an example of the typical magnitude of spray drift onto adjacent areas, Bird *et al.* (1996) reported that median values of pesticide deposition from spray drift with aerial application dropped from the order of 5 % of the nominal application rate at 30 m downwind to approximately 0.5 % at 150 m during low-flight applications. However, the amount of non-target drift relative to conventional application equipment could be reduced by a factor of ten by application of a relatively coarse spray, or increased by a factor of ten with application of a fine spray, as with ultra-low volume (ULV) application. In contrast, drift values from ground application indicated by Ganzelmeier and Rautmann (2000) are of the order of 1 % or less at 30 m downwind, although air blast sprayers used early in the season in orchards (i.e. before full canopy development) may generate around 12 % drift at 10 m downwind.

In summary, spray drift is an important factor in the off-target contamination of both food and, in particular, drinking water, and it is important to follow practices that minimize the extent this occurs.

TRANSPORT AND PARTITIONING OF CHEMICALS IN THE ENVIRONMENT

Once released to the environment, chemicals do not generally stay on the target area or initial surface reached. On release, chemicals will tend to partition into air, water, soils, sediments, biota, etc., with the extent of movement into the individual compartments depending critically on their physicochemical properties. This concept is often called *fugacity*, which can be conceived of as the 'escaping tendency' of a substance from any given compartment or phase (Mackay and Paterson, 1981). Chemicals tend to partition from phases in which they have a high fugacity to those where their fugacity is low. This has led to the generation of a series of models which allow with increasing sophistication the prediction of the global distribution of chemicals, in reality a mass balance of a chemical once released into the environment. As an example, the global chemical fate has recently been modelled for α-hexachlorocyclohexane (Wania et al., 1999).

TRANSPORT IN AIR

The atmosphere is a very important compartment through which pesticides may be transported in a variety of forms. These include volatilization into the air in the vapour form from plants, water and soils, and transport both in dissolved forms in fog, rain, etc. or adsorbed to particles such as dusts (Majewski and Capel, 1995; Unsworth et al., 1999). Transport may be medium or long-range (inter-regional, intercontinental or throughout the globe). Concentrations of pesticides in air can be expressed per volume of air, or in terms of the transport medium if on dust or fog. Transport to the cold, seasonally-low sunlight conditions of polar regions may allow substances to become persistent, both because further volatilization is not favoured and because they may degrade much more slowly than under the warm conditions in agricultural areas where they were applied (Bidleman, 1999).

Volatility

Relatively volatile pesticides may be released directly into air, e.g. methyl bromide (MeBr) during fumigation treatments or dichlorvos from pest strips. However, pesticides may also move into air indirectly after application. As noted above, many chemicals may move into the atmosphere after the deposition of spray droplets onto the target surface by virtue of their volatility through evaporation or sublimation. Active ingredient properties, meteorological conditions, management practices and the nature of the soil or crop interact to affect the volatilization rate in the days to weeks over which volatilization may continue (Bedos *et al.*, 2002). Volatilization of some persistent pesticides from old residues in soil and water may still be contributing to their transport in air long after use has ceased in an area (Bidleman, 1999).

Endosulfan is an example of a pesticide where significant amounts volatilize from soil and more so from foliage or leaf surfaces, particularly soon after application. The α-isomer is more volatile than the β-isomer (see structures in Figure 2.1), which in turn is more volatile than endosulfan sulfate, the product of their oxidation.

Sampling of ambient air in the US during the early 1970s found α-endosulfan (mean 0.11 μg/m^3, maximum 2.26 μg/m^3) in about 2% of samples and β-endosulfan (mean 0.02 μg/m^3, maximum 0.06 μg/m^3) in about 0.3% (IPCS, 1984). More recently, sampling at Bloomington, Indiana, over a 14 month period found an average concentration for α-endosulfan of 86 pg/m^3, with a summer maximum of 890 pg/m^3 (10^6 pg = 1 μg). Results, including those summarized from earlier studies, indicate that temperature is the major significant predictor of atmospheric concentration, with wind direction playing an important but secondary role (Burgoyne and Hites, 1993).

Endosulfan is clearly a mobile chemical with the potential to contaminate non-target areas even when used according to label. This potential has been investigated on a global scale through analysis of tree bark samples, which accumulate airborne lipophilic pollutants because of the bark's relatively high lipid content. In general, the β-isomer is present in tree bark around the world at slightly higher levels than the α-isomer, with endosulfan sulfate being the dominant residue (Simonich and Hites, 1997).

However, while it is found at great distances from likely sources, the distribution of endosulfan residues is more regional than global in nature. Simonich and

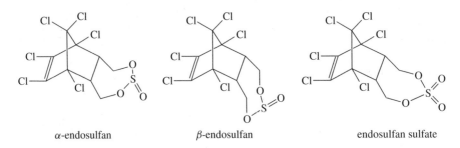

α-endosulfan   β-endosulfan   endosulfan sulfate

Figure 2.1   Structures of the α- and β-isomers of endosulfan and endosulfan sulfate

Hites (1995) noted that more volatile persistent organochlorines than endosulfan were found in higher concentrations in bark with increasing latitude (i.e. relatively low in bark in equatorial regions compared to boreal regions), thus making them truly global pollutants. In contrast, there was no correlation with latitude with β-endosulfan, for which concentrations in bark tended to be higher near the original region of use. For example, total endosulfan residues, normalized to lipid content, were high (100–1000 ng/g lipid) in samples of bark from New South Wales and SW Western Australia, but relatively low (10–100 ng/g lipid) in less agriculturally intensive areas, such as Tasmania. Only very low residues (below 10 ng/g lipid) were found at remote sites such as the Marshall Islands, while the highest residues (1000–10 000 ng/g lipid or more) were found in the Pacific Rim and India (thought to be from use in rice production) and agriculturally intensive regions of the USA and Europe.

Chlorpyrifos (Figure 2.2) is also mobile in the environment by virtue of its volatility. Volatilization from foliage is particularly pronounced, with around 80 % lost within 24–48 h, compared with up to 25 % from soil surfaces. The Henry's law constant is high enough that volatilization should also occur from water.

Movement of chlorpyrifos vapours has recently been studied in California's Central Valley where it finds widespread use on a range of orchard, vineyard and row crops, and prevailing daytime winds carry contaminated air masses into the adjacent Sierra Nevada mountain range (Aston and Seiber, 1997). Chlorpyrifos vapours are diluted as they disperse, with further declines in concentration through such processes as deposition to soil, water and vegetation, partitioning to airborne particles, washout by rain, and degradation. High-volume air and pine needle samples were taken throughout the summer of 1994 at three stations, situated at elevations of 114, 533 and 1920 m, in order to measure the rate of this decline. The lowest site was situated on the eastern edge of the valley and was surrounded by large areas of commercial citrus. The second station was located in Sequoia National Park in the southern Sierras, some 22 km east of the nearest agriculture, and the highest station at an exposed rocky outcrop, some 10 km to the north-east.

Chlorpyrifos and chlorpyrifos oxon (see Figure 2.2) were consistently found on vegetation (pine needles) at the site within the valley, each at concentrations ranging up to about 100 μg/kg. Only occasional detections occurred at the two higher

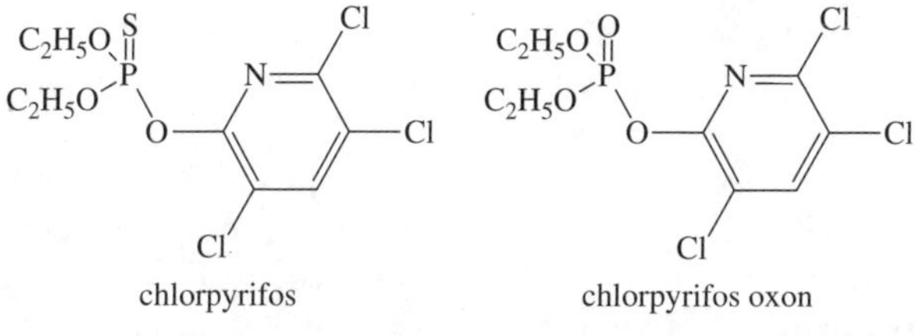

chlorpyrifos   chlorpyrifos oxon

Figure 2.2   Structures of chlorpyrifos and chlorpyrifos oxon

Table 2.1 Chlorpyrifos and chlorpyrifos oxon residues in air (Aston and Seiber, 1997)

Sample elevation (m)	Mean concentration (range) (ng/m^3)	
	Chlorpyrifos	Chlorpyrifos oxon
114	63 (3.9–180)	27 (2.4–63)
533	0.31 (0–0.49)	1.3 (0.25–3.6)
1920	0.19 (0–0.13)	0.33 (0.11–0.65)

sites, with chlorpyrifos reaching about 30 μg/kg and its oxon 60 μg/kg at 533 m, falling to the 5–15 μg/kg range at 1920 m. Residues were more frequently found in air samples with mean concentrations as shown in Table 2.1.

Airborne residues along the Mississippi were also taken from a moving research vessel during the first 10 days of June 1994 (Majewski *et al.*, 1998). Chlorpyrifos was found in all samples, peaking at 1.6 ng/m^3 near the town of St Louis. The median concentration was 0.29 ng/m^3. Samples were also analysed for another 42 pesticides and 3 transformation products. Among the pesticides, 15 of 25 herbicides and 7 of 17 insecticides were detected. There was no obvious relationship with such parameters as application rate or vapour pressure. Concentrations were most closely correlated to use on crop land within 40 km of the river, or to local uses in urban areas.

Transport in Fog

Fog is a lesser recognized transport medium for pesticides which, however, do seem to become very enriched in fog–water and may then be transported for long distances through the air. Rice (1996) has recently reviewed the literature on this phenomenon. Again, the bulk of the data were generated in Californian studies close to the area of pesticide use, although there are also some data for levels in fog distant from agricultural activity. Concentrations for the former are generally in the low ppb range with a high close to 100 ppb, whereas those in non-agricultural areas do not exceed 5 ppb. A noteworthy aspect of the data for thion-containing organophosphorus compounds was the apparent facilitated conversion in fogs to their more toxic oxon forms (cf. chlorpyrifos and its oxon – see Figure 2.2); the proportion of oxon to thion present was evidently time-related and increased with distance from application sources.

The pesticide levels in fog from the Bering and Chukchi Seas (as well as in air, ice, sea water and the surface microlayer), referred to in the review by Chernyak *et al.* (1996), were several thousand kilometres from likely usage areas. Fog appeared to be an efficient scavenger of airborne pesticides, with up to 5 ng/l chlorpyrifos found in fog condensate samples, but no detections in air samples. Chlorpyrifos was found at trace levels in six of nine water samples collected from 0.5–1 m depth, with maximum concentrations (0.046–0.067 ng/l) in northern and western areas receiving ice melt, and no detectable contamination in the central Bering Sea. Two microlayer samples from the eleven analysed contained chlorpyrifos at higher levels (above 0.10 ng/l). Both were from near-shore locations. A single integrated sample from an ice flow contained 0.17 ng/l chlorpyrifos. This is a clear example of how chemicals may move far distant from their areas of use; not only were organophosphorus (OP) compounds measured, but also some herbicides and fungicides such as metolachlor and chlorothalonil.

Transport on Particles, Including Dust

Movement on dusts or on particulate matter is another pathway for aerial transport of pesticides, although generally considered to be less important than volatilization (Majewski and Capel, 1995; Bidleman, 1999). This includes pesticides that enter the atmosphere adsorbed to eroded dust particles per medium of the wind, or volatilized chemical that then sorbs onto suspended particulate matter. Far less attention seems to have been given in the literature to this aspect, although it is suspected that in many instances measurements of chemicals in air have combined the gaseous and particulate forms.

MECHANISMS FOR REMOVAL OF PESTICIDES FROM AIR

How long a pesticide stays in the atmosphere depends on how rapidly it is removed by deposition or chemical transformation. Both gaseous pesticides and those sorbed to particulate matter are removed by closely related processes, namely wet and dry deposition (i.e. those involving precipitation or no precipitation, respectively), but likely at very different rates. Removal involving fog, mist or dew lies somewhere in between these processes, but is believed to be more closely related to dry deposition. Again, the effectiveness of the various removal processes depends on the physicochemical characteristics of the particular chemical, along with meteorological processes (Majewski and Capel, 1995).

Wet Deposition

Regional studies of atmospheric deposition of chlorpyrifos to Chesapeake Bay, Maryland/Virginia, USA, have been reported by McConnell *et al.* (1997). Chlorpyrifos is widely used in agriculture between April and June in catchments to the bay, as well as being used in urban areas for termite control and turf care. The main channel for Chesapeake Bay runs for some 200 km north to south with an average width of about 10 km and numerous sounds and tributaries to either side. Water and air samples were collected during cruises down the main channel in March, April, June and September of 1993.

The highest levels in water samples were generally found in the north near the inflow from the Susquehanna River. Decreasing concentrations north to south correlate with increasing salinity in the bay. Peak concentrations of 1.67 ng/l were recorded in March, and 1.60 ng/l in April, with the latter occurring halfway down the bay near the inflow from the Potomac River. Such results were unexpected as the March samples were intended as pre-season controls, and are difficult to explain except that riverine flows are highest in March and April. June was expected on the basis of use patterns to provide the highest residues, but peak residues during the month were only 0.55 ng/l, suggesting that warmer temperatures favour more rapid dissipation of chlorpyrifos residues. The highest concentration found in September was lower still at 0.25 ng/l.

Results from air monitoring appear contradictory in that highest levels (0.097 ng/m^3) were recorded in June, with only very low levels found in March. It is thought that this reflects increased volatilization inputs from local uses. Air concentrations are much lower than observed in the Californian studies described above and below. The intensity of use appears higher in California, and high foliar volatilization rates from citrus, grown across extensive areas in California's Central Valley, may further account for the differing observations.

A basic model focusing on interactions between the atmosphere and surface water suggested that, notwithstanding the low water temperatures, volatilization from water prevails during March and April, when most inputs of chlorpyrifos are from rivers. The model estimated that some 145 g/d volatilize from the bay, or about 10 % per month. In June and September when concentrations in the water are relatively low and there are higher vapour concentrations from use on crops during the warmer weather, the model predicts deposition of about 85 and 56 g/d, respectively. This illustrates very well the dynamic processes that are occurring, with pesticides moving in and out of the atmosphere depending on meteorological conditions and their concentrations.

More recent studies in California examined wet deposition (rain and snow) at the same sites in the southern Sierras and at Lake Tahoe (2200 m) in the northern Sierras during the winter and spring season when most precipitation occurs. There was substantial use of chlorpyrifos during the sampling period (December to April).

Chlorpyrifos was a pervasive contaminant of rain and snow samples, being present at 1.3–4.4 ng/l at 533 m, 1.1–13 ng/l at 1920 m and 0.3–3.4 ng/l at 2200 m. Chlorpyrifos was also ubiquitous in water samples taken from various depths to 350 m in Lake Tahoe during June. Levels detected (0.18–4.2 ng/l) correspond well with those found in snow, but this is likely to be coincidental as residential and commercial development around the lake provides a number of local sources of chlorpyrifos. Contamination levels in the Sierra Nevada were much lower than had been recorded in Central Valley fog–water (900–14200 ng/l) and rain (< 1.3–180 ng/l) or in the San Joaquin River (< 10–220 ng/l) which is mainly contaminated through run-off (McConnell et al., 1998).

While the emphasis above has been on chlorpyrifos and endosulfan, it should be made clear that there are a number of other pesticides found in the atmosphere or in rain water. If a pesticide is reasonably persistent and commonly used, the chances of detection in rain or remote situations also depend on the detection capabilities of the analytical methods. Hence, we can expect more detections in future as the methods continue to improve and more efforts are made to monitor different pesticides. Examples from the recent literature include atrazine, metolachlor and trifluralin measured in rain deposition in coastal waters of the Atlantic Bight (Alegria and Shaw, 1999), estimated to represent between 1–10 % of the yearly riverine deposits. In a highly developed agricultural river basin in Greece, a whole range of pesticides have been measured in rainfall including atrazine, metolachlor, lindane, molinate and a number of organophosphates, with concentrations in the range of 0.002 to 6.82 μg/l (Charizopoulos and Papadopoulou-Mourkidou, 1999). In spite of its unfavourable properties, atrazine is widely detected in rainfall, and was present in precipitation at levels of 0.1–0.4 μg/l in rain falling on Lake Michigan during 1991–1995, with atmospheric input estimated at nearly 25 % of the total loading of this herbicide to the lake (Miller *et al.*, 2000).

The work of Kreuger and Staffas (1999) provides some interesting insights into pesticide deposition in rainfall in northern Europe. Rainwater was collected over the years 1990 to 1992 at three locations in Finland, including sites in the south and far south near areas treated with agricultural pesticides, and in the far north, well away from land treated with pesticides. Seventeen pesticides and one metabolite were detected at the southern sites, with the most frequently detected pesticide being the organochlorine insecticide, lindane (γ-hexachlorocyclohexane and its other isomers – detection frequency 95–100 %). Phenoxy acid herbicides (2,4-D, dichlorprop, MCPA and mecoprop), triazine herbicides (atrazine and its metabolite *N*-deethylatrazine, cyanazine, simazine and terbuthylazine), and the herbicides bentazone and triallate were also detected on many occasions at one or both of the southern sites. Maximum concentrations detected in rainfall for several of the pesticides were at the 100–200 ng/l level, up to 240 ng/l for MCPA, with median concentrations often approximately an order of magnitude lower. These resulted in calculated deposition rates of up to about 10 μg/m^2 during May to September for the major pesticides detected.

Lindane isomers were also detected at low levels (1–5 ng/l) in all samples from the northernmost site, but the only other pesticides detected in rainwater at that site were traces of atrazine, MCPA and dichlorprop, which were detected on isolated occasions. There were also differences in the ratio of the isomers of lindane present compared to the southern sites (an increasing α/γ ratio from south to north), consistent with the pattern expected with increasing transport distances from input sources.

In many cases, pesticides detected in the south were also being used to a substantial extent in nearby agricultural production (MCPA, dichlorprop, mecoprop and bentazone were those most heavily used). However, some pesticides where there was little or no local use were also detected frequently in rainfall, in some cases also at relatively high concentration ranges (e.g. there were no sales of lindane or atrazine, low sales figures for simazine and terbuthylazine, and 2,4-D had been recently withdrawn from the Swedish market). Kreuger and Staffas (1999) noted that estimated atmospheric residence times (half-lives) for these pesticides range from less than one day for atrazine to about one day for 2,4-D, to about one week for lindane, from which potential transport distances can be estimated based on average wind speed. Transport in air from neighbouring countries presumably accounted for much of the residues found (e.g. 2,4-D was in common use in the Baltic states and transport of ester formulations would have been favoured by their volatility). However, transport may also have occurred over very much longer distances, particularly in the case of lindane.

Dry Deposition

This phenomenon includes deposition to the earth's surface by impact with surfaces such as vegetation, soil and water of airborne pesticide vapours as well as particle-bound pesticides, and by gravitational settling of the latter. Dry deposition is a continuous but slow process and its contribution to the total deposition burden is largely unknown (Majewski and Capel, 1995). This reference notes that the relative importance of wet versus dry deposition depends on the frequency of occurrence of precipitation and fog events, as well as the concentration of pesticides in air, the particle size distribution and concentration, and the efficiency of the removal process.

Majewski and Capel (1995) also note that direct measurement of dry deposition rates of air pollutants is difficult, and that the results have a high degree of uncertainty associated with them. As a result, there are far fewer literature results, although it should be noted that pesticides which move in air largely associated with particles can be scavenged by the forming rain drop and thus can be deposited in precipitation. This may be the case for many of the chemicals discussed above.

In the study in Finland by Kreuger and Staffas (1999) discussed above, pesticide recovered in rinsings from collection funnels during one to two week periods without precipitation made little contribution to overall atmospheric deposition of pesticides, even though samples were collected during the main pesticide application season. Transport on dust may be more significant in drier environments, such as inland Australia.

Chemical Transformation

Another important removal process for chemicals in air is their photochemical transformation, either by the direct effect of sunlight or indirectly by their reaction with free radicals and other reactive species produced by sunlight (Majewski and Capel, 1995; Unsworth et al., 1999). With direct phototransformation, excitation of a molecule through absorption of a photon (ultraviolet or visible light) is followed by a chemical reaction, usually oxidation through reaction with oxygen (OECD, 1993). For this to occur, the pesticide molecule must have an unsaturated or aromatic structure which absorbs photon energy at the UV–visible wavelengths present in sunlight (longer than 290 nm, because of upper atmospheric absorption of lower UV wavelengths). The indirect transformation processes include reaction with ˙OH (hydroxyl) radicals, ozone and other photochemically generated species.

Of the direct and indirect photochemical processes possible in the atmosphere, reaction with ˙OH radicals is generally the most important because it is the most rapid phototransformation process for the majority of chemicals (OECD, 1993). This reference notes that organic chemicals that do not react or react only slowly with ˙OH radicals do not tend to react with other photochemically derived species. While ozone concentrations are relatively high when compared with other photochemically formed reactive species, only unsaturated aliphatics, amines, polycyclic aromatic hydrocarbons, phenols and some sulfur compounds undergo ozonolysis easily, and of these only the first named would react with ozone faster than with ˙OH radicals. OECD (1993) provides methods to allow estimation of half-lives in air for a variety of organic compounds. (Half-life – time taken for the concentration to decline by one half.)

Thurman and Cromwell (2000) found trace concentrations of the triazine herbicides atrazine and cyanazine in rainfall in national parks near Lake Superior. Based on the predominant wind direction, these residues arose hundreds of kilometres away in the mid-western United States. The triazine metabolites deethylatrazine and deisopropylatrazine were also detected, at levels suggesting that their origin was primarily photodegradation in the atmosphere, rather than volatilization of residues formed in the soil where the herbicides had been applied. Deposition of these herbicides in rain was seasonal, with the highest rainfall concentrations and total deposition occurring during June, which corresponds to the time of herbicide application, and a few weeks thereafter in part of the US Corn Belt. The chloroacetanilide herbicides, alachlor and metolachlor, are also used extensively in the US Corn Belt. Although these herbicides are also volatile, they were not detected in rainfall in the test area (detection limit 5 ng/l). The suggested explanation was that they degrade in the atmosphere before they can be transported long distances.

CONCLUSIONS REGARDING TRANSPORT AND FATE IN AIR

In summary, the atmosphere is a very important compartment for the transport of pesticides off-target. The extent of the various processes, and their importance depends greatly on the physicochemical properties of the chemical. For example, although volatilization is the main route by which off-target movement of endosulfan occurs, it occurs gradually and off-target deposition via this route over short intervals is some 200 times lower than can occur from spray drift. Off-site contamination by dust movement is also relatively insignificant. Processes that can move large quantities of endosulfan in a short time, namely spray drift and especially storm run-off, appear to be the main contributors to major aquatic contamination incidents involving endosulfan (Muschal, 1998).

A number of chemicals have been identified as persistent organic pollutants (POPs) based on their semi-volatility, environmental persistence and tendency to associate with lipids. These properties enable the global distillation effect whereby substances used in tropical regions can contaminate cooler regions of the globe through long-range atmospheric transport, with residues accumulating in Arctic wildlife and indigenous inhabitants (Webster *et al.*, 1998; UNECE, 1996). Drinking water will also be contaminated by these processes, but possibly not to the same extent, given the lipophilicity and bioaccumulation potential of the pollutants in question.

Thus, pesticides and their degradation products may be transported in the atmosphere up to very long distances, and may then reach food and drinking water at remote locations where they may no longer be, or were never, used. Because of dispersion and loss over the distances involved, the resulting residue levels in food and drinking water from long-distance transport are likely to be insignificant when compared to direct treatment or neighbouring uses. Volatile, persistent organic pollutants (POPs) are of greater concern, both because they may move long distances and because residues accumulated by organisms low in the food web may reach significant levels through biomagnification up the food web to species eaten by humans. While possibly not a concern in the general human diet, such contamination could be highly significant to the Inuit (Eskimos), whose traditional diet is high in fat from Arctic fish and meat (e.g. bear and seal).

MOBILITY, TRANSPORT AND DEGRADATION IN SOIL AND WATER

PATHWAYS TO SOIL AND WATER

Pesticides may be applied directly onto or into soil (e.g. for pre-emergence weed control and pest and disease control in soil) and are also likely to reach soil directly during application to vegetation through spray missing foliage. During or immediately after application, pesticides may also reach soil through spray running off foliage or other surfaces and through spray drift. Pesticides and their degradation products may subsequently reach soil in rain or irrigation water washing off or leaching from treated surfaces and by residues in decaying material from treated crops.

Similarly, pesticides may reach water by direct treatment (e.g. for weed control in irrigation channels or semi-aquatic crops such as rice, and for pest control in fish production) and may also reach water incidentally during spraying (e.g. of weeds on a channel bank). More likely is indirect contamination from application on land through spray drift or in surface run-off and drainage of hills or beds. Movement of a pesticide in water may occur either dissolved in the water or adsorbed to soil and organic matter particles carried in the water. Ground water or sub-surface drainage water may also be contaminated if pesticides or their metabolites leach sufficiently deeply in soil.

Soil and water may also be contaminated accidentally or if operations associated with pesticide application are inappropriately managed, such as spray preparation, cleaning of equipment, disposal of waste and storage.

Pesticide residues in urine or manure from treated animals may reach soil and water, either where it falls in the field, through drainage from affected areas, or after collection and transport, e.g. of manure from poultry sheds or cattle feedlots. Pesticide residues may also be released during processing of produce. For example, residues of ectoparasiticides in wool may be released during scouring, after transport over long distances from where the sheep were treated. Depending on the chemical characteristics of the particular pesticide, residues released during scouring may partition either to the wool wax or to the scouring plant effluent. In Australia, most scour plant effluent passes through sewage plants to ocean outfalls and therefore would not reach drinking water. However, in other countries with different practices, effluent may reach rivers after relatively little processing.

Pesticide residues in treated crops may also be transported away from the original application site when used for animal feed. In addition to potentially contaminating soil and water, such residues may accumulate in the animals and contaminate milk or meat, as occurred in Australian cattle fed cotton trash containing residues of the insecticide chlorfluazuron when other feeds were short in supply during a drought period.

Pesticides may also reach soil and water on a more global scale, through residues depositing from air, as discussed above.

The introduction to plants through genetic engineering of genes to produce toxins such as those from *Bacillus thuringiensis* may also be perceived to be a means by which pesticides are ultimately released, e.g. through toxin residues in trash reaching soil.

TRANSFORMATION IN SOIL AND WATER

Degradation and Mineralization

The fate of pesticides once they reach soil or water depends on their chemical and physical characteristics and susceptibility to various transformation and transport processes (Kookana *et al.*, 1998; Schnoor, 1996; Lyman, 1995; Mill, 1993; Wolfe *et al.*, 1990; Bollag and Liu, 1990). Environmental conditions, biota, soil and sediment characteristics and water composition also influence fate. Degradation rates after release to the environment vary widely between substances, with half-lives from minutes to many years.

The extent to which degradation proceeds also varies widely, from minor alterations of the pesticide molecule to complete mineralization to carbon dioxide, ammonia, water and inorganic salts. Degradation may occur through both abiotic (hydrolysis, photolysis and other physicochemical processes) and biotic (aerobic and anaerobic metabolism) processes, as discussed below.

Hydrolysis

Hydrolysis refers to the cleavage of a bond and formation of a new bond with the oxygen atom of water, i.e. hence introducing HOH or OH into the molecule. Mill (1993) gives the generalization that hydrolysis may be important in any molecule where alkyl, carbonyl or imino carbon atoms are linked to halogen, oxygen or nitrogen atoms or groups through σ-bonds. This author instances the conversion of alkyl halides to alcohols, esters to acids and epoxides to diols.

Hydrolysis may occur abiotically or biotically and is a major means of chemical alteration in the degradation pathways of many pesticides. Abiotic hydrolysis may be the principal means of pesticide degradation where biological activity is low. These reactions may be strongly pH-dependent, occurring in the presence of H_2O, H_3O^+ and OH^- to varying degrees (respectively, neutral, acid and base hydrolysis), and related to the acid–base dissociation characteristics (pK_a) of the molecule. The rate of hydrolysis increases with increasing temperature, and may be affected by other environmental factors, such as whether the pesticide is present in solution or adsorbed to particles. In general, hydrolysis products are more polar than the molecules from which they are derived and may be significantly more water soluble and less subject to bioaccumulation.

For example, hydrolysis of the insecticide carbaryl to 1-naphthol (Figure 2.3) occurs rapidly in neutral or basic waters and is then likely to be the principal initial step in the degradation pathway of this substance (WHO, 1994). However, whereas at 20–25 °C the half-life in water is 1.3–1.9 d at pH 8 and approximately 0.1 d at pH 9, hydrolysis is markedly slower at pH 7 (half-life of 10.5–16.5 d). At acid pH levels (pH values below 7), carbaryl is effectively stable to hydrolysis, with half-lives exceeding 70 d at ~20–25 °C.

Photolysis

Degradation of substances in water or on exposed surfaces such as foliage or soil may be facilitated through the action of sunlight. As for pesticides in air, direct photolysis may occur when the absorption spectrum of the pesticide molecule overlaps the spectral distribution of sunlight in the UV–visible range. Absorption of a photon then leads to direct transformation of the molecule. Similarly,

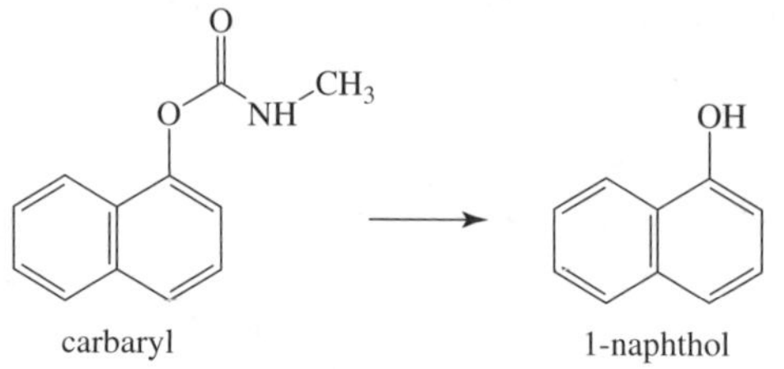

carbaryl   1-naphthol

Figure 2.3   Hydrolysis of carbaryl

photochemical transformation may occur following photon absorption by some other molecule (sensitizer), indirectly leading to reactions affecting the pesticide molecule.

At UV wavelengths reaching the earth's surface, sunlight has sufficient energy to cause direct photochemical reactions by rearranging or cleaving carbonyl double bonds, carbon–halogen, carbon–nitrogen, some carbon–carbon, and peroxide O–O bonds, but not enough to cleave most carbon–oxygen or carbon–hydrogen bonds (Mill, 1993). Lyman (1995) states that the end result of photolysis may include such reactions as dissociation or fragmentation, rearrangement or isomerization, cyclization, photoreduction by hydrogen-ion extraction from other molecules, dimerization and related addition reactions, photoionization and electron transfer reactions.

Photolysis is relatively insensitive to temperature and pH effects compared to hydrolysis (Mill, 1993; Haag and Holgnè, 1986). However, as would be expected, photolysis is strongly affected by factors influencing the spectral distribution, intensity and duration of sunlight. Such factors include latitude, time and date, cloud cover, dust, etc. and the extent of absorption of UV–B radiation by atmospheric ozone. In water, sunlight penetration varies with different wavelengths and is affected by reflection from the water surface and attenuation in the water by absorption and scatter. The angle of incidence of the light and movement of the water alter reflection, while attenuation is influenced by water depth, turbidity and dissolved substances. Thus, aqueous photolysis may be very limited in turbid water.

Humic substances (organic matter) dissolved in water are well known to be photosensitizers, as they lead to the formation of oxidants such as singlet oxygen (i.e. $^1O_2$ or atomic oxygen), superoxide anion ($^{\bullet}O_2{}^-$), alkyl peroxy radicals ($ROO^{\bullet}$), hydrogen peroxide (H_2O_2) and hydroxyl radicals ($^{\bullet}OH$) in reactions following photon absorption. These oxidants may also be produced by various other organic and inorganic substances in water. Such radicals are likely to occur transiently and only at very low concentrations due to rapid interactions with water and dissolved substances. For example, singlet oxygen is only present at very low steady-state concentrations ($< 10^{-12}$ mol/l) and only while exposure to sunlight continues (Haag and Holgnè, 1986). Hydroxyl radical concentrations may be lower still (10^{-15}–10^{-18} mol/l), but nonetheless may contribute significantly to the degradation of pesticides in aquatic situations, most importantly shallow water bodies with relatively strong sunlight penetration (Armbrust, 2000).

In general, substances such as phenols, furans, aromatic amines, sulfides and nitro-aromatics may undergo indirect photochemical transformation in water. In air, photo-oxidation of less reactive substances may occur, including alkanes, olefins, alcohols and simple aromatics (Lyman, 1995; Mill, 1993).

Laboratory studies have generally found that direct or indirect photolysis occurs more slowly on soil surfaces than in water. Only a thin layer of soil is either reached directly by photons or indirectly by diffusion of reaction products such as singlet oxygen. Hence, the extent to which photolysis occurs is affected by the amount of exposure of the soil surface to sunlight and the amount of pesticide available at the soil surface. Once incorporated into the soil by cultivation or leached in by rain or irrigation, a large proportion of pesticide is likely to be unavailable for photolysis, unless returned to the surface by volatilization or in water by capillary action.

Particularly in clear water and in the atmosphere, direct or indirect photolysis may be an important means of initial degradation of some pesticides, or may assist mineralization of pesticide degradation products. For example, Wolfe *et al.* (1990) note that photolytic reactions may assist the degradation of surface-applied pesticides containing sulfide linkages or thiocarbonyl groups (e.g. the fungicide thiram), and Tomlin (1997) indicates that photodegradation is an important means for loss from soil and water of the herbicide napropamide (Figure 2.4).

Photolytic degradation is generally not an important mechanism of pesticide loss from soil. However, there are several examples where photo-induced transformations appear to occur and some examples where it is the major means of dissipation. Photolysis in soil may be facilitated by photo-induction of oxidizing species from organic matter or by catalysis on clay minerals (Racke *et al.*, 1997). Joseph (1999) reported clear evidence from identified metabolites in field dissipation studies that soil photolysis is the major initial means of degradation for the fungicide azoxystrobin in the field (Figure 2.5).

As Klöpffer (1992) comments, one question which still needs much further examination is the extent to which a molecule is available for photolytic reactions when adsorbed to soil, rather than in solution (e.g. in interstitial water). This is also relevant to residues adsorbed to particles suspended in stream or pond water rather than dissolved in the water.

It is often the case that photolysis produces oxidation products which are more water soluble, less volatile and less subject to bio-uptake than their parent molecules (Lyman, 1995), but this is not always the case. For example, photolysis of the organophosphate insecticide parathion-methyl (Figure 2.6) occurs through oxidative desulfuration to form paraoxon-methyl, which is more toxic – the same reaction occurs by metabolism in the liver to form the active acetylcholinesterase inhibitor. However, paraoxon-methyl is more susceptible to hydrolysis than parathion-methyl, and hence photolysis may hasten the

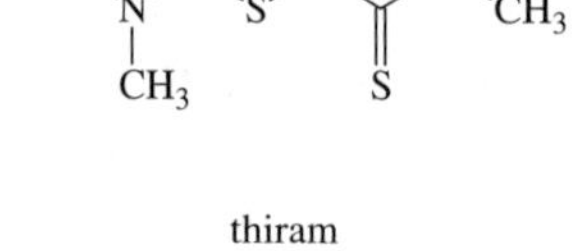

thiram 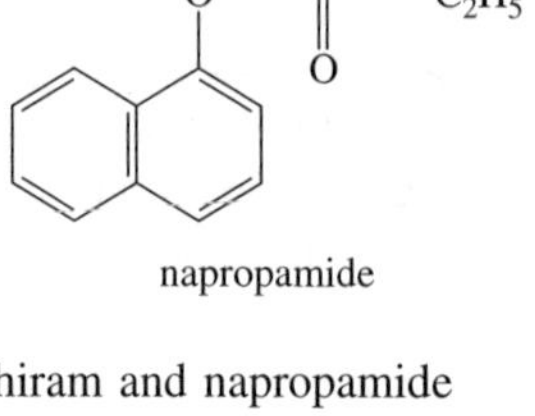   napropamide

Figure 2.4   Structures of thiram and napropamide

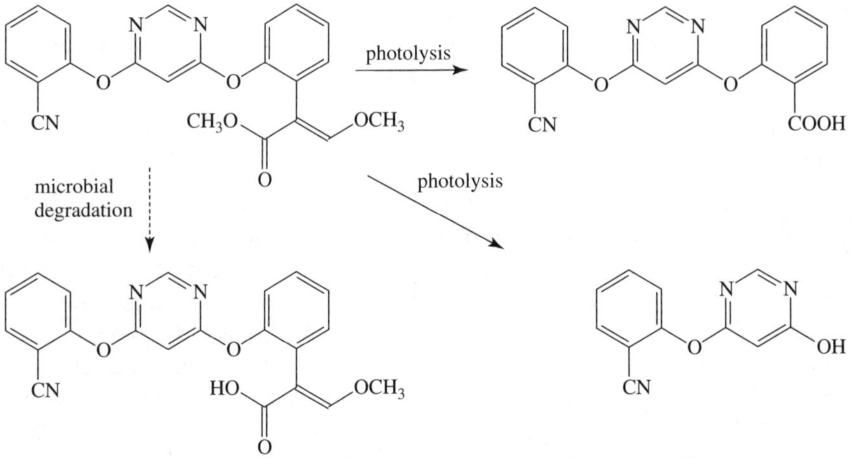

Figure 2.5 Degradation of azoxystrobin

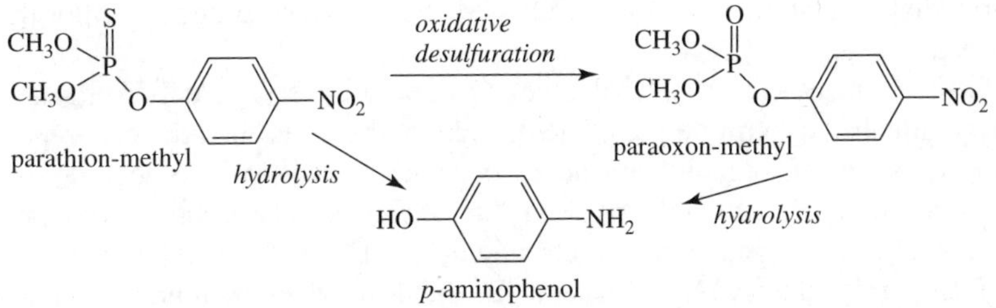

Figure 2.6 Parathion-methyl photolytic degradation and hydrolysis. Direct hydrolysis from parathion-methyl to *p*-aminophenol occurs at pH 9, whereas there is no such qualification for hydrolysis of the oxon

degradation of parathion-methyl, although microbial degradation is thought to predominate in aquatic situations (WHO, 1993).

Other Physical and Chemical Processes

Some pesticides may undergo non-biological reactions with ions or radicals present in water or in the soil solution, or may undergo catalytic reactions on the surfaces of clay minerals, organic matter or metal oxides. Reactions with free radicals produced by sunlight were described under 'photolysis'. Under reducing conditions, e.g. in anaerobic situations, reactions with ions such as Fe^{2+} may occur. Wolfe *et al.* (1990) discussed these and other non-biological chemical reactions such as nitrosation of the herbicide glyphosate by nitrite ion, displacement of the chlorine atom in chlorotriazine herbicides by nucleophiles (e.g. sulfhydryl groups – Lippa and Roberts, 2002), and rearrangement and hydrolysis reactions of organophosphates such as parathion when adsorbed on clay minerals such as kaolinite.

Microbial Degradation

Abiotic degradation processes may be significant in the dissipation of pesticides from air, soil and water. However, in many cases pesticides or their initial degradation products are relatively stable to abiotic degradation processes. Pesticide residues may also reach environments where conditions are unfavourable for abiotic degradation to occur (e.g. unsuitable pH for hydrolysis or protection from sunlight). Fortunately, biological processes, primarily microbial metabolism, are often highly effective in assisting the dissipation of pesticides once they reach the environment.

Under appropriate conditions, micro-organisms may be able to utilize certain synthetic organic compounds as nutrients, enabling ultimate biodegradation to CO_2 and inorganic components (i.e. mineralization). For example, in natural waters and soil the organophosphate dichlorvos is rapidly mineralized via intermediate metabolites to form CO_2 and the inorganic products, chloride and phosphate (Figure 2.7).

However, many pesticide molecules have structures (e.g. polycyclic aromatics) or attached groups (e.g. halides) making them sufficiently different from naturally occurring organic substances that they are difficult to degrade or cannot be assimilated, although they may be altered at any reactive sites present. For example, the organochlorine insecticide aldrin undergoes epoxidation readily, forming the closely related insecticide dieldrin, which then breaks down very slowly (WHO, 1989). Scheunert et al. (1992) noted that the metabolite dihydrochlordene dicarboxylic acid (Figure 2.8) was still detectable bound to soil and still slowly releasing in leachate 15 years after application of ^{14}C-aldrin to the topsoil. The environmental persistence of organochlorine pesticides or their degradates is one reason that most are no longer widely used, but these substances or their metabolites may still be found in soil many years after their use has ceased.

Much microbial metabolism affecting pesticides in the environment occurs through co-metabolism, i.e. where metabolic reactions transform the pesticide molecule incidentally, without the organism deriving energy or useful metabolites

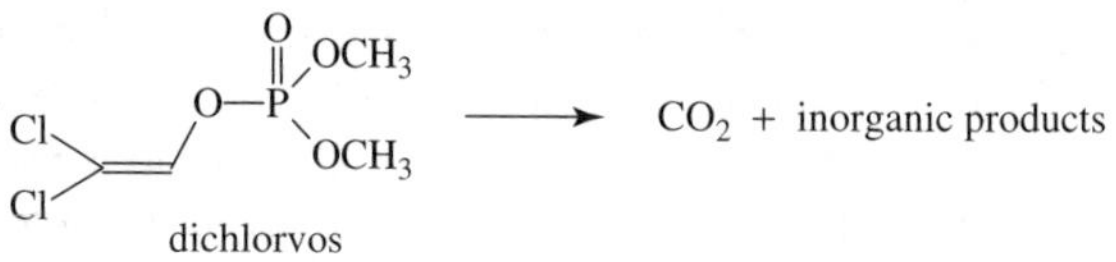

Figure 2.7   Mineralization of dichlorvos

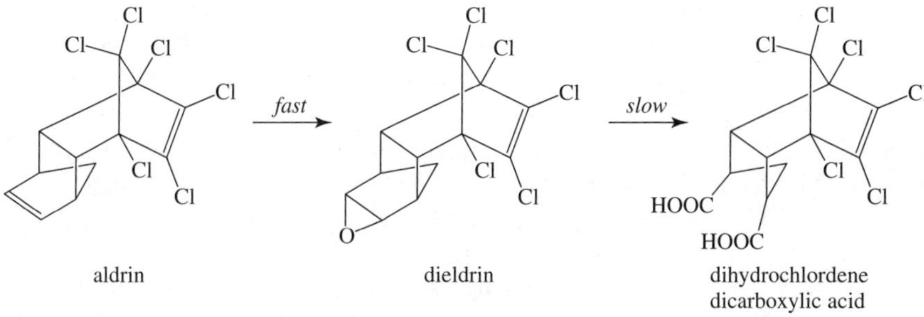

Figure 2.8 Metabolism of aldrin in soil

for cell growth or division, or only using a portion of the molecule. Microbial activity may also lead to polymerization involving pesticide or metabolite molecules, oxidation and additions such as acetylation or methylation, or conjugation with endogenous substrates such as glycosides or amino acids. Changes to the original molecule through these processes may assist in detoxification and elimination of the pesticide.

Metabolites formed by organisms initially taking up a pesticide may be amenable to assimilation by other organisms, so enabling degradation to proceed. However, mineralization of pesticide molecules often occurs only slowly or to a very limited extent, although the parent molecule may be significantly altered. Mineralization may be indefinitely delayed by incorporation of residues into soil organic matter. Assimilation of useful portions of the original molecule (e.g. as protein) may also delay release of carbon as CO_2, nitrogen as NH_3, etc., with those components again being assimilated by higher organisms in the food chain and potentially being incorporated into human foods without being completely mineralized. Residues of unchanged pesticide or metabolites may persist and may accumulate with repeated use. The extent to which such residues are bioavailable depends on their water solubility and the strength of adsorption or binding to soil or sediment.

Various organochlorine pesticides or their degradates are well known to be persistent, and although bound to soil may be slowly released into soil water if sufficiently soluble, as in the case of the aldrin metabolite discussed above.

An example where the parent molecule is altered more significantly is the fungicide cyprodinil. Dec et al. (1997a,b,c) examined the degradation in soil of cyprodinil in a series of studies, using material with a radiolabelled carbon in either the phenyl or the pyrimidyl ring. Metabolites will only be radiolabelled if they contain that part of the parent molecule with the ^{14}C; cyprodinil requires labelling in two places to produce ^{14}C in each major fragment. Some formation of $^{14}CO_2$ occurred over a six month incubation period, but only a small amount, and more so from the phenyl than the pyrimidyl ring. Several metabolites were identified in methanol extracts from incubated soil, including some where both rings were present and others where the pyrimidyl, but not the phenyl ring was present. A high proportion of the applied radioactivity remained bound to the soil and was identified as unchanged or slightly changed cyprodinil, sequestered or entrapped into the humus and soil micropores. Also present were phenyl and pyrimidyl cleavage products which had been covalently bound to become an integral part of the humus. Plant uptake studies showed that ^{14}C remaining in soil after six months incubation could be taken up to a small extent by plants, particularly from ^{14}C which was in the pyrimidyl rather than the phenyl ring. While reduced, some uptake still occurred when the plants were grown in soil which had been extracted with solvent leaving only bound residues, evidently through the activity of micro-organisms re-releasing material from the bound residues (Figure 2.9).

Pesticide degradation through microbial activity is not limited to intracellular metabolism. For example, microbes may significantly alter pH and reducing–oxidizing conditions in their immediate environment, indirectly assisting non-enzymatic reactions. Cells (e.g. algae) may also release enzymes or other reactive substances, either while living or when they die and decompose. Living or dead cells may also passively absorb and retain pesticides or their metabolites without metabolizing them, removing them from the surrounding medium, but potentially introducing them into the food web (Bollag and Liu, 1990).

Many interacting factors influence overall microbial activity and the ability of microbes to degrade a particular substance; the species composition and characteristics of the organisms, the immediate environmental conditions, the soil characteristics, and the concentration and properties of the substance are all influential. In some cases, appropriate organisms may need to be fostered or

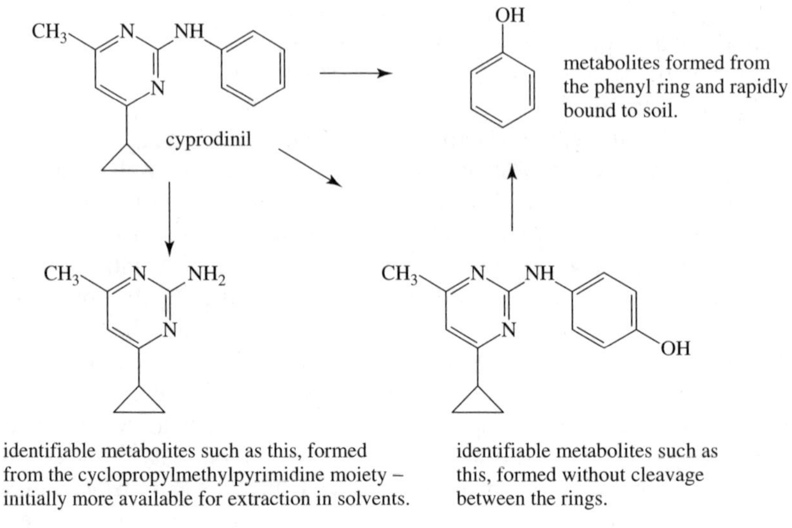

identifiable metabolites such as this, formed from the cyclopropylmethylpyrimidine moiety – initially more available for extraction in solvents.

identifiable metabolites such as this, formed without cleavage between the rings.

metabolites formed from the phenyl ring and rapidly bound to soil.

Figure 2.9 Metabolism of cyprodinil in soil

necessary enzymes may need to be induced, hence resulting in a lag phase before degradation can proceed. Thus, prior exposure of a site to a particular substance may usefully increase the rate of biodegradation of a pesticide added subsequently.

In some cases 'enhanced biodegradation' may become sufficiently rapid that the effectiveness of the pesticide is reduced, e.g. as has been observed with the carbamate soil insecticide carbofuran (Racke, 1990). On the other hand, it may be possible to enhance natural degradation of relatively intractable substances at a contaminated site by strategies such as inoculation with suitable organisms (perhaps genetically engineered), treatment with microbial enzymes and adjustment of conditions to foster the growth of appropriate micro-organisms (Bollag and Liu, 1990).

Degradation Under Anaerobic Conditions

Low oxygen content, or anaerobic conditions, may arise in situations such as flooded soil (e.g. rice fields), perched water tables and ground water, in stagnant or eutrophic water bodies and in the lower layers of sediment in lakes and ponds. In addition to depleted oxygen levels, flooding may lead to low pH and reducing conditions (e.g. the presence of Fe^{2+} and S^- ions).

Some substances are amenable to anaerobic as well as aerobic biodegradation. Flooding may directly facilitate hydrolysis and may alter pH, e.g. favouring acid hydrolysis. Reducing conditions are also likely to develop, favouring abiotic or biologically mediated reactions such as dehalogenation which may both reduce the toxicity of a pesticide molecule and facilitate further degradation. However, anaerobic or reducing conditions may also greatly slow primary degradation, or may inhibit the degradation of metabolites. They may also lead to different metabolites from those formed under aerobic conditions, e.g. the formation of 4-amino-parathion-methyl from parathion-methyl (Coats, 1991) (Figure 2.10).

In some cases, oxidative reactions which have already occurred may be reversed, e.g. phorate is rapidly oxidized to the sulfoxide under aerobic conditions in soil, but the sulfoxide may be reduced back to phorate under flooded anaerobic conditions (Coats, 1991). However, both phorate and its sulfoxides have pesticide activity (Figure 2.11).

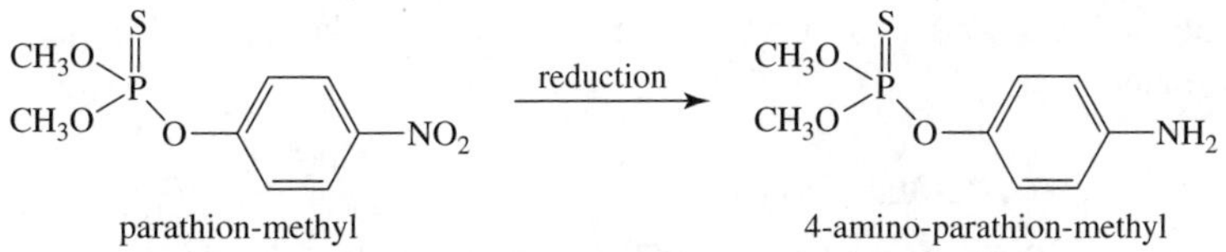

Figure 2.10   Anaerobic degradation of parathion-methyl

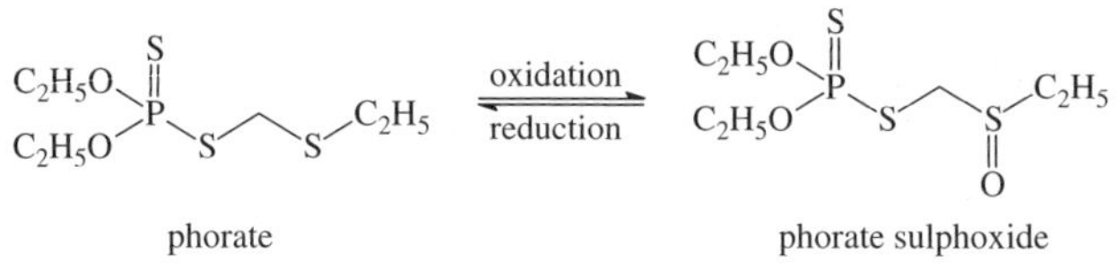

phorate   phorate sulphoxide

Figure 2.11 Phorate and phorate sulfoxide interconversion in aerobic and anaerobic systems

Evaluating Pesticide Behaviour

Regulators generally rely on a suite of standard tests for evaluating the degradation of pesticides through various mechanisms, with a view to estimating the rate of degradation, understanding the metabolic pathway and determining whether there are any metabolites reaching significant levels (typically considered to be $\geq 10\,\%$ of the amount of pesticide originally applied, in which case further evaluation of the metabolite itself may be conducted). Degradation tests conducted in the laboratory include hydrolysis at various pH levels, aqueous photolysis, soil photolysis, aerobic and anaerobic soil metabolism and aerobic and anaerobic aquatic metabolism (usually conducted in water–sediment systems). In general, these tests are conducted at 20–$25\,°C$ and in the dark, except for photolysis studies, where natural sunlight or a light source simulating sunlight is used. Test soils and sediments may be incubated at cooler temperatures where a chemical is to be used in areas such as northern Europe.

Standard laboratory tests are also used to evaluate the mobility of pesticides and metabolites (including adsorption/desorption and leaching of freshly applied pesticide and aged residues) and bioaccumulation (usually in fish). Various field, lysimeter and glasshouse tests may also be used to augment knowledge of a pesticide's behaviour under field or semi-field conditions and in plants. These include field dissipation, soil accumulation and rotational crop studies.

Guidelines for various tests have been published by the Organization for Economic Cooperation and Development (OECD), US Environmental Protection Agency (EPA) and similar agencies in Germany, Japan and some other countries, and efforts to produce 'harmonized' guidelines between the US EPA and OECD are well advanced (US EPA (website)). The book by Leng *et al.* (1995) provides a critical evaluation of laboratory and field methods used to assess the fate of pesticides.

In addition to such standardized tests, there is, of course, a large body of scientific knowledge on pesticides which is accumulating through manifold studies of pesticide behaviour and surveys of levels in the general environment, foods and humans.

MOBILITY IN SOIL AND WATER

Pesticides and their degradation products range widely in their physical and chemical characteristics, affecting their tendency to move to soil, water or air and mobility in those media (Lyman, 1995; Koskinen and Harper, 1990; Calder-bank, 1989).

Adsorption to Soil

Various interactions between soil particles and the molecules of pesticides or their degradation products affect the degree to which these molecules are retained in the soil or partition to water and air and are therefore potentially mobile. At one extreme, chemicals reaching the soil in water may continue to move relatively freely with the wetting front and may leach right through the soil profile to ground water, or wash out of soil beds into irrigation furrows and hence reach surface water in run-off. At the other extreme, substances may readily become attached to soil particles and remain attached to them, even if the particles are dislodged and carried elsewhere by flowing water. Most substances fall somewhere in between. Similarly, substances vary in the extent to which they partition to sediment or suspended matter or remain in water if they are added directly or indirectly to water (e.g. via pesticide drift to a pond).

Retention of organic substances in soil occurs largely through their accumulation on soil particle surfaces ('adsorption'), although absorption processes also contribute and precipitation as insoluble material may also occur (hence the wider term 'sorption' may be used). Molecules near the surface of soil particles may be attracted and held by various physical and chemical means which vary in their strength and reversibility. These include mechanisms such as Van der Waals forces, hydrogen bonding, ion exchange, co-ordination through attached metal ions, the formation of charge-transfer complexes, and most stable of all, covalent bonding (Koskinen and Harper, 1990).

The extent to which adsorption may occur is strongly dependent on the physicochemical characteristics of the substance (e.g. ionic or non-ionic, polarity, presence of functional groups, water solubility, hydrophobicity, molecule size and shape) and may be affected by factors such as the amount of water present, pH and presence of other solutes, as well as soil characteristics. In addition to affecting mobility, adsorption may also affect bioavailability to plant roots, animals and micro-organisms. With some substances, the rate of degradation of a substance slows with time because of adsorption in a fashion reducing availability to micro-organisms, while in some cases adsorption may increase the rate of degradation, e.g. through catalysis by clay minerals (Wolfe *et al.*, 1990).

Of overwhelming importance in determining the adsorptive capacity of soil are the amounts and composition of clay and organic colloids present, because of both their large surface area and the presence of sites assisting adsorption. Humic material contains numerous chemically reactive functional groups such as carboxyl and phenolic hydroxyl groups, leading to a high cation-exchange capacity and favouring various adsorption mechanisms. Humic acid, fulvic acid and humin fractions of humus may differ in the extent to which they adsorb a particular substance, and differences in the age and origin of organic matter may also affect its properties. Clay minerals contain reactive sites such as siloxane and inorganic hydroxyl groups, varying in their abundance and availability from swelling clays such as montmorillinite, with a high cation-exchange capacity, to non-swelling clays such as kaolin, low in cation-exchange capacity. Adsorption to other soil minerals may also be significant, e.g. due to hydrophobicity effects, some substances may sorb significantly even in low-organic-matter-content sands, and in some cases adsorption may occur to metallic hydrous oxides.

Characterizing Mobility of a Substance

One of the key indicators of the tendency for adsorption to occur with a substance is the distribution coefficient, based on the ratio of pesticide in solution to that adsorbed after equilibrium is established. Provided that the substance is sufficiently stable under the test conditions, this is usually determined for various soils and sediments by measuring the relative amounts of chemical partitioning to soil and water after shaking the soil with water containing a range of concentrations of the chemical (batch equilibrium method – Weber, 1995; US EPA (website)). The tendency of the chemical to partition back into water from soil (desorption) can then be determined by shaking the soil containing the residues with fresh water.

Partitioning is usually adequately modelled by the 'Freundlich' equation to give a distribution coefficient (K_d) for adsorption or desorption. For many pesticides, organic matter is the most important soil component. Hence, the partition is often expressed in terms of the organic carbon content of the soil as K_{OC}, the soil-organic-carbon distribution coefficient.

However, it is by no means always true that sorption is related directly to the soil organic matter content, and hence K_{OC} values for a substance may still vary widely between soils, e.g. because particular clay or other soil minerals contribute strongly, or because of soil pH effects on dissociation of the molecule.

Based on numerous measurements with various chemicals and different soil types, McCall et al. (1980) suggested a scale of mobility classes based on K_{OC}. This is shown in Table 2.2, together with some representative examples drawn from published data.

Pesticide mobility may also be assessed in the laboratory by techniques such as soil thin layer chromatography or measurement of leaching using soil columns. Regression equations based on known data have also been developed to enable these parameters to be predicted for various classes of compounds based on other physicochemical characteristics of the substance, such as the n-octanol–water partition coefficient (P_{OW}), water solubility or 'parachor' value (Gawlik et al., 1997; Lyman, 1990; Green and Karickoff, 1990).

Extent of Leaching in the Field

Provided the hydraulic characteristics of the soil profile permit through drainage, when sufficient rainfall or irrigation occurs substances present in the surface soil may be leached deeper into the profile, ultimately permeating to sub-surface

Table 2.2 Scale of soil mobility classes (McCall *et al.*, 1980) and examples

K_{OC}	Mobility class	Examples
>5000	Immobile	dieldrin, DDT, trifluralin, paraquat, deltamethrin, chlorpyrifos, abamectin
2000–5000	Low mobility	endosulfan, glyphosate, parathion-methyl
500–2000	Slightly mobile	chlorothalonil, malathion, linuron, diallate
150–500	Medium mobility	carbaryl, diuron, monuron, propachlor, diazinon
50–150	Highly mobile	atrazine, simazine, bromacil, 2,4,5-T
0–50	Very highly mobile	acrolein, aldicarb, 2,4-D, dimethoate, mevinphos, dichlorvos

drains or ground water, from where they may reach surface drains, re-emerge in springs, or be pumped out for irrigation, stock or domestic water purposes. However, leaching to ground water is likely to be a gradual process, unless high volumes of water are applied over a short period and preferential pathways are present or the soil profile is highly permeable. Hence, even for mobile pesticides there may be a limited period of opportunity for this to occur before the substance degrades or dissipates to air.

On the other hand, slow degradation because of cold conditions may enable deeper leaching to occur. Even if a pesticide degrades relatively rapidly in surface soil, leaching may move it to a depth where the degradation rate slows markedly because conditions are no longer favourable. For example, alkaline conditions do not favour hydrolysis of sulfonylurea herbicides, and hence once moved to depths where biological activity is low, these herbicides are likely to become more persistent in soils with alkaline pH at depth (Sarmah *et al.*, 2000). Similarly, Wells and Waldman (1995) determined that there is a positive correlation between detections of the carbamate insecticide aldicarb in ground water in the US and vulnerable soils, usage and cold temperatures.

In many cases, degradates of the applied substance may be more mobile, both because they are relatively more polar and water soluble than the parent substance and because they are formed before sufficient water is applied to the soil to cause significant leaching. Degradates may also form after the parent has moved. Hence, when leaching is evaluated in the laboratory in soil columns, movement of degradates as well as the parent substance is usually considered. Movement of 'aged residues' (i.e. in soil treated with the substance and incubated for a period to allow ~50 % degradation to proceed) may be evaluated, as well as from substance freshly applied to the top of the column. While accessions to ground water would be expected to be restricted to relatively mobile pesticides, the presence of hydrophobic substances in some ground waters has led to the suggestion that substances may move through soil via attachment to colloidal particles or dissolved organic matter.

A screening indicator of whether or not a pesticide applied to the soil surface is likely to reach ground water in field situations is the *Gustafson Ubiquity Score*

(GUS) (Gustafson, 1989). This was derived from data for compounds which had been categorized as to their leachability from field experience for which consistent sets of soil degradation half-life and K_{OC} data were available:

$$GUS = [\log_{10}(\text{soil half-life})][4 - \log_{10}(K_{OC})] \qquad (2.1)$$

From the data available, Gustafson (1989) concluded that where the GUS > 2.8 the compound is unlikely to leach to ground water, while if the GUS < 1.8 ('leacher') or intermediate between these values ('transitional leacher') closer consideration of leaching behaviour is warranted.

For the reasons discussed, the profile of pesticide-related substances found in surface waters may differ significantly from that for ground waters. Persistent chemicals could still remain in ground water when they are no longer used and therefore no longer likely to enter surface water, e.g. Kookana et al. (1998) noted the presence in ground water in Australia of the pesticides dieldrin, lindane and alachlor, none of which had been registered in agriculture for more than 10 years at the time of testing.

Surface Drainage

Rather than leaching deeper into the soil, residues dissolved in water or adsorbed to soil particles may be transported in surface water (run-off), flowing into drains and from there potentially into ponds or streams. Kookana et al. (1998) note that particularly in dispersive soils, losses in surface run-off may in fact be greater for pesticides strongly adsorbed to soil particles than those which are relatively mobile in soil, as movement into the soil makes a substance less prone to loss by surface run-off. Steps which can assist in reducing run-off losses include timing application of the pesticide to provide a delay allowing adsorption or movement into soil before heavy rainfall or irrigation which might trigger run-off, avoiding use of problem pesticides on areas with significant risk of run-off (e.g. sloping ground), and where appropriate, incorporation of a pesticide into the soil. Downstream areas can be protected from residues in run-off by not cultivating adjacent to aquatic areas, by providing a vegetative filter strip and by capturing drainage in a tailwater recirculation dam.

Studies with sulfonylurea herbicides (relatively mobile in soil) exemplify the extent of run-off which might be expected under practical conditions. Wauchope et al. (1990) examined the loss of chemical from bare soil and grassy plots (loamy sand, 3 % slope) treated with two formulations of the sulfonylurea herbicide sulfometuron-methyl at 400 g ai/ha[1] and given simulated rain at 69 mm/h until 2 mm of run-off occurred. Losses ranged from 0.4–2.3 % of the applied amount, with the average concentrations in run-off water being 0.05–0.3 mg/l (ppm). Afyuni et al. (1997) looked at two sulfonylureas, nicosulfuron and chlorimuron,

---

[1] ai/ha, active ingredient per hectare.

in conventional tillage and no-tillage systems on a sandy loam and a sandy clay loam. Artificial rain was applied at a low and high rate of 1.27 and 5.08 cm/h, respectively, 24 h after application. One week later, the high rainfall rate was again used. The average herbicide loss was around 1 % and 2 %, respectively, for the conventional tillage and no-tillage systems, irrespective of the soil type. The loss in run-off declined to < 0.2 % of the applied rate one week later. Pesticide losses in run-off of the order of 2 % of the applied amount may also be expected with other pesticides (Leonard, 1990). However, Kookana et al. (1998) indicate situations where losses greater than 10 % have been encountered

Loss by Volatilization and Wind Erosion

As discussed earlier, pesticide residues may be lost from soil and water by volatilization and from soil in dust eroded by wind. These are complex processes depending on the physical properties of the substance (e.g. vapour pressure, water solubility, Henry's Law constant, diffusivity, etc.), physical, chemical and structural properties of the water body or soil, interactions with other substances and soil components (hence K_d), and atmospheric conditions (primarily wind).

PESTICIDE FATE ON PLANT SURFACES AND IN PLANTS

Pesticide landing on plant surfaces may dissipate through volatilization, photolysis and by microbial activity on the leaf surface, as well as by washing off in rain or spray irrigation. Concentrations in or on plant tissue may also decline through growth dilution effects. Many pesticides adsorb to the leaf surface, move into the waxy surface of the cuticle or are absorbed into plant cells, reducing the amount of residues which might wash off and enabling degradation by plant enzymes to occur. Depending on its physical and chemical properties, a pesticide or altered product may become 'systemic,' entering the sap stream of the plant and moving either acropetally (towards the apex, in the xylem) or basipetally (towards the base, in the phloem). Again depending on their chemical and physical properties, pesticides or their residues in soil beneath plants may be taken up by roots and be moved upwards into the plant in the sap stream. With persistent substances, this may occur well after the original application (hence investigations may include rotational crop studies to examine residue uptake by subsequent crops). Thus, pesticides may move in the sap stream to parts of the plant not directly sprayed, and may leave residues within growing tissues, not just on the plant surface. Pesticide residues remaining in plants may be removed in harvested produce, or may be released in the soil or retained in organic matter as the plant tissue decays.

FIELD DISSIPATION

As indicated in the preceding sections, scientists use various laboratory studies to evaluate the degradation and mobility of pesticides under standard conditions.

While these may explain how a pesticide degrades, predict its mobility and identify the metabolites formed, the best indication of a pesticide's behaviour is to
evaluate it in a field situation similar to that where it will actually be used. Field
studies may be very comprehensive, examining the dissipation rate and mobility of a pesticide and its major metabolites in soil, often at more than one site
and sometimes with repeated spraying over several years to determine the extent
of soil accumulation. When field lysimeters are used, drainage water may be
evaluated, which is clearly of value when significant leaching is expected.

Dissipation mechanisms which are tested independently in the laboratory may
act together to enhance the degradation rate in the field. Hence, the dissipation
rate in the field is sometimes found to be significantly faster than that indicated
in the laboratory, given similar average temperature conditions. Most importantly, degradation in the field may be enhanced by photolysis, whereas standard
microbial degradation studies in the laboratory are conducted in the dark: a clear
example is azoxystrobin, discussed above. Similarly, microbial activity in field
soils may be greater than in soils removed, prepared and stored for laboratory
use, and the microbial viability of samples undergoing incubation may be difficult
to maintain under the unnatural conditions of prolonged (e.g. 12 months) incubation in the laboratory. On the other hand, in the field there is limited ability to
control conditions such as temperature and moisture, and hence the degradation
rate may vary with weather conditions and the seasons.

CONCLUSIONS REGARDING PESTICIDES IN SOIL AND WATER

Pesticides and their breakdown products directly or indirectly reaching soil may
dissipate by various means, including mobilization to the atmosphere through
volatilization or dust, transport in water though leaching and run-off, and degradation through abiotic or biotic processes. Similarly, pesticides and their breakdown
products directly or indirectly reaching water (surface water such as streams,
ponds, dams and wetlands, or ground water), may be dissipated from the water
column by processes including adsorption to sediment, degradation in the water
column or sediment, and volatilization. Before dissipating, they may be transported elsewhere, e.g. downstream or back to the land through irrigation or
flooding. Ground water may ultimately return to the surface through pumping
or in springs. Residues of more persistent substances may remain indefinitely in
soil and sediment, adsorbed to organic matter or clay particles. Provided adsorption is not too strong, prior to degradation, pesticide residues may be taken up
by plants or animals and may therefore enter the food web, even if not directly
applied to the produce.

BIOACCUMULATION AND BIOMAGNIFICATION

Some substances present in water may be absorbed by an organism through the
gills and epithelial tissue at a faster rate than they can be degraded or eliminated,

resulting in 'bioconcentration'. The wider term 'bioaccumulation' is used when accumulation of chemicals also involves food consumption, while 'biomagnification' occurs when bioconcentration and bioaccumulation take place through several trophic levels, i.e. up the food chain. Hence, some substances may be present in fish or other aquatic organisms used for food at a significantly greater concentration than they are found in the surrounding water, and similar effects may occur in terrestrial animals. Bysshe (1990) noted that such residues can accumulate to levels that are harmful to consumers of such organisms, or even to the organisms themselves.

Typically, such substances are hydrophobic, with a low solubility in water and an affinity for lipids or non-polar solvents. The solubility, octanol–water partition coefficient (P_{OW}) and organic-carbon distribution coefficient (K_{OC}) give some indication of this tendency, but tests with living organisms are preferable as bioconcentration and bioaccumulation are affected by various factors, such as the ways in which uptake and elimination may occur, toxicity, degradability of the parent substance and characteristics of the metabolites.

In such tests, species such as bluegill sunfish are exposed for a prolonged period (e.g. one to two months) to concentrations of the chemical below chronic toxicity levels, followed by a depuration period (e.g. one to two weeks) in water free of the test substance. Concentrations of the ^{14}C-labelled test substance (and often its degradates) in the water and in various tissues of the organism are monitored and a bioconcentration factor (BCF) calculated based on the ratio of these concentrations once an equilibrium concentration has been reached. While not necessarily accurate, various equations have been determined based on measured data to allow BCF values to be estimated, as discussed by Bysshe (1990).

Among more notorious examples of bioaccumulators are organochlorine insecticides such as DDT and chlordane, with BCF values of the order of 30 000 or higher.

SUMMARY AND CONCLUSIONS

This discussion of the fate of pesticides makes it clear that we need to be aware of relevant metabolites and degradation products as well as the parent pesticide when assessing residues in food and drinking water: such products may sometimes retain significant toxicity, and may be more persistent than the parent substance itself.

In is important to consider not just residues obviously associated with the particular treatment used, such as those on the skin of fruit and vegetables which have been sprayed or in meat of dosed animals. Residues from an applied substance may also reach food crops through indirect routes, such as root uptake of pesticide applied to soil or via translocation within the plant from sprayed leaves into developing grains or fruit. Contamination can also arise from previous use in the same field. Pesticide residues may also be transported into the field where food is being produced by a number of mechanisms, ranging from spray drift from nearby areas, to introduction in irrigation or drainage water or on imported material such as manure, to long-range transport mechanisms such as in rain or dust.

Residues may be found in animals as a result of consumption of feed containing residues. This could arise from treatments used on the crop or directly on the feed (e.g. stored grain), or from contamination such as spray drift onto pasture (e.g. to protect beef producers, there is now a large downwind buffer specified for use of the insecticide endosulfan on Australian cotton). Fish and aquatic foods may take up pesticide residues in water and food sources, and biomagnification may then occur up the food chain.

Drinking water supplies in surface streams and water bodies may be contaminated by direct exposure, spray drift, run-off and drainage. Downstream flow may then carry residues large distances from the original site of contamination. Similarly, deep leaching may carry residues to ground water which may subsequently be accessed for drinking water, stock water or irrigation purposes. All residues in ground water will have resulted from environmental contamination rather than direct use, and ground water may also have moved some distance from the original source of contamination.

A thorough knowledge of the environmental fate and behaviour of a pesticide is an essential component of assessing its potential for causing residues in food and drinking water. Most times, especially for food, the assessment will conclude that the environmental pathway for contamination is very minor compared with direct application. If any residues appear in drinking water, the environmental pathway is the more likely, so the environmental assessment will provide information on the nature of the residue and the levels to be expected.

REFERENCES

Afyuni, M. M., Wagger, M. G. and Leidy, R. B. (1997). Runoff of two sulfonylurea herbicides in relation to tillage system and rainfall intensity, *J. Environ. Qual.*, **26**, 1318–1326.

Alegria, H. and Shaw, T. J. (1999). Rain deposition of pesticides in coastal waters of the South Atlantic Bight, *Environ. Sci. Technol.*, **33**, 850–856.

Armbrust, K. L. (2000). Pesticide hydroxyl radical rate constants: measurements and estimates of their importance in aquatic environments, *Environ. Toxicol. Chem.*, **19**, 2175–2180.

Aston, L. S. and Seiber, J. N. (1997). Fate of summertime airborne organophosphate residues in the Sierra Nevada Mountains, *J. Environ. Qual.*, **26**, 1483–1492.

Bedos, C., Cellier, P., Calvet, R., Barriuso, E. and Gabrielle, B. (2002). Mass transfer of pesticides into the atmosphere by volatilization from soils and plants: overview, *Agronomie*, **22**, 21–33.

Bidleman, T. F. (1999). Atmospheric transport and air-surface exchange of pesticides, *Water Air Soil Pollut.*, **115**, 115–166.

Bird, S. L., Esterly, D. M. and Perry, S. G. (1996). Off-target deposition of pesticides from agricultural aerial spray applications, *J. Environ. Qual.*, **25**, 1095–1104.

Bollag, J.-M. and Liu, S.-Y. (1990). Biological transformation processes of pesticides, in *Pesticides in the Soil Environment: Processes, Impacts and Modeling*, Cheng, H. H. (Ed.), Soil Science Society of America, Inc., Madison, WI, USA, Ch. 6, pp. 169–212.

Burgoyne, T. W. and Hites, R. A. (1993). Effects of temperature and wind direction on the atmospheric concentrations of α-endosulfan, *Environ. Sci. Technol.*, **27**, 910–914.

Bysshe, S. E. (1990). Bioconcentration factor in aquatic organisms, in *Handbook of Chemical Property Estimation Methods*, Lyman, W. J., Reehl, W. F. and Rosenblatt, D. H. (Eds), American Chemical Society, Washington, DC, USA, Ch. 5, pp. 1–30.

Calderbank, A. (1989). The occurrence and significance of bound pesticide residues in soil, *Rev. Environ. Contam. Toxicol.*, **108**, 71–103.

Charizopoulos, E. and Papadopoulou-Mourkidou, E. (1999). Occurrence of pesticides in rain of the Axios River Basin, Greece, *Environ. Sci. Technol.*, **33**, 2363–2368.

Chernyak, S. M., Rice, C. P. and McConnell, L. L. (1996). Evidence of currently used pesticides in air, ice, fog, seawater and surface microlayer in the Bering and Chukchi Seas, *Mar. Poll. Bull.*, **32**, 410–419.

Coats, J. R. (1991). Pesticide degradation mechanisms and environmental activation, in *Pesticide Transformation Products: Fate and Significance in the Environment*, Somasundaram, L. and Coats, J. R. (Eds), American Chemical Society, Washington, DC, Ch. 2, pp. 10–30.

Dec, J., Haider, K., Rangaswamy, V., Schäffer, A., Fernandes, E. and Bollag, J.-M. (1997a). Formation of soil-bound residues of cyprodinil and their plant uptake, *J. Agric. Food Chem.*, **45**, 514–520.

Dec, J., Haider, K., Benesi, A., Rangaswamy, V., Schäffer, A., Plücken, U. and Bollag, J.-M. (1997b). Analysis of soil-bound residues of the ^{13}C-labeled fungicide cyprodinil by NMR spectroscopy, *Environ. Sci. Technol.*, **31**, 1128–1135.

Dec, J., Schäffer, A., Haider, K. and Bollag, J.-M. (1997c). Use of the silylation procedure and ^{13}C-labeled fungicide cyprodinil by NMR spectroscopy to characterize bound and sequestered residues of cyprodinil in soil, *Environ. Sci. Technol.*, **31**, 2991–2997.

Ganzelmeier, H. and Rautmann, D. (2000). Drift, drift reducing sprayers and sprayer testing, *Aspects Appl. Biol.*, **57**, 1–10.

Gawlik, B. M., Sotirou, N., Feicht, E. A., Schulte-Hostede, S. and Kettrup, A. (1997). Alternatives for the determination of the soil adsorption coefficient, K_{OC}, of non-ionic organic compounds – a review, *Chemosphere*, **34**, 2525–2551.

Green, R. E. and Karickoff, S. W. (1990). Sorption estimates for modeling, in *Pesticides in the Soil Environment: Processes, Impacts and Modeling*, Cheng, H. H. (Ed.), Soil Science Society of America, Inc., Madison, WI, USA, Ch. 4, pp. 79–102.

Gustafson, D. I. (1989). Groundwater ubiquity score: a simple method for assessing pesticide leachability, *Environ. Toxicol. Chem.*, **8**, 339–357.

Haag, W. R. and Holgnè, J. (1986). Singlet oxygen in surface waters. 3. Photochemical formation and steady-state concentrations in various types of waters, *Environ. Sci. Technol.*, **20**, 341–348.

Hewitt, A. J. (2000). Spray drift modelling, labelling and management in the US, *Aspects Appl. Biol.*, **57**, 11–19.

Hewitt, A. J., Johnson, D. R., Fish, J. D., Hermansky, C. G. and Valcore, D. L. (2002). Development of the Spray Drift Task Force database for aerial applications, *Environ. Toxicol. Chem.*, **21**, 648–658.

IPCS (1984). *Environmental Health Criteria 40: Endosulfan*, International Program on Chemical Safety, World Health Organization, Geneva, Switzerland.

Joseph, R. S. I. (1999). Metabolism of azoxystrobin in plants and animals, in *Pesticide Chemistry and Bioscience: The Food–Environment Challenge*, Brooks, G. T. and Roberts, T. R. (Eds), The Royal Society of Chemistry, Cambridge, UK, pp. 265–278.

Klöpffer, W. (1992). Photochemical degradation of pesticides and other chemicals in the environment: a critical assessment of the state of the art, *Sci. Total Environ.*, **123/124**, 145–159.

Kookana, R. S., Baskaran, S. and Naidu, R. (1998). Pesticide fate and behaviour in Australian soils in relation to contamination and management of soil and water: a review, *Aust. J. Soil Res.*, **36**, 715–764.

Koskinen, W. C. and Harper, S. S. (1990). The retention process: mechanisms, in *Pesticides in the Soil Environment: Processes, Impacts and Modeling*, Cheng, H. H. (Ed.), Soil Science Society of America, Inc., Madison, WI, USA, Ch. 3, pp. 51–78.

Kreuger, J. and Staffas, A. (1999). Atmospheric deposition of pesticides in rainfall in Sweden (manuscript), in Kreuger, J., *Pesticides in the Environment – Atmospheric Deposition and Transport to Surface Waters,* Doctoral Thesis, Swedish University of Agricultural Sciences, Uppsala, Sweden (Acta Universitatis Agriculturae Sueciae, Agraria 162).

Leng, M. L., Leovey, E. M. K. and Zubkoff, P. L. (1995). *Agrochemical Environmental Fate: State of the Art*, CRC Press/Lewis Publishers, Boca Raton, FL, USA.

Leonard, R. A. (1990). Movement of pesticides into surface waters, in *Pesticides in the Soil Environment: Processes, Impacts and Modeling*, Cheng, H. H. (Ed.), Soil Science Society of America, Inc., Madison, WI, USA, Ch. 9, pp. 303–350.

Lippa, K. A. and Roberts, A. L. (2002). Nucleophilic aromatic substitution reactions of chloroazines with bisulfide (HS^-) and polysulfides (S_n^{2-}), *Environ. Sci. Technol.*, **36**, 2008–2018.

Longley, M. and Sotherton, N. W. (1997). Measurements of pesticide spray drift deposition into field boundaries and hedgerows: 2. Autumn applications, *Environ. Toxicol. Chem.*, **16**, 173–178.

Longley, M., Cigli, T., Jepson, P. C. and Sotherton, N. W. (1997). Measurements of pesticide spray drift deposition into field boundaries and hedgerows: 1. Summer applications, *Environ. Toxicol. Chem.*, **16**, 165–172.

Lyman, W. J. (1990). Adsorption coefficient for soils and sediments, in *Handbook of Chemical Property Estimation Methods: Environmental Behaviour of Organic Compounds*, Lyman, W. J., Reehl, W. F. and Rosenblatt, D. H. (Eds), American Chemical Society, Washington, DC, USA, Ch. 4, pp. 1–33.

Lyman, W. J. (1995). Transport and transformation processes, in *Fundamentals of Aquatic Toxicology*, Rand, G. M. (Ed.), 2nd Edn, Taylor and Francis, Washington, DC, USA, Ch. 15, pp. 449–492.

Mackay, D. and Paterson, S. (1981). Calculating fugacity, *Environ. Sci. Technol.*, **15**, 1006–1014.

Majewski, M. S. and Capel, P. D. (1995). *Pesticides in the Atmosphere: Distribution, Trends and Other Governing Factors*, Ann Arbor Press, Ann Arbor, MI, USA.

Majewski, M. S., Foreman, W. T., Goolsby, D. A. and Nakagaki, N. (1998). Airborne pesticide residues along the Mississippi River, *Environ. Sci. Technol.*, **32**, 3689–3698.

Marrs, R. H. and Frost, A. J. (1995). Minimising the effects of herbicide spray drift, *Pesticide Outlook*, **6**(4), 28–31.

Mathews, G. A. (1992). *Pesticide Application Methods*, 2nd Edn, Longmans, Singapore.

Mathews, G. and Piggott, S. (1999). Can farmers reduce spray drift?, *Pesticide Outlook*, **10**(2), 31–33.

McCall, P. J., Swann, R. L., Laskowski, D. A., Unger, S. M., Vrona, S. A. and Dishburger, H. J. (1980). Estimation of chemical mobility in soil from liquid chromatographic retention times, *Bull. Environ. Contam. Toxicol.*, **24**, 190–195.

McConnell, L. L., Nelson, E., Rice, C. P., Baker, J. E., Johnson, W. E., Harman, J. A. and Bialek, K. (1997). Chlorpyrifos in the air and surface water of Chesapeake Bay: prediction of atmospheric deposition fluxes, *Environ. Sci. Technol.*, **31**, 1390–1398.

McConnell, L. L., LeNoir, J. S., Datta, S. and Seiber, J. N. (1998). Wet deposition of current-use pesticides in the Sierra Nevada Mountain Range, California, USA, *Environ. Toxicol. Chem.*, **17**, 1908–1916.

Mill, T. (1993). Environmental chemistry, in *Ecological Risk Assessment*, Suter, G. W. II (Ed.), Lewis Publishers, Boca Raton, FL, USA, Ch. 4, pp. 91–127.

Miller, M. M., Sweet, C. W., Depinto, J. V. and Hornbuckle, K. C. (2000). Atrazine and nutrients in precipitation: results from the Lake Michigan mass balance study, *Environ. Sci. Technol.*, **34**, 55–61.

Muschal, M. (1998). Central and North-West Regions Water Quality Program, 1996/97 report on pesticides monitoring, CNR98.038, Ecosystem Management, Centre for Natural Resources, Department of Land and Water Conservation, Parramatta NSW, Australia.

OECD (1993). The Rate of Photochemical Transformation of Gaseous Organic Compounds in Air under Tropospheric Conditions, Environment Monograph No. 61, Organization for Economic Co-operation and Development, Paris.

Preston, M. (2000). *BNA Chem. Reg. Rep.*, **24**(37), 1851 (Bureau of National Affairs, Inc., Washington, DC, USA).

Racke, K. D. (1990). Pesticides in the soil microbial ecosystem, in *Enhanced Biodegradation of Pesticides in the Environment*, Racke, K. D. and Coats, J. R. (Eds), ACS Symposium Series 426, American Chemical Society, Washington, DC, USA, Ch. 1, pp. 1–12.

Racke, K. D., Skidmore, M. W., Hamilton, D. J., Unsworth, J. B., Miyamoto, J. and Cohen, S. Z. (1997). Pesticide fate in tropical soils, *Pure Appl. Chem.*, **69**, 1349–1371.

Rice, C. P. (1996). Pesticides in fogwater, *Pesticide Outlook*, **7**(4), 31–36.

Salyani, M. and Cromwell, R. P. (1992). Spray drift from ground and aerial applications, *Trans. Am. Soc. Agric. Eng.*, **35**, 1113–1120.

Sarmah, A. K., Kookana, R. S. and Alston, A. M. (2000). Leaching and degradation of triasulfuron, metsulfuron-methyl, and chlorsulfuron in alkaline soil profiles under field conditions, *Aust. J. Soil Res.*, **38**, 617–631.

Scheunert, I., Mansour, M. and Andreux, F. (1992). Binding of organic pollutants to soil organic matter, *Int. J. Environ. Anal. Chem.*, **46**, 189–199.

Schnoor, J. L. (1996). Toxic Organic Chemicals, in *Environmental modeling: fate and transport of pollutants in water, air and soil*, Wiley, New York, Ch. 7, pp. 305–380.

Simonich, S. L. and Hites, R. A. (1995). Global distribution of persistent organochlorine compounds, *Science*, **269**, 1851–1854.

Simonich, S. L. and Hites, R. A. (1997). Relationship between socioeconomic indicators and concentrations of organochlorine pesticides in tree bark, *Environ. Sci. Technol.*, **31**, 999–1003.

Spray Drift Task Force (1997). A Summary of Aerial Application Studies, Airblast Application Studies and Ground Application Studies (available from David Johnson at Stewart Agricultural Research Services, Inc., PO Box 509, Macon, MO 63552, USA).

Teske, M. E., Bird, S. L., Esterly, D. M., Curbishley, T. B., Ray, S. L. and Perry, S. G. (2002). AgDRIFT®: a model for estimating near-field spray drift from aerial applications, *Environ. Toxicol. Chem.*, **21**, 659–671.

Thurman, E. M. and Cromwell, A. E. (2000). Atmospheric transport, deposition, and fate of triazine herbicides and their metabolites in pristine areas at Isle Royale National Park, *Environ. Sci. Technol.*, **34**, 3079–3085.

Tomlin, C. (Ed.) (1997). *The Pesticide Manual*, 11th Edn, British Crop Protection Council, Farnham, Surrey, UK.

UNECE (1996). Review of risk characterisation information on selected persistent organic pollutants, paper presented by the UK to the *United Nations Economic Commission for Europe Convention on Long Range Transboundary Air Pollutants Preparatory Working*

Group on Persistent Organic Pollutants, 4th Session, Ottawa, Quebec, Canada, 21–23 October 1996.

Unsworth, J. B., Wauchope, R. D., Klein, A. W., Dorn, E., Zeeh, B., Yeh, S. M., Akerblom, M., Racke, K. D. and Rubin, B. (1999). Significance of the long range transport of pesticides in the atmosphere, *Pure. Appl. Chem.*, **71**, 1359–1383.

US EPA (1999). Spray Drift of Pesticides, Fact sheet: [http://www.epa.gov/pesticides/citizens/spraydrift.htm].

US EPA (website). [http://www.epa.gov/docs/OPPTS_Harmonized/]

Wania, F., Mackay, D., Li, Y.-F., Bidleman, T. F. and Strand, A. (1999). Global chemical fate of α-hexachlorocyclohexane, 1. Evaluation of a global distribution model, *Environ. Toxicol. Chem.*, **18**, 1390–1399.

Wauchope, R. D., Williams, R. G. and Marti, L. R. (1990). Runoff of sulfometuron-methyl and cyanazine from small plots: effects of formulation and grass cover, *J. Environ. Qual.*, **19**, 119–124.

Weber, J. B. (1995). Physicochemical and mobility studies with pesticides, in *Agrochemical Environmental Fate: State of the Art*, Leng, M. L., Leovey, E. M. K. and Zubkoff, P. L. (Eds), CRC Press/Lewis Publishers, Boca Raton, FL, USA, Ch. 10, pp. 99–115.

Webster, E., Mackay, D. and Wania, F. (1998). Evaluating environmental persistence, *Environ. Toxicol. Chem.*, **17**, 2148–2158.

Wells, D. A. and Waldman, E. (1995). Analysis of aldicarb leaching potential, in *Agrochemical Environmental Fate: State of the Art*, Leng, M. L., Leovey, E. M. K. and Zubkoff, P. L. (Eds), CRC Press/Lewis Publishers, Boca Raton, FL, USA, Ch. 23, pp. 253–257.

WHO (1989). *Aldrin and Dieldrin*, Environmental Health Criteria 91, International Programme on Chemical Safety, World Health Organization, Geneva, Switzerland.

WHO (1993). *Methyl Parathion*, Environmental Health Criteria 145, International Programme on Chemical Safety, World Health Organization, Geneva, Switzerland.

WHO (1994). *Carbaryl*, Environmental Health Criteria 153, International Programme on Chemical Safety, World Health Organization, Geneva, Switzerland.

Wolfe, N. L., Mingelgrin, U. and Miller, G. C. (1990). Abiotic transformations in water, sediments and soil, in *Pesticides in the Soil Environment: Processes, Impacts and Modeling*, Cheng, H. H. (Ed.), Soil Science Society of America, Inc., Madison, WI, USA, Ch. 5, pp. 103–168.

3 Pesticide Metabolism in Crops and Livestock

[1]**MICHAEL W. SKIDMORE** and [2]**ÁRPÁD AMBRUS**
[1] *Syngenta AG, Bracknell, UK*
[2] *Joint FAO/IAEA Division of Nuclear Techniques in Food and Agriculture, Vienna, Austria*

Pesticide Residues in Food and Drinking Water: Human Exposure and Risks.   Edited by Denis Hamilton and Stephen Crossley
© 2004 John Wiley & Sons, Ltd   ISBN: 0-471-48991-3

INTRODUCTION

When a pesticide is applied it enters a hostile environment and is subjected to a wide range of biological (enzymes), chemical (hydrolysis) and physical (photolysis) reactions, which may change its chemical nature. These new chemical structures, called *metabolites* or *degradation products*, have different inherent properties to the parent pesticide and the effects of these changes need to be assessed both in terms of environmental and human safety. These studies are generically referred to as *metabolism* or, in the case of crops and livestock, nature of the residue studies.

Metabolism studies in pesticide science are conducted to provide a detailed understanding of the fate and behaviour of the chemical in environmental systems, which in the context of this chapter is limited to plants and animals. This understanding is crucial throughout the development life cycle of the pesticide. In the early invention stages simple, usually *in vitro*, studies are employed to assess the relative rates of metabolism, or to identify metabolic differences between species. These are usually high-throughput studies and are used where candidates are pro-pesticides, or are being developed for their selective activity. In the latter stages of development detailed, usually *in vivo* studies, are carried out to define the nature of the residues in commodities used for human food or animal feed. These data are used to support a registration and to provide critical endpoints for human dietary risk assessment. In this latter case, the metabolism studies are the first and most crucial step in the assessment, i.e. until the nature of the residues, to which humans will be exposed, is appreciated neither the hazard nor the level of exposure can be defined. These studies are therefore the foundation for understanding the fate and behaviour of the pesticide and of any subsequent risk assessment.

Residues resulting from the use of the pesticide may enter the human food chain either directly – through the consumption of treated foods, e.g. grain or fruit, or indirectly – through the transfer of residues into milk, eggs and meat products from treated feed items. Three study types form the backbone of the registration package and the data used to define the residue definition (Figure 3.1).

The studies outlined in Figure 3.1 are a core regulatory requirement and as such have descriptive guidelines on their design and conduct. The guidelines are, with perhaps the exception of rotational crop studies, fairly well harmonized

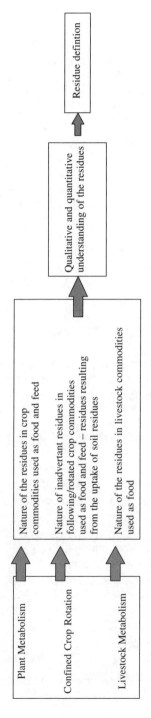

Figure 3.1 Flow diagram showing the process for the interpretation of study data

internationally, having a common objective of defining the nature of the residue in food and feed items and of understanding the behaviour of the chemical in biological systems.

These studies are, by their very nature, complex and not insignificant in terms of time and cost. Considerable effort should go into the planning and interpretation of these studies to fully appreciate the behaviour of the pesticide and its metabolites (the residue), when the product is used as recommended on the label. A major emphasis in the conduct of the study is the identification of the metabolites found in food and feed commodities and, to facilitate the isolation and identification, a tracer is incorporated. This usually takes the form of a radio-atom incorporated into the pesticide molecule, a practice which has implications for the study design resulting from local radiochemical safety laws.

The objective of this chapter is to discuss and highlight some of the complexities and considerations associated with metabolism studies, namely:

- factors influencing the nature of the residues in food and feed items
- examples of metabolic reactions reported for pesticides in plants and livestock
- test substances used in metabolism studies
- achieving reality
- design and conduct of core regulatory studies
- interpretation and significance of metabolites

Throughout the chapter, the term *xenobiotic* is used to describe a foreign compound within a biological system. In the case of pesticides, this could be either the parent compound or a metabolite or degradation product.

FACTORS INFLUENCING THE NATURE OF THE RESIDUES IN FOOD AND FEED ITEMS

The qualitative and quantitative nature of residues in a biological system following its exposure to a pesticide or its metabolites is a function of the following processes:

- **Absorption** – movement across a biological cell wall or membrane
- **Distribution** – transport within the system
- **Metabolism** – biological or chemical modification of the pesticide
- **Elimination** – the pesticide or the products of metabolism are eliminated from active cell processes

In turn, the factors that influence each of the processes include the physicochemical properties of the xenobiotic, the nature of the biological system, species, sex and dose rate. These influences and relationships are discussed below.

ABSORPTION

In the context of this chapter, *absorption* is considered as the movement of the xenobiotic across membranes. This includes absorption from the site of exposure and subsequent absorption from the systemic circulation. Xenobiotics can pass into and out of cells by passive diffusion, osmosis or active transport mechanisms. These factors are particularly important to the metabolism chemist in gaining an understanding of the likely nature and extent of the residues in the commodities being analysed.

Livestock

The exposure of livestock species to pesticides can effectively be limited to oral ingestion, i.e. through ingestion of residues on feed items, or via direct application, when used in animal health. Irrespective of the route of exposure, a xenobiotic must cross biological membranes before it can be absorbed and distributed around the organism. These membranes comprise phospholipids, proteins and polysaccharides.

There are generally accepted to be three main processes whereby molecules can cross the membrane barrier. These are passive diffusion, facilitated diffusion and active transport.

In the case of passive diffusion, movement across the membrane is fundamentally dictated by the ability of a xenobiotic molecule to penetrate the lipophilic core of the membrane. Ionized molecules are too polar in nature to penetrate this region but neutral or un-ionized molecules are able to penetrate.

The driving force for passive diffusion is movement down the concentration gradient of neutral or un-ionized species until equilibrium is achieved either side of the membrane (i.e. a driving force of zero).

Key factors that affect the rate of passive diffusion of xenobiotic molecules therefore include:

(a) Relative concentrations of neutral or un-ionized species either side of the membrane.
(b) Factors that dictate the equilibrium concentrations of ionized and non-ionized species, i.e. the pH either side of the membrane, with the ionization potential of the xenobiotic molecule expressed as pK_a.
(c) Lipophilicity of the xenobiotic molecule expressed as log P_{OW} (the logarithm of its octanol–water partition coefficient).

Where passive diffusion alone applies, lipophilic, non-ionic compounds will generally diffuse more rapidly across membranes than hydrophilic, ionic compounds.

However it is also known that small (ca. < 200 molecular weight) ionic or water-soluble compounds can also diffuse rapidly via water-filled pores present in membranes. Under these circumstances, the molecular weight, charge and configuration of the compound will influence the rate of diffusion. Mass flow of water, dictated by cellular osmotic potential, is also thought to influence the rate at which qualifying molecules transfer through these pores.

Other polar molecules and ions may be rapidly transported across cell membranes as a result of facilitated diffusion. This involves a highly selective transport mechanism where membrane proteins act as carriers for specific endogenous substrates, which would otherwise find difficulty in crossing the membrane. The result is an increase in orders of magnitude in the rate of trans-membrane passage of the substrate when compared with that of simple passive diffusion. Where a xenobiotic sufficiently resembles the natural substrate it may also be transported by this mechanism.

In both facilitated and passive diffusion the passage of molecules always occurs down the concentration or electrochemical gradient and is not continued beyond the point of equilibrium. Active transport, by contrast, utilizes transport proteins to enable accumulation of selected polar and ionic substrates beyond the point of equilibrium, i.e. against the concentration gradient. This transport process is thermodynamically unfavourable and occurs only when coupled to an energy releasing process such as the breakdown of adenosine triphosphate (ATP) to adenosine diphosphate (ADP). Very considerable mass transfer across membranes is achieved in this way. Examples of processes that utilize active transport in vertebrates are the movement of amino acids and glucose from the intestinal lumen into blood. A schematic representation of these diffusion processes is shown in Figure 3.2.

Absorption through the skin following dermal treatments comprises a series of diffusion and partitioning processes through the stratum corneum, epidermis and dermis (Guy *et al.*, 1987).

Plants

The processes of absorption across plant cell walls and membranes are basically the same as those described above. In a review of the absorption of herbicides, Sterling (1994) concludes that most herbicides are absorbed by plant cells by passive diffusion; however, examples are also given where the evidence suggests that some herbicides, dalapon, 2,4-D, glyphosate and paraquat are absorbed through active transport mechanisms. It is also concluded that weakly acidic herbicides can reach a higher concentration in the cell than that on the outside due to ion trapping in the alkaline components of the cell – the cytoplasm has a pH of approximately 7.5. Weakly basic herbicides are described as accumulating in the more acidic cell compartments such as the vacuole, pH 5.5. The physico-chemical properties influencing absorption are defined as lipophilicity, acidity, the cell membrane and the electrochemical potential in the cell.

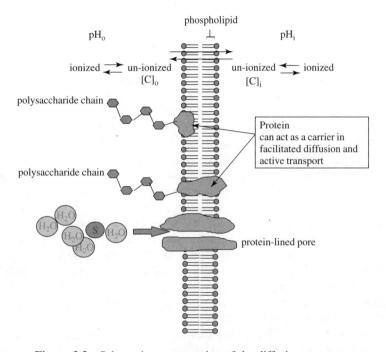

Figure 3.2 Schematic representation of the diffusion processes

The uptake of a xenobiotic by the crop, following a pesticide treatment, depends on the degree of exposure from both the roots and aerial parts of the plant. Pesticides on the surface of the plant or in the soil will be subject to a range of environmental factors, e.g. photolysis and microbial activity, which can result in degradation of the pesticide. These degradation products and the parent pesticide are therefore available for absorption. A further factor influencing the uptake of xenobiotics from the soil is the interaction of chemicals with the soil. To be absorbed into the roots the xenobiotic needs to be bioavailable or present in the soil–water compartment, a function of the interactions of the chemical with soil organic matter or clay particles. This interaction is measured as the adsorption coefficient, K_d or K_{oc}, and in most cases shows some reversibility.

The uptake of compounds by the roots is therefore is a factor of the soil adsorption coefficient of the xenobiotic, the concentration gradient between the soil solution and that inside the root, lipophilicity, degree of ionization, and on the mass flow of water.

Aerial parts of the plant, i.e. the outer surfaces of the leaf and stem, have layers of cuticular wax above the cell walls which serves as a barrier to water loss and to the entry of xenobiotic compounds.

DISTRIBUTION

Livestock

Once in the systemic circulation or blood stream, xenobiotics are carried through-out the body. Xenobiotic distribution in tissues will be dependent on the blood-tissue dynamics and the propensity of the compounds to bind with plasma protein. The mechanistic processes for transfer into tissues are similar to those described above.

One of the most important food items from livestock is milk that forms a significant part of the diet for children and is usually subject to special atten-tion. The significant mass flow of water into mammary tissues would suggest that small water-soluble xenobiotic compounds could be transferred into milk; indeed, Levine (1983) reported that ionized compounds with a molecular weight of less than 200 can transfer into milk. Two possible examples of this mechanism are flusilazole and oxamyl. In the case of flusilazole (a triazole fungicide – see Figure 3.3), the metabolite 1,2,4-triazole is found in milk following dosing of the parent pesticide (Anderson *et al.*, 1999). 1,2,4-Triazole (molecular weight 69) has a low lipophilicity, a pK_a of 2.3 and would be ionized at physiological pH levels and therefore passive diffusion from the blood stream into tissues or milk is not to be expected. The likely mechanism is therefore mass flow, albeit active transport cannot be ruled out. In the second example, thiocyanate, a significant metabolite of oxamyl (Li *et al.*, 1997), accounted for up to 49 % of the total radioactive residue in milk following oral dosing. In the same study report, the presence of thiocyanate was also a prominent metabolite in eggs.

Plants

Plants have an equally effective transport system that uses a collection of vascu-lar conduits, xylem and phloem, to distribute nutrients, water and assimilates. The xylem system is primarily concerned with the transport of water from the roots, while the phloem is concerned with the transport of assimilates from the 'sources', mature leaves and roots, to 'sinks', young growing leaves and roots

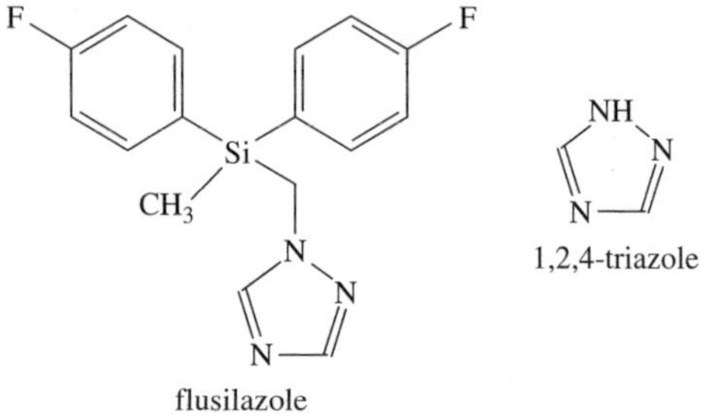

Figure 3.3   Structures of flusilazole and 1,2,4-triazole

and developing fruits and seeds. The flow within these vascular tissues is driven by differences in water potential (the mass flow of water from areas of high water potential such as the root–soil interface and those of low water potential such as the leaf–air interface) and osmotic gradients (the flow of water from areas of low solute concentration to those of high solute concentration). A schematic representation of the transport system in plants is shown in Figure 3.4.

The translocation of xenobiotics within these vascular systems depends on the entry into the system and retention of the compound in the conduits for a sufficient time for them to be transported to other plant tissues.

The direction of xylem movement is the same throughout the plant's life and represents a conduit for flow of water, inorganic ions, amino acids and absorbed xenobiotics from the roots to the leaves. The rate of flow of water in the xylem is between 50 and 100 times that of the flow in the phloem system, and has a pH similar to that in most of the living plant tissue, i.e., pH 5–6. Transport in the xylem depends on the ability of the xenobiotic to be absorbed and on its partitioning behaviour, i.e. the physicochemical properties, log P and pK_a. The optimum log P for xylem transport has been described as 1.8 for non-ionized compounds (Briggs *et al.*, 1982). Xenobiotic compounds transported in the xylem will be accumulated at the leaf tip and leaf margins, the extremities of the water flow.

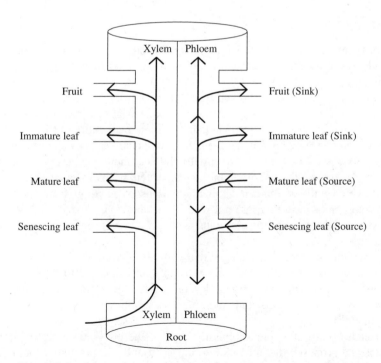

Figure 3.4 Schematic representation of the transport system in plants

The same properties that affect the absorption of xenobiotics into cells also affect the passage and retention in the phloem. This infers that most compounds will be capable of entering the phloem; however, because of the greater flow in the xylem and the commensurate reduction in concentration there will be a net diffusion back into the xylem. The pH of the phloem is ca. 8 and so compounds such as weak acids which change from a non-ionized state at pH 5 will be ionized at pH 8 and are therefore trapped in the phloem system. In general, weakly acid or zwitterionic compounds, e.g. glyphosate, are likely to undergo long-range transport in the phloem while non-ionized lipophilic compounds will have a low phloem mobility. Models that attempt to predict systemic behaviour based on pK_a and log P have been proposed by Bromilow *et al.* (1990).

It is important to recognize that a pesticide may be chemically changed during absorption or distribution by enzymes within the plant cells. Any changes are likely to impact the physicochemical characteristics and hence its distribution throughout the plant, e.g. a non-ionized lipophilic compound may enter a plant cell and be conjugated, thus increasing its polarity and hence its ability to diffuse out of the cell. Weak acids which are inherently phloem-mobile may also be conjugated (a reaction where the xenobiotic is linked to an endogenous compound, e.g. to a sugar or amino acid (Holland, 1996)), which will facilitate their diffusion from the phloem stream.

METABOLISM

The term 'metabolism' generally refers to the chemical transformation of the pesticide resulting from natural processes in the environmental system under investigation.

Biological systems have evolved to survive against a wide range of environmental influences. To survive, the system needs to absorb nourishment and to defend itself against threats from chemical or biological entities. To achieve this, it has developed a wide range of enzyme and chemical defence mechanisms which will assert themselves on any foreign compound entering the system. Therefore, a xenobiotic entering a biological system is likely to suffer a chemical change which will facilitate its utilization or its elimination from the system. Some compounds, however, are not readily metabolized, e.g. sterically hindered carboxylic acids, strong organic acids with $pK_a < 2$, strong bases and highly lipophilic compounds.

The plethora of chemical changes that can be imparted within biological systems is impressive by any chemist's standards and can lead to highly complex biotransformation pathways. Examples of the biotransformation of the fungicide azoxystrobin in plants and rotated crops are shown in Figures 3.5 and 3.6 (Joseph, 1999).

Dorough (1980) attempted a classification of the possible nature of pesticide residues and proposed that a residue resulting from the use of the pesticide could be fundamentally characterized into one of four categories, as follows:

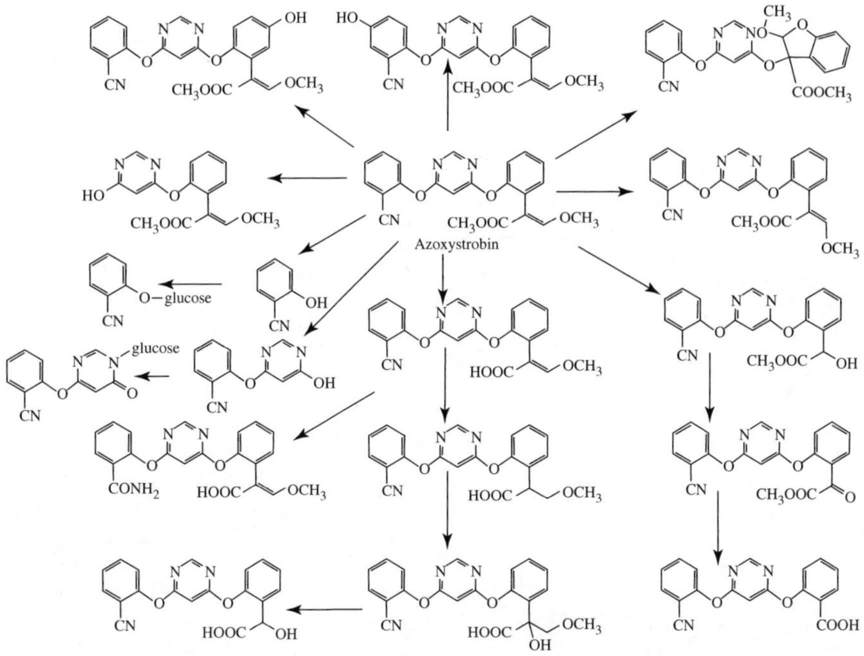

Figure 3.5 Biotransformation pathways of azoxystrobin in plants (Joseph, 1999)

- Phase I – free metabolites resulting from the functionalization of the pesticide
- Phase II – conjugated residues
- Phase III – bound or compartmentalized residues
- Phase IV – naturally incorporated

In terms of dietary risk considerations, it is essential to understand the extent and type of metabolism occurring and the distribution of the residues before the inherent hazard and exposure is assessed. Phase I and II residues are generally extractable and can be readily characterized or identified and their significance assessed, based on their concentration and toxicity or by structure–activity relationships. Residues, which are unambiguously shown to result from the incorporation of the radio-label from the pesticide into naturally occurring compounds, e.g. proteins and sugars, are of no toxicological concern. The significance of bound or unextracted residues is more complex since the nature of the residue is frequently unknown and for many years it was believed that unextractability was synonymous with non-bioavailability, i.e. no exposure – no risk. Recent investigations have, however, shown that the bioavailability of a bound residue upon oral ingestion is dependent on the xenobiotic and the nature of the binding (Sandermann *et al.*, 1990). The critical questions for phase III metabolism are its definition and the rigour of the extraction procedure. Guidance on these

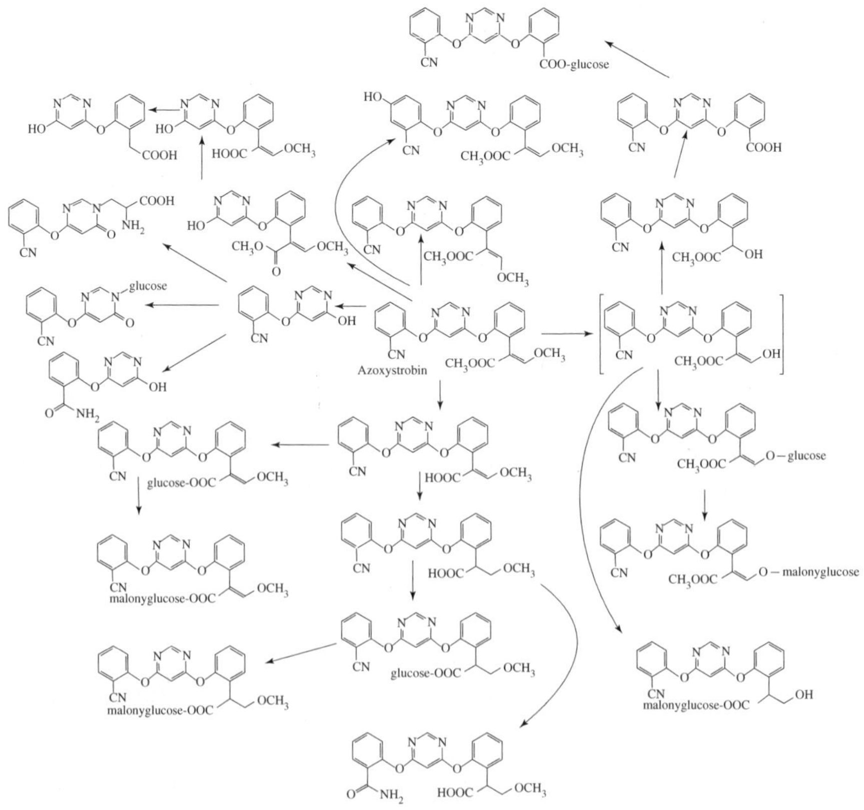

Figure 3.6   Biotransformation pathways for azoxystrobin in rotated crops (Joseph, 1999)

questions has been given by the IUPAC Commission on Agrochemicals and the Environment (Skidmore *et al.*, 1998).

Livestock

Although the metabolism in biological systems is described as similar, some species have a particular diversity which can add complexity to the metabolism. In all animals, macromolecules in food and feed items are digested into simpler compounds, which can be absorbed into the blood stream through the gastrointestinal tract. This digestion occurs through fermentation, by microorganisms in the gut, by hydrolysis and enzymatic reactions. The ruminant is, however, unique in being highly specialized in terms of fermentation, which takes place in the rumen. The latter acts as a batch fermenter and has been defined as an intact ecosystem comprising bacteria, protozoa and anaerobic fungi. In return for the appropriate conditions for survival, the microbial population assist in the digestion of fibrous feeds which would otherwise be underutilized, which is a classical example of a symbiotic relationship (Hungate, 1988). The environment in the rumen is reducing (the opposite of oxidizing). The gases above the rumen comprise 65 % CO_2, 27 % methane, 7 % nitrogen and 0.6 % oxygen, while the fluid has a pH of 6–7 and a negative redox potential (E_h) of −250 mV (Church, 1993). In this environment, microbes have little oxygen and therefore any metabolism is limited; however, reductive processes are abundant, e.g. conversion of CO_2 to CH_4, and unsaturated acids are converted to saturated acids. In the rumen, it is therefore possible that the xenobiotic may be chemically modified prior to being introduced into the gastric stomach.

Although the major organ of metabolism is the liver, the areas of administration of the pesticide also have significant metabolizing ability, i.e. the gastrointestinal tract (GIT) and the skin. In the GIT, the intestinal microflora operate in an oxygen-free environment which can favour reductive processes, e.g. reduction of a nitro moiety to an amine. In contrast to the liver, which forms conjugates to assist elimination, the gut flora hydrolyse conjugates excreted in the bile such as glucuronides and sulfate esters (Goldman, 1982). For dermal applications, reactions include bacterial degradation on the skin surface and enzymatic reactions in the epidermis and dermis (Hotchkiss and Caldwell, 1989). The functions of the enzyme systems in the skin were reported as including all those found in the liver, including conjugating ability.

The qualitative and quantitative nature of the metabolism can be influenced by species, gender and dose rate, e.g. when tebufenozide was fed to rats, 4 % of the 250 mg/kg single dose was metabolized, while about 46 % of the 3 mg/kg dose could be metabolized (FAO, 1997a). In many cases, however, an approximately proportional relationship was found between the dose rate and the level of residues.

Plants

In plant studies, the term 'metabolism' is used in a wider context and includes processes forming products from chemical reactions on the plant surfaces (hydrolysis and photolysis), biological processes which occur outside of the plant (e.g. microbiological degradation in soil) and biotransformations of the pesticide or any of the degradation products. The formation of photo-products can be a significant issue in the definition of the residue of concern; the products can be unique and unlikely to be animal metabolites. They may therefore require specific toxicology testing to assess their relevance and analysis of the crop commodities to assess their significance, e.g. 8,9-Z-avermectin B_{1a}, a photo-product of abamectin B_{1a}.

The metabolizing capability of plants has been compared to that in animals on many occasions and has been shown to be similar, with some differences around the catabolism of specific conjugates or the nature of the conjugating endogenous material. Sandermann (1994) went so far as to describe the xenobiotic metabolism in plants as resembling that found in the animal liver based on metabolic patterns and enzyme classes and proposed the 'green liver' concept. The capability of a plant to metabolize a xenobiotic can be influenced by its growth stage at the time of application. Hatton *et al.* (1996) showed that the levels of glutathione enzymes were higher in young leaves, which would facilitate the levels of metabolism.

ELIMINATION

Livestock

The excretory pattern for a xenobiotic in livestock is a function of the absorption, distribution and metabolism. Poorly absorbed compounds will be excreted in the faeces while absorbed compounds are excreted either in the urine or the bile. It is generally accepted that urinary excretion occurs for compounds having molecular weights less than 400–500, depending on species, and biliary excretion occurs for compounds having higher molecular weights. Biliary excreted products enter the GIT and are generally excreted in the faeces; in some cases, however, biliary metabolites are further modified in the gut, e.g. hydrolysed, and are re-absorbed. To fully define the extent of absorption, the amount eliminated in the bile must therefore be measured.

In the case of poultry, the contents of the digestive tract and the urinary tract empty into the cloaca, a chamber which has a single external opening, the vent; there is therefore no direct separation of urine and faeces.

Examples of the excretion patterns of some pesticides from livestock are shown in Table 3.1.

In the case of ruminants, minor routes of elimination also include elimination through bodily secretions, e.g. milk and expired air. The latter can sometimes be significant, e.g. lactating goats eliminated 39–48 % of the total thiram (Figure 3.7) dose by the expired air (FAO, 1998i).

Plants

Although plants do not have a classical excretory system, they do effectively eliminate xenobiotic compounds by removing them from active cell processes by storage in the cell vacuole, which is a large fluid filled cavity within the cell bounded by a membrane called the tonoplast, with the fluid containing sugars, salts, pigments and waste products dissolved in water (Cole, 1994), or by reaction with the structural compartments of the plant, e.g. lignin and cellulose, or by exudation from the roots into the soil (Walker *et al.*, 1994), or through volatilization.

Pesticide residues eliminated from cell processes by reaction with lignin and cellulose represent bound residues and require specific assessment of their relevance during risk assessment.

Table 3.1 Excretion patterns of some pesticides in livestock

Compound (molecular weight)	Animal and dosage rate[a]	Excretion pattern[b]				Reference
		Faeces			Urine	
		T	P	M	T	
Glufosinate-ammonium (223.2)	Lactating goats: 3 mg/kg bw/day for 4 days	$69 + 11^c$	–	–	2.9	FAO, 1999a
	Lactating goats: 3 mg N-acetyl-L-glufosinate/kg bw/day for 3 days	$68 + 18$	34^d	52^e	7.3	
Teflubenzuron (381.1)	Lactating goats: 7 mg/kg bw/day	99^f	76.9	9.5^g	–	FAO, 1997b
Kresoxim-methyl (313.4)	Lactating goats: equivalent to 7.1 ppm in feed for 5 days	18.1	nc^h	nc^h	69.5	FAO, 1999b
	Lactating goats: equivalent to 450 ppm in feed	24.5	0.67	99.3^i	59.3	

[a] bw, body weight.
[b] T, total residue; P, parent compound; M, conjugated or free metabolites.
[c] In two feeding studies, 11–18 % remained in the gastrointestinal tract of the animal 15–16 h after the last dose.
[d] Excreted as glufosinate, the administered metabolite was converted back to the parent compound.
[e] The administered material, N-acetyl-L-glufosinate, is the metabolite of the pesticide.
[f] Including intestinal contents *post mortem*.
[g] One metabolite, the meta-hydroxybenzoyl derivative of teflubenzuron, was present in 3.6 %, an unknown metabolite was found in 5.9 %, while the reminder consisted of unidentified minor components.
[h] nc, not characterized.
[i] Seven polar metabolites, including glycoside conjugate, were identified. The extractable unidentified metabolites amounted to 9.6 %, while the unextractable residues were present in 3.1 %.

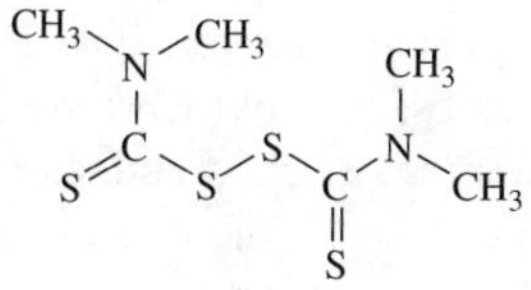

thiram

Figure 3.7 Structure of thiram

EXAMPLES OF PESTICIDE METABOLIC REACTIONS IN PLANTS AND LIVESTOCK

In this section, the focus will be on phases I–IV metabolic reactions, occurring in both livestock and plants. Although the reactions are broadly similar, some specific differences exist and will be highlighted in the text. Bounds and Hutson (2000) have described a generalized comparison of the features of the metabolic reactions in mammals and plants, as follows:

- both have a similar range of oxidation reactions
- both hydrolyse esters
- plants make infrequent use of sulfate conjugation
- plants use glucose for conjugation rather than glucuronic acid which is mainly used by livestock
- both use glutathione conjugation but catabolize the product differently

It is not our intention to provide an exhaustive set of references to all reactions since several very comprehensive treatises are now available (e.g. Roberts and Hutson, 1999; Roberts, 1998). The aim of this section is to demonstrate that the metabolism of a xenobiotic can be extremely complex, resulting from a wide range of reactions some of which are likely to be unpredictable. Indeed, a 'holy grail' for the metabolism chemist would be a reliable predictive model for pesticide metabolism; however, this has so far been an elusive goal.

PHASE I METABOLISM

Phase I metabolism mostly involves oxidation, reduction and hydrolytic reactions, introducing functional groups into the xenobiotic compound, and hence generating functionalities such as hydroxy, carboxylic acid and amine groups. These reactions can be sequential, leading to a complex mixture of phase I metabolites which may in turn act as substrates for phase II metabolic reactions.

Oxidation

Oxidation reactions are arguably the most important metabolic reactions and are found in both plants and animals. Although the enzymology of oxidative reactions is beyond the scope of this present chapter, it is evident that oxidation can be mediated by a range of enzymes, e.g. cytochrome P450s and peroxidases. The former enzyme has been widely researched and shown to consist of a wide range of different isoforms (Gonzalez, 1992), the relative proportions and types present being dependent on the species, e.g. chlortoluron undergoes benzylic oxidation in maize and *N*-demethylation in wheat. Oxidation is also an important mechanism in the de-activation and metabolic selectivity of the sulfonylurea herbicides. A range of the possible oxidative reactions is shown in Table 3.2.

Hydrolysis

Hydrolysis reactions are limited by the chemistry and so appear to be less common than for oxidative processes. They nevertheless have significant importance in the detoxification of many pesticides and can be both chemical- or

Table 3.2 Typical oxidative reactions occurring in plants and livestock

Reaction	Example	Reference
Aliphatic hydroxylation	Hexaconazole	Skidmore et al. (1990)
Alicyclic hydroxylation	Carbofuran	Schlagbauer and Schlagbauer (1972)
Aromatic hydroxylation	Fonophos metabolites	Subba-Rao et al. (1997)

(continued overleaf)

Table 3.2 (*continued*)

Reaction	Example	Reference
Benzylic oxidation	Chlortoluron	Gross et al. (1979)
Epoxidation	Carbendazim	Roberts and Hutson (1999)
Oxidative cleavage	Cycloxydim	Huber et al. (1988)
Alcohol oxidation	Prochloraz (2,4,6-trichlorophenoxyethanol metabolite) $OCH_2CH_2OH \xrightarrow{[O]} OCH_2COOH$	Laignelet et al. (1992)
Aldehyde oxidation	Ethylene dichloride (chloroacetaldehyde metabolite) $ClCH_2-CHO \xrightarrow{[O]} ClCH_2-COOH$	McCall et al. (1983)

O-, *N*-dealkylation

Metoxuron

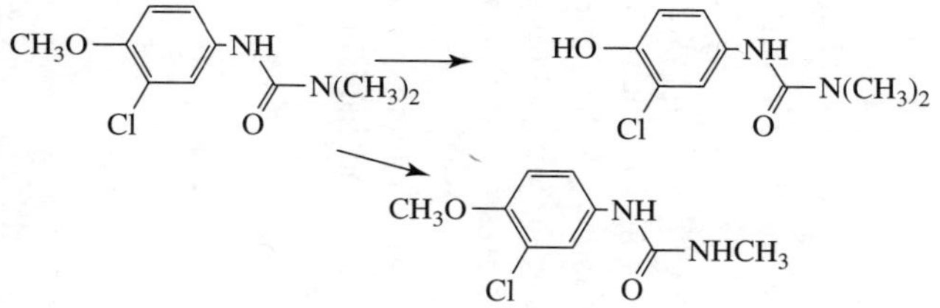

Owen (1987)

N-, *S*-oxidation

Sethoxydim

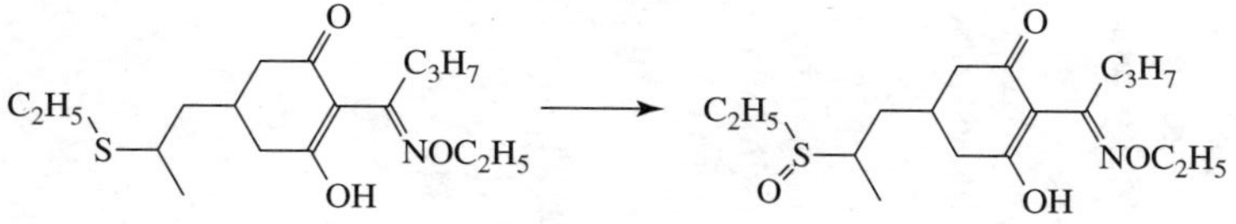

Ishihara *et al.* (1988)

Disulfoton – includes oxidative desulfurization

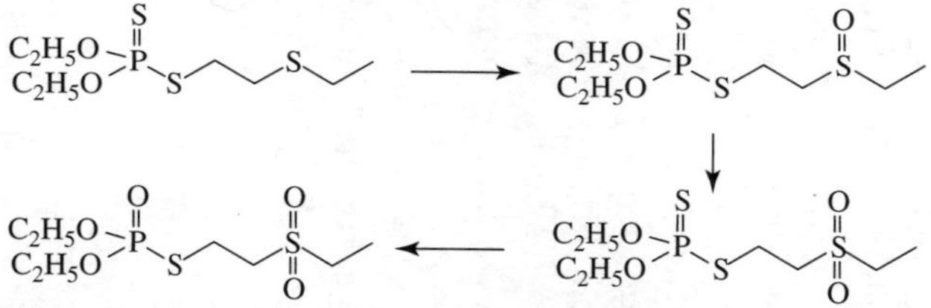

Metcalf *et al.* (1957)

enzyme-mediated. Ester hydrolysis is particularly important in the case of the arylphenoxypropionic acid herbicides, e.g. fluazifop and fenoxaprop, where the herbicide is formulated and used as the alkyl ester to facilitate foliar absorption. Once in the plant, the ester is rapidly hydrolysed to the free acid, which is the active moiety. Ester hydrolysis is also a key metabolic step in the detoxification of the pyrethroid insecticides.

Esterase activity is particularly important in the toxicity of the phosphorothionate insecticides. In an *in vitro* study investigating the relative toxicity of diazinon to mammalian and avian species (Machin *et al.*, 1975), it was demonstrated that the oxon metabolites (active moieties) were generated in the liver at essentially an equivalent level. However, studies investigating the stability of the oxon in blood showed that mammalian blood (cow, sheep, pig and rat) readily hydrolysed the ester, whereas avian blood had virtually no hydrolytic activity. This finding is consistent with the observed toxicity which shows that avian species are highly susceptible to diazinon toxicity. It was therefore concluded that the extrahepatic metabolism of diazinon is more toxicologically important than liver metabolism. These data also demonstrate that esterases in the blood are important metabolically and are species-dependent.

Hydrolysis reactions resulting in the opening of heterocyclic ring systems have also been reported, e.g. cleavage of the oxazolidone ring in vinclozolin, and the heterocyclic ring in some of the sulfonylurea herbicides. In the poultry metabolism of the fungicide vinclozolin (Dean *et al.*, 1988), the major component of the residue in fat, liver and muscle resulted from the hydrolytic cleavage of the oxazolidone ring (Figure 3.8).

Other dicarboximide fungicides, e.g. procymidone and chlozolinate, also readily undergo opening of the heterocyclic ring by hydrolysis, a mechanism for which is described by Villedieu *et al.* (1994).

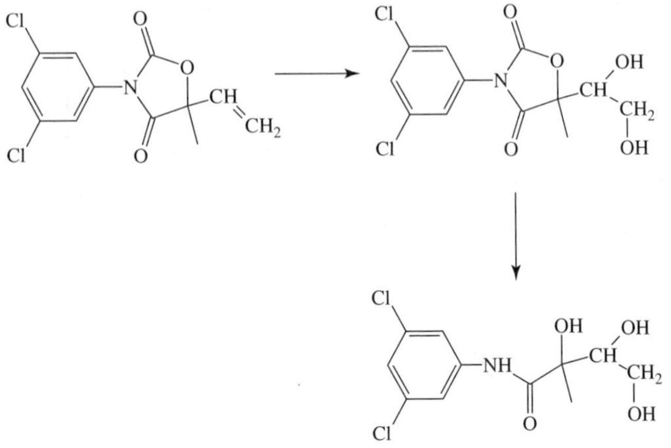

Figure 3.8   Biotransformation of vinclozolin in poultry

In the case of the sulfonylurea herbicides containing triazine or pyrimidine as the heterocyclic ring, mechanisms for the hydrolytic ring cleavage have been proposed. In the cases of chlorsulfuron (Reiser *et al.*, 1991) and prosulfuron (Bray *et al.*, 1997), the product of the opening of the triazine ring is a triuret (Figure 3.9(a)), while the pyrimidine ring of bensulfuron-methyl and halosulfuron-methyl hydrolyses to give a guanidine (Dubelman *et al.*, 1997) (Figure 3.9(b)).

The classic example of amidase activity is propanil, a rice herbicide, which is rapidly metabolized in rice (Still, 1967) and animals to dichloroaniline.

REDUCTIVE PROCESSES

Reduction reactions have been reported in both livestock and plants. It is probable that these reactions are more common, but the reaction products may be subsequently re-oxidized within other compartments of the biological system.

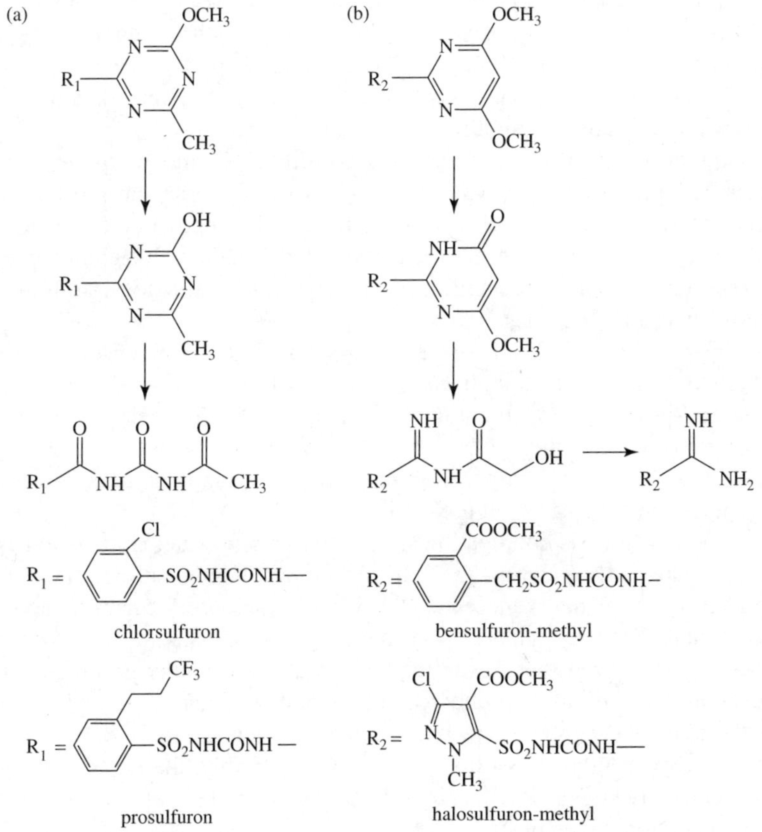

Figure 3.9 Ring-opening reactions observed for sulfonylureas containing (a) triazine, and (b) pyrimidine heterocycles

Typical reactions include the reduction of nitro groups, aldehydes, ketones and alkenes. In animals, the reactions are characteristically found in the liver and GIT and would be expected to occur in ruminants. Examples of some reduction reactions representing the above groups are included in Table 3.3.

PHASE II METABOLISM

Phase II or conjugation reactions represent chemical synthesis where the xenobiotic (exocon) is chemically bonded to an endogenous substrate (endocon).

Glutathione

Glutathione (GSH) is a tripeptide found in both plants and animals and is important as a source of endogenous thiols, which act as scavengers of free radicals and active electrophiles. The reaction results from nucleophilic attack of the thiolate anion on an electrophilic centre and is catalysed by the enzyme, glutathione-*S*-transferase. In some plant species, e.g. soybean and other leguminous species, homoglutathione (HGSH) (Figure 3.10) is found in preference to glutathione. It should also be noted that non-enzymatic reactions of an electrophile with glutathione can occur at a significant rate.

Glutathione chemistry has attracted a considerable amount of research and multiple isoforms of the enzyme glutathione-*S*-transferase have been isolated and characterized. These isoforms provide for a wide range of electrophiles to act as substrates for the enzyme, e.g. reactions of pesticides with glutathione have been reported for chloroacetanilides, diphenyl ethers, triazines, sulfonylureas, thiocarbamates and triazines.

The initial GSH conjugate is subsequently catabolized to the cysteine conjugate which, in turn, is further catabolized to a complex mixture of products. The nature of these end-products is different in plants and animals. A special feature of GSH conjugation is that the conjugates formed are subject to catabolism, thus resulting in a complex mixture of components. The different catabolic reactions are shown as a generalized scheme in Figure 3.11.

The products of GSH conjugation are generally more water soluble and as such will not readily diffuse across the cell membranes. Kreuz and Martinoia (1999) concluded that in plants xenobiotic–GSH conjugates are formed in the cytosol and are transported by an energy-dependent process into the cell vacuole. It has been hypothesized that glutathione conjugates are inhibitors of glutathione-*S*-transferase and of glutathione reductases and as such need to be removed from the cytosol (Schroeder, 1998). Within the vacuole, the GSH conjugate is catabolized by peptidases (Wolf *et al.*, 1996) to, possibly, the cysteine conjugate. It is not currently known where subsequent catabolism, i.e. oxidative deamination and malonylation, takes place.

In animals, the formation of mercapturates facilitates the elimination of the xenobiotic in the urine. The transport of the conjugates between cells and to

Table 3.3 Typical reduction reactions found in plants and livestock

Species/type	Pesticide or metabolite	Reaction	Reference
Nitro moiety	Dicloran		FAO (1999c)
	Quintozene		FAO (1995)
Aldehyde	Deltamethrin metabolite		Ruzo and Casida (1979)
Ketone	Triadimefon		Clark *et al.* (1978)

(continued overleaf)

Table 3.3 (*continued*)

Species/Type	Pesticide or metabolite	Reaction	Reference
Reductive dehalogenation	Fluoroimide metabolite		Ohori and Aizawa (1983)
	Bromuconazole		PSD (1996)
Alkene	Fluoroimide metabolite		Ohori and Aizawa (1983)

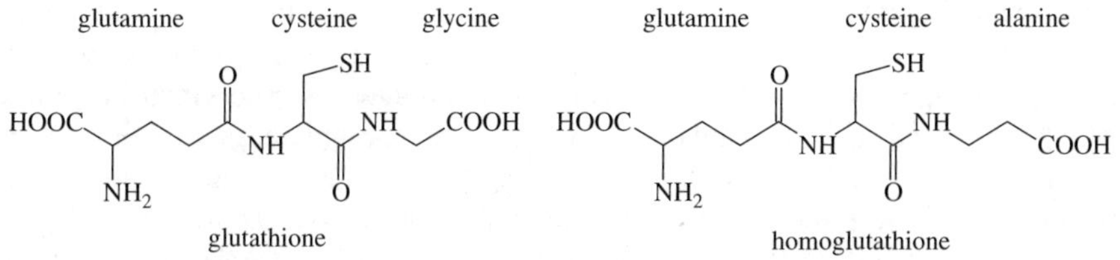

Figure 3.10 Structures of glutathione and homoglutathione

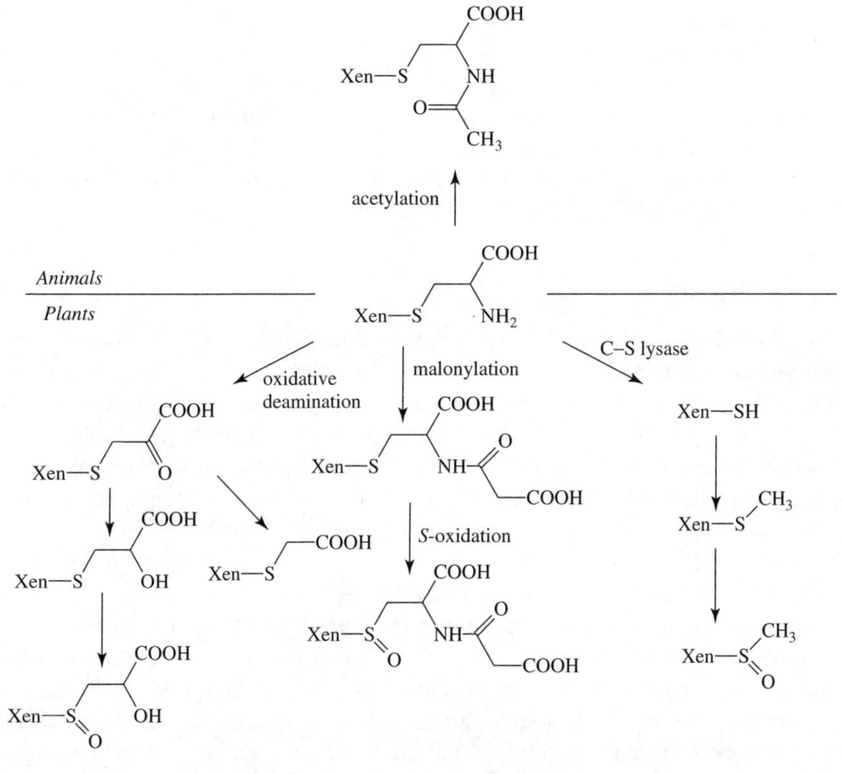

Figure 3.11 Catabolism of glutathione conjugates

various organs is complex and has been described (Stevens and Jones, 1989) as involving the formation of the glutathione conjugate inside the cell which is then released for conversion to the S-cysteine conjugate. This must then be re-absorbed into the cell to be N-acetylated, and finally the mercapturate is released from the cell and excreted. These authors cited research demonstrating that the kidney and liver were both involved in these transformations prior to excretion and hence the presence of catabolites in these tissues.

A typical biotransformation pathway of a xenobiotic resulting from glutathione conjugation is shown in Figure 3.11. This figure is a generalization and starts from the xenobiotic cysteine conjugate. This conjugate is a common catabolite which is formed in both plants and animals from the glutathione conjugate via specific peptidases. The complexity of the pathway is made even more complex by parallel phase I metabolism. Although the figure only shows the acetylation of the cysteine to the mercapturic acid in animals, several other reactions can also occur (Stevens and Jones, 1989), i.e. (a) deconjugation in which the GSH is removed intact, (b) formation of di-GSH conjugates, (c) sulfur oxidations and reductions, (d) deacetylations of mercapturates, and (e) deamination and elimination.

Glutathione conjugates of the metabolites of various pesticides were identified in different organs or excreta of livestock species, e.g. benomyl/carbendazim (S-[4,5-dihydro-5-hydroxy-2-(methoxycarbonylamino)-1H-benzimidazol-5-yl]glutathione) in cow kidney and poultry tissues (FAO, 1999e), and dicloran (2,6-dichloro-4-nitro-3-glutathionylaniline and 4-amino-3-chloro-5-glutathionyl-acetanilide) in goat liver and milk (FAO, 1999f).

Sugar Conjugates

Conjugation of xenobiotic chemicals with endogenous sugar units is common in both plants and animals and is most frequently observed with alcohols, amines, mercapto moieties and carboxylic acids. In plants, sugar conjugates are usually in the form of glucosides and in animals as glucuronides (Figure 3.12), although there are exceptions, e.g. glucuronic acid conjugates of 4-nonylphenol found from wheat suspension cultures and the use of glucose conjugates in mammals to conjugate endogenous steroids and bilirubin (Bounds and Hutson, 2000). Examples for O-, S- and N-glycosides (FAO, 1997n, 1998c and 1999g) are given in Figure 3.13 and some glucuronides in Figure 3.14.

In plants, sugar conjugates may undergo further conjugation with extra sugar units, or with malonic acid, e.g. the malonyl glucoside of a metabolite of fenbuconazole was identified in wheat, peanut and peach (FAO, 1998c). In the case of animals, xenobiotic chemicals are usually conjugated with glucuronic acid, which can be further conjugated by sulfation. The existence of sugar conjugates is inferred on the basis of a reaction of the metabolite with enzymes or after chemical hydrolysis, e.g. it has been a common procedure to cleave plant conjugates with β-glucosidase and to identify the nature of the aglycone. This can, however, lead to a failure in the appreciation of the extent of the sugar conjugation, as in the case of polyglucosides, or incorrect characterization in the case of malonylglucosides which do not react with β-glucuronidase but are readily cleaved by cellulase. In recent years, the advent of liquid chromatography linked to mass spectrometry has greatly facilitated the identification of sugar conjugates and it is likely that a greater number of these conjugates will be reported in future research papers.

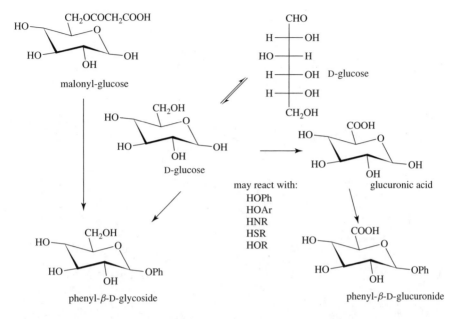

Figure 3.12   Structures of glucose and glucuronic acid

The most extensive reports of sugar conjugates in plants are from the pyrethroid insecticides, e.g. cypermethrin, permethrin and fenvalerate (Figure 3.15).

The primary metabolism of these compounds is ester hydrolysis and both the acid and the alcohol can be conjugated with endogenous sugars. The phenoxy-benzyl alcohol is also readily oxidized to 3-phenoxybenzoic acid (3-PBA). The sugar conjugation of 3-PBA has been extensively investigated in a range of plant species (Mikami *et al.*, 1984). In this investigation, abscised leaves of cabbage, cotton, cucumber, kidney bean and tomato were placed in a radiolabelled solution of 3-PBA. Analysis of the leaves showed that 3-PBA had been extensively metabolized to a range of sugar conjugates which included esters of glucose, glucosylxylose, cellobiose, gentiobiose, malonylglucose and triglucose.

In a similar investigation, the formation of sugar conjugates of 2-(4-chlorophenyl)-3-methylbutyric acid, resulting from the hydrolysis of fenvalerate, was investigated using abscised leaves from a range of plant species (Mikami *et al.*, 1985a). In this study, the authors demonstrated some plant specificity for the type of sugar conjugate; in cabbage, bean and cucumber, the malonylglucoside predominated while the triglucose ester was found only in the tomato. This finding was further supported by an investigation into the metabolism of fenpropathrin where the 2,2,3,3-tetramethylcyclopropanecarboxylic acid was conjugated predominantly to the gentiobioside in tomato, and to the malonylglucoside in cabbage and bean (Mikamia *et al.*, 1985b).

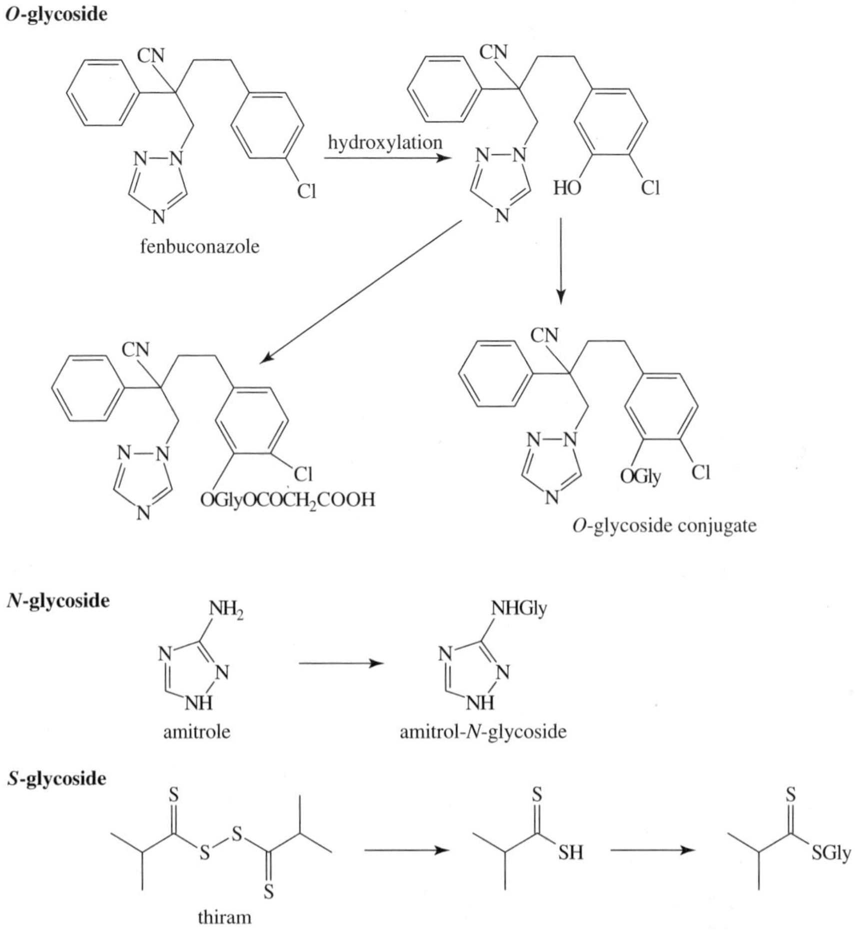

O-glycoside

N-glycoside

S-glycoside

Figure 3.13   Examples of *O*-, *N*- and *S*-glycosides from pesticides

In animals, glucuronic acid is an important conjugation partner for foreign and endogenous substances in preparation for excretion, especially phenols and acids which are frequently excreted in urine as glucuronides. An additional complexity of glucuronide conjugation was described by Sidelmann *et al.* (1995) who showed that glucuronide conjugates of 4-fluorobenzoic acid could exist as isomers through internal acyl migration and rotation. The glucuronides rearrange to the 2-, 3- and 4-acyl derivatives, each of which exists in α- and β-forms.

Glucuronide conjugates of metabolites are very common and have been reported in goat bile (teflubenzuron – FAO, 1997c), goat and hen liver (6-hydroxybentazone and 8-hydroxy bentazone – FAO, 1996a), cow kidney and liver (flumethrin – FAO,

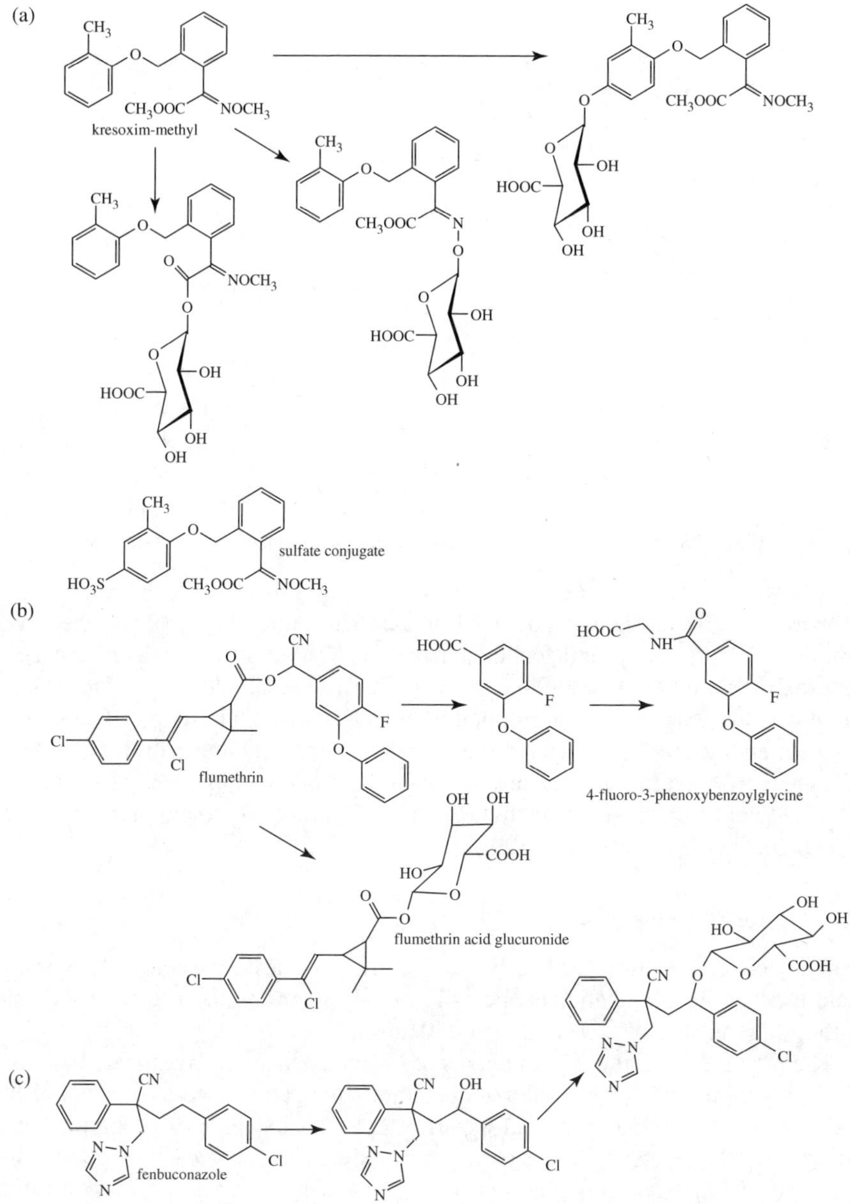

Figure 3.14 Glucuronide conjugates of (a) kresoxim-methyl, (b) flumethrin, and (c) fenbuconazole

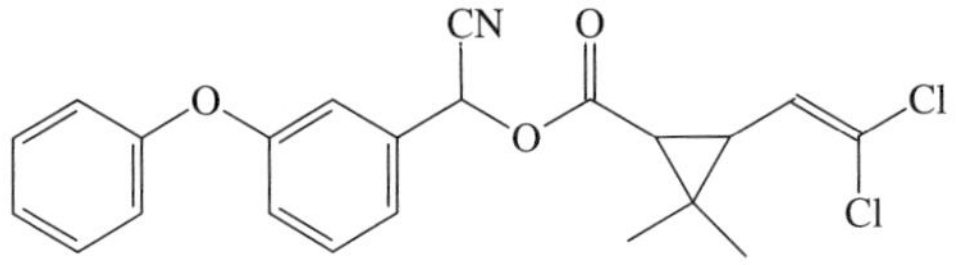

Figure 3.15   Structures of cypermethrin, permethrin and fenvalerate

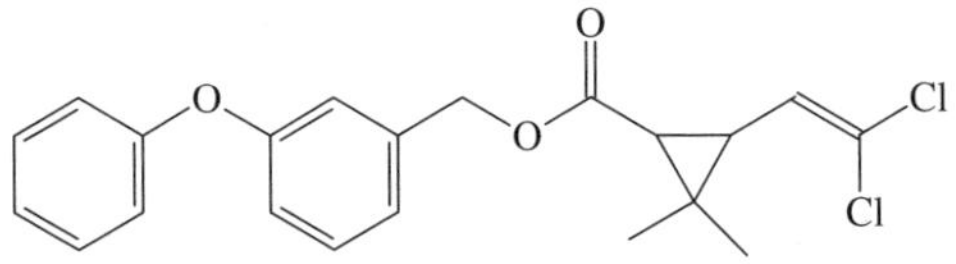

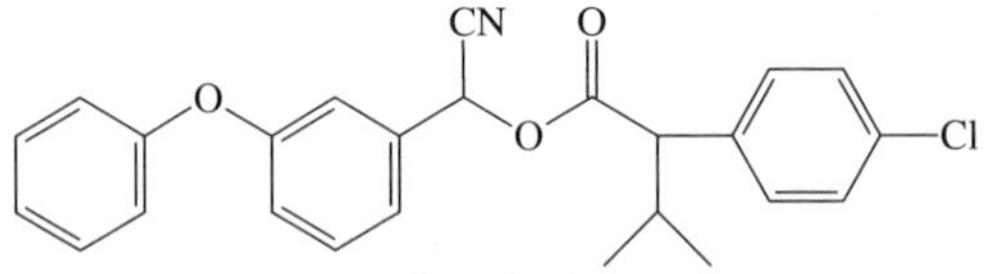

1997d) cow, goat and hen kidney (benomyl/carbendazim – FAO, 1999h), hen liver
(thiram – FAO (1997e) and fenbuconazole – FAO (1998d)), hen liver and eggs
(kresoxim-methyl – FAO, 1999i), to name but a few. In addition to the usually
conjugated primary hydroxyl group, the secondary hydroxyl group of fenbuconazole metabolite (Figure 3.14) and the carboxyl group of flumethrin (FAO, 1997f)
(Figure 3.14) also formed glucuronides. N-glucuronide conjugates of bentazone
(FAO, 1996a) and kresoxim-methyl (Figure 3.14) were also found in the liver and
excreta of laying hens (FAO, 1999j).

Amino Acid Conjugation

Amino acid conjugation has been observed in all species; the nature of the amino
acid used in the conjugations appears to be dependent on the nature of the acid
rather than on the species (Bounds and Hutson, 2000).

The nature of amino acid conjugates has been extensively investigated for 2,4-
D. Feung *et al.* (1973) reported that the major conjugates found in soybean callus
cultures were glutamic acid and aspartic acid; the longer the incubation, then
the greater the proportion of aspartic acid conjugate. These data suggested that
interconversion of the conjugates was occurring. This observation was confirmed
by observing the fate of the glutamic acid conjugate after its introduction into
the cultures. Other amino acid conjugates found in the cultures were alanine,
leucine, phenylalanine, tryptophan and valine. It is particularly interesting to
note that these metabolites retained the biological activity of 2,4-D.

Amino acid conjugates have also been observed in the metabolism of MCPA,
triclopyr, and the pyrethroid metabolites, 3-PBA and DCVA, of MCPA with

aspartic acid in peas, rape and red campion, of triclopyr with aspartate and glutamate in soya cell suspension, and of the pyrethroid metabolites 3-PBA with glycine, glutamic, acetylornithine and DCVA with glycine, taurine (DCVA is 3-(2,2-dichlorovinyl)-2,2-dimethylcyclopropanecarboxylic acid).

In the case of thiram, the initial fission of the S–S bond generates the dimethyldithiocarbamic acid, which is subsequently conjugated with alanine (Figure 3.16). It has been reported that the reaction with alanine is reversible and results in an interconversion with a glucose conjugate (Roberts and Hutson, 1999).

One of the best known examples of amino acid conjugation is the reaction of 1,2,4-triazole with serine to give triazolylalanine. The metabolite has been reported as a significant plant metabolite of all of the triazole fungicides and of the herbicide amitrole. The reaction mechanism is thought to result from the conjugation of triazole with glycerol–phosphate which is replaced in the presence of the enzyme with serine. In the reaction, triazole is thought to be acting as a precursor for the enzyme tryptophan synthetase (Smith and Chang, 1973). In plants, triazolylalanine is usually found in the seeds and fruit. Subsequent catabolism of the triazolylalanine through oxidative de-amination and oxidation results in the formation of 1,2,4-triazolylacetic acid (Figure 3.17).

It has been generally accepted that triazolylalanine is a unique plant metabolite and can also be detected in rotational crops (e.g. fenbuconazole (FAO, 1998a)) but recent reports have shown that it is also found in the goat and hen following dosing with fenbuconazole. Following feeding at 100 mg/kg in the diet, the

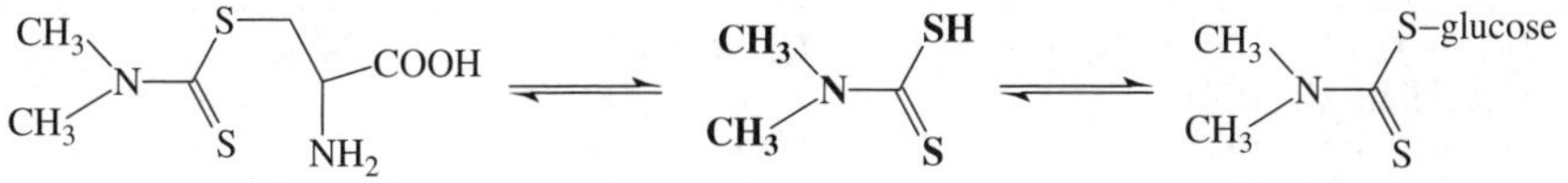

Figure 3.16   The reaction of the dimethyldithiocarbamic acid metabolite of thiram with alanine and glucose (Roberts and Hutson, 1999)

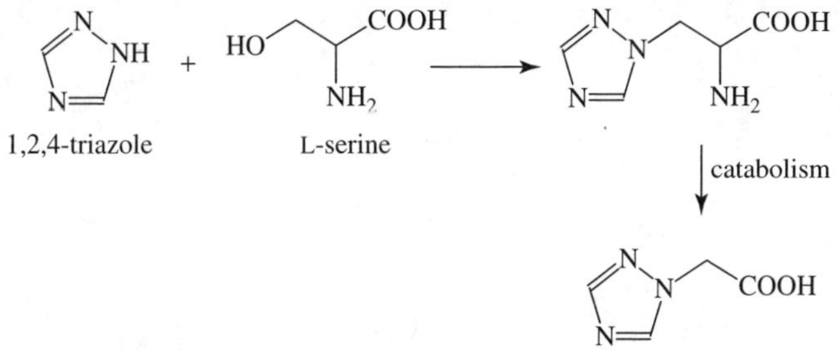

Figure 3.17   Production of triazolylalanine and triazolylacetic acid in plants

metabolite was identified in liver, milk, kidney and muscle of the goat and in hen muscle (FAO, 1998e; PSD, 1995). Triazolylalanine was not found in the rat following dosing with fenbuconazole.

Lipophilic Conjugation

Some conjugation reactions increase the lipophilicity of the xenobiotic with the implication that it will be retained within the system and not be as readily eliminated. Some of the best known examples of these conjugates are found in the metabolism of the pyrethroid insecticides (Table 3.4).

The metabolism of the miticide cycloprate in the cow resulted in 52–76% of the radio-label becoming associated with triacylglycerols in the milk (Quistad et al., 1978). These were identified as comprising a mixture of cyclopropyl fatty acids resulting from the metabolism of cycloprate to cyclopropanecarboxylic acid (CPCA) and subsequent chain elongation of the carboxylic acid group in a similar manner to that found in fatty acid biosynthesis. It was particularly interesting that a CPCA conjugate of carnitine was present in milk at a level of 39% of the total radioactive residue in milk 12 h after dosing (Figure 3.18). The significance of this finding is related to the entry of fatty acids into the mitochondria where fatty acid synthesis and oxidation take place. In this instance, the carnitine acts as a carrier facilitating the transfer of CPCA across the membrane.

Table 3.4 Some examples of lipid conjugates

Pesticide (and metabolite)	Lipid conjugate	Reference
Fenvalerate (2-(4-chlorophenyl)-3-methylbutyric acid)	Cholesterol	Kaneko et al. (1986)
Fluvalinate (anilino acid)	Cholic acids	–
Cypermethrin (3-PBA)	Fatty acids, glycerol	–
Tefluthrin	Fatty acid	–
Haloxyfop	Triglycerol	FAO (1997g)
Tebufenozide	Triglycerol	FAO (1997h)

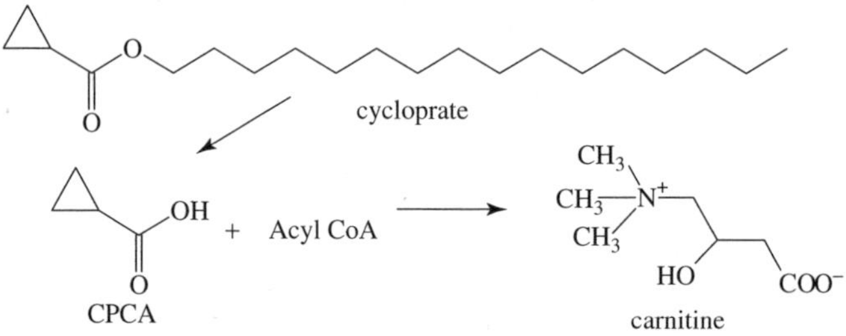

Figure 3.18 Metabolism of cycloprate in ruminants

Based on the above findings it is likely that other xenobiotic carboxylic acids can be incorporated into triglycerides (Caldwell and Marsh, 1983) and can be expected to be present in high-fat commodities such as milk.

Chain elongation was also observed in the alfalfa metabolism of 2,4-D where 2,4-DB and 2,4-dichlorophenoxycaproic acids were formed (Figure 3.19).

An unusual angelic acid ((Z)-2-methyl-2-butenoic acid) conjugate of the 3-hydroxycarbofuran metabolite was found in carrots, and although not lipophilic in the true sense it did serve to increase the lipophilicity of the metabolite. This metabolite appeared to be unique to carrots and represented the major component (61 %) of the residues (Sonobe et al., 1981).

Sulfate Conjugates

Sulfate conjugates are commonly observed in animals, and are formed from substrates similar to those which readily form sugar conjugates. Indeed, in many cases these conjugates are competitive with glucuronidation and it has been shown that sulfate conjugation can be promoted by administration of sulfate precursors, e.g. cysteine (Scheline, 1978). The formation of sulfate conjugates is proposed as being an enzyme-catalysed transfer of sulfate from 3-phosphoadenosine-5-phosphosulfate to the substrate and has been widely reported for phenolic xenobiotics. A recent example of the formation of sulfate conjugation is the livestock metabolism of kresoxim-methyl where the complex metabolic pathway generated a number of phenolic metabolites which were identified as the sulfate conjugates in the excreta and in the eggs and muscle of hens (FAO, 1999i). Similar conjugations are observed with thiabendazole which was rapidly metabolized by both goats and hens to 5-hydroxythiabendazole. This major metabolite was found in the excreta, edible tissues, milk and eggs as the sulfate conjugate (Figure 3.20) (Chukwudebe et al., 1994). Similarly, in the hen metabolism of orally dosed deltamethrin and 3-phenoxybenzoic acid (3-PBA), the 3-PBA, the primary metabolite of deltamethrin, was readily metabolized into 3-hydroxybenzoic acid which is conjugated with a variety of endogenous substances, including sulfates (Akhtar et al., 1994).

An example of sulfate conjugates in plants was found in the cotton metabolism of profenofos which involves cleavage of the phosphorothioate ester to yield

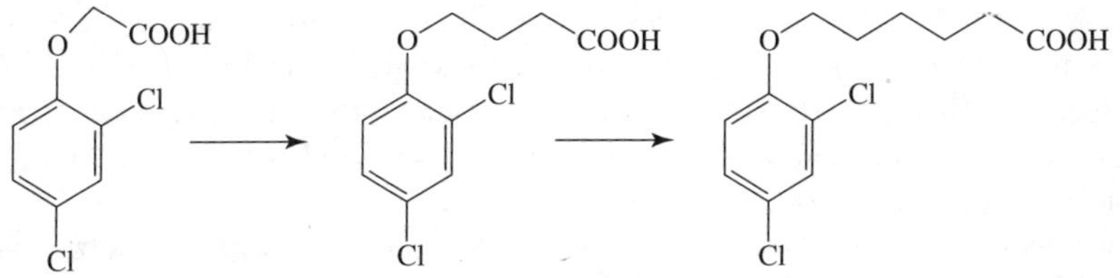

Figure 3.19 Chain elongation of 2,4-D by alfalfa

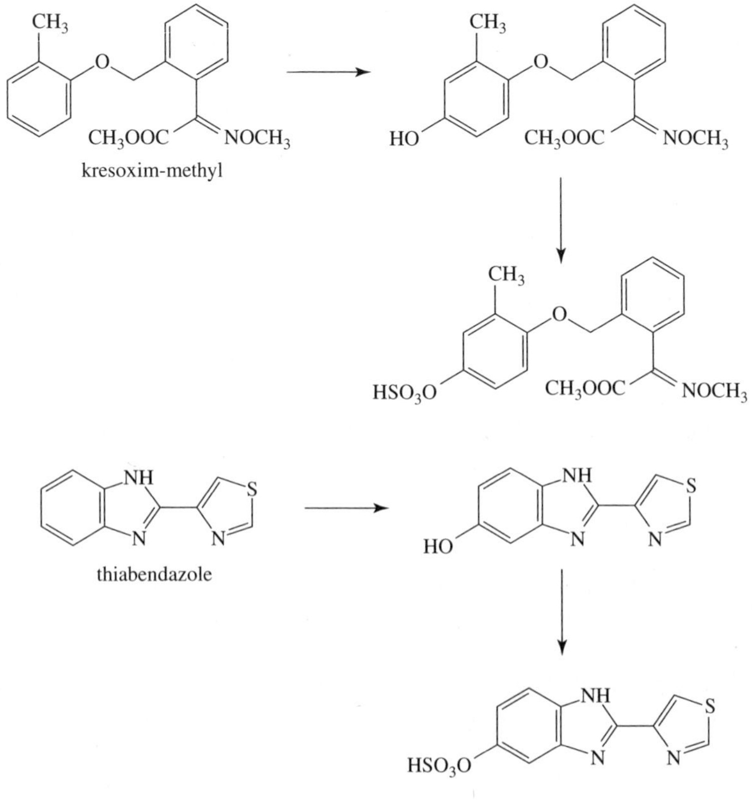

Figure 3.20 Sulfation of the phenolic metabolites of kresoxim-methyl and thiabendazole in animals

4-bromo-2-chlorophenol. This is followed by conjugation with sugars and the formation of the glucosylsulfate conjugate which was found in the stalks and the seeds (Capps *et al.*, 1996).

PHASE III METABOLISM

Phase III metabolism refers to pesticide residues that are associated with endogenous materials; they can be either covalently bound, or in some way physically encapsulated within the macromolecular matrix. The significance of these residues is difficult to assess and requires that they are characterized in terms of bioavailability (Skidmore *et al.*, 1998) and differentiated from the natural incorporation of the radio-atom.

In animals, the most common endogenous materials to which xenobiotics are bound are protein or nucleic acids. In the case of the triazine herbicide ametryn, 61 and 79 % of the liver residue in goat and hen, respectively, were unextracted, but following protease hydrolysis could be solubilized. After a further hydrolysis using conditions typical of those used to degrade proteins, several metabolites containing the intact triazine ring were released. The study concluded that conjugations with protein represented a major pathway in the biotransformation pathway of ametryn.

Monson (1991) also reported that, following dosing with benomyl or carbendazim, the majority of the radioactive residue in the liver remained unextracted. After a Raney nickel desulfurization treatment, the residues were released. The nature of the metabolites recovered supported the conclusion that the xenobiotic was conjugated with glutathione and either this conjugate or a catabolite was incorporated into the protein through the glutathione derived sulfide bond.

In plants, associations of the xenobiotic with natural macromolecules in the cell wall have been demonstrated by Langebartels and Harms (1985) who investigated the binding of pentachlorophenol (PCP). In the investigation, ^{14}C-pentachlorophenol was introduced into cell suspension cultures of wheat, soya and lupins, during late logarithmic growth. The cells were harvested and extracted to fractionate the cell wall components. Radioactivity was found in all fractions but levels were significant in proteins, pectin, lignin and hemicellulose, which suggests that PCP residues were bound by several different mechanisms. Radioactivity associated with hemicellulose was purified by dialysis and molecular sieve chromatography and shown to be congruent with high-molecular-weight material. The radioactivity could only be released using hemicellulase, whereas no radioactivity was released with sodium dodecylsulfate (SDS) or urea solutions. These data suggested a covalent association. The authors showed that PCP was released following hemicellulase digestion and although the exact nature of the association was not defined it was noted in a pulse chase experiment that the metabolism of PCP was by the initial conjugation with glucose followed by malonylation. The levels of the glucose conjugate increased up to 12 h after a 30 min pulse treatment. After subsequent intervals, the levels decreased with a commensurate increase in unextracted material; since no other metabolite changed significantly it was assumed that the glucose conjugate was the substrate for cell wall incorporation.

In separate studies, Schmidt *et al.* (1994) investigated the metabolism of 4-nitrophenol and 3,4-dichloroaniline in carrot cell suspension cultures and demonstrated that the formation of unextractable residues coincided with the initiation of cell aggregation and lignification. These authors explained that from biochemical processes phenolic secondary plant products were known to be stored as glycosides in the vacuole. These conjugates are hydrolysed to liberate the aglycones which are subsequently polymerized to produce insoluble structures. Based on this information, the authors proposed that 4-nitrophenol as the gentiobioside could follow this same route whereby the glucose units are hydrolysed and the phenol incorporated into natural cell wall components such as lignin.

Lignin is the structural unit of plants and forms the 'secondary wall' around the plant cell and is the basic component of the xylem structure. It comprises

coumaryl alcohol   coniferyl alcohol   sinapyl alcohol

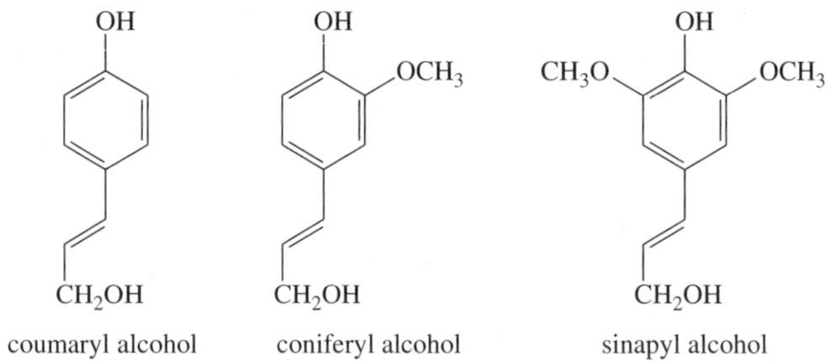

Figure 3.21   Structures of coumaryl, coniferyl and sinapyl alcohols

a polymeric, undefined structure which varies depending on the components available in the cytoplasm and the environment. The three most important starting materials are coniferyl, sinapyl and coumaryl alcohols (Figure 3.21). In the lignin reaction, the alcohols are secreted into the cell wall where they are oxidized by peroxide to form free radicals. These readily interconvert to different isomers which react to form dimers. The latter react further to eventually form a random and complex polymer. It is evident that xenobiotic compounds may become associated with this biosynthesis and thus form bound residues.

Sandermann *et al.* (1983) pointed out that there were some major uncertainties in the reaction of xenobiotic chemicals with lignin since not all associations can be accounted for by covalent bonds and hypothesized the formation of inclusion complexes with the lignin matrix. This was supported by data from several studies where xenobiotics in their free form had been recovered from an isolated lignin fraction. In the case of these xenobiotics, i.e. carboxin and buturon, the intact molecules were released from isolated lignin after dissolution in dimethylformamide (DMF). Since these molecules have no functional group that could be covalently bonded with lignin, it is therefore concluded that these associations are due to inclusion. In addition, these associations can be released by using DMF as a solvent for lignin.

An example of the diversity of the associations of residues with cell wall components is seen in the plant metabolism of carbofuran. Metabolites of the latter were identified in soybean hay and forage (FAO, 1998f). The plants were grown in carbofuran-treated soil (5.5 kg ai/ha equivalent of phenyl-ring labelled and 0.5 kg ai/ha equivalent of ^{13}C geminal-methyl-group-labelled carbofuran).[1] The samples were extracted with methanol/water (4/1 v/v), refluxed with 0.25 N HCl, hydrolysed with cellulase, β-glucosidase, amyloglucosidase, pectinase and protease enzymes, and finally hydrolysed with 6 N HCl and 2 N NaOH. After each hydrolysis, the aqueous products were adjusted to pH 2 and extracted to recover the organosoluble residues. The distribution of the residues is shown in Table 3.5.

---

[1] ai/ha, active ingredient per hectare.

Table 3.5 Fractionation of carbofuran residues in soya

Fraction	Total radioactive residue (%)		
	Forage	Beans	Hay
Total residue (mg/kg)	63 (49)[a]	0.32 (139)[a]	36 (139)[a]
Methanol/water extract	80	59	35
0.25 N HCl treatment	2.3	9.3	15
Cellulase-released organosoluble	0.18	3.6	1.1
β-Glucosidase-released organosoluble	0.16	0.92	0.72
Amyloglucosidase-released organosoluble	0.38	1.8	0.65
Pectinase-released organosoluble	0.22	4.3	0.36
Protease-released organosoluble	1.3	6.9	0.48
6.0 N HCl-released organosoluble	1.3	4.0	0.97
2.0 N NaOH-released organosoluble	6.5	5.9	1.1
Final residual solid	6.9	5.9	43[b]

[a] Entries in parentheses represent days elapsed between soil treatment/sowing and sampling.
[b] Released as lignin.

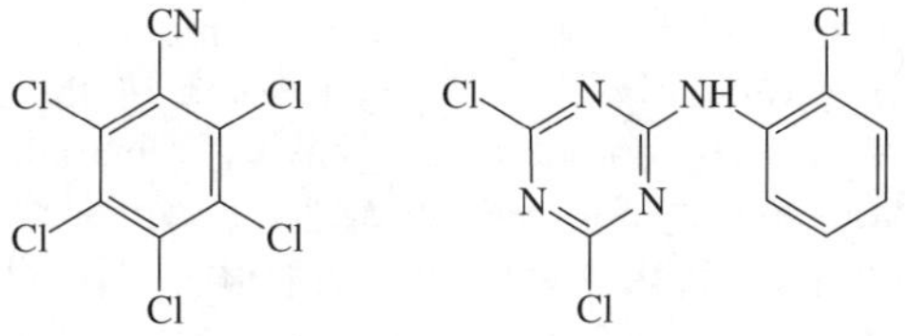

Figure 3.22 Structures of chlorothalonil and anilazine

As stated earlier, the nature of the residues in plants results not only from the uptake and metabolism of the pesticide but also as a result of abiotic changes on the leaf surface. These changes can involve reaction with plant material resulting in the formation of unextractable residues, e.g. photochemical-induced binding to plant cuticles has been reported (Breithaupt et al., 1998). These authors concluded that the plant cuticle is a significant binding site for chlorinated pesticides, e.g. chlorothalonil and anilazine (Figure 3.22). The mechanism for this reaction was reported to be the high affinity of the pesticide for olefinic compounds in the cuticle.

PHASE IV METABOLISM

A phase IV residue is essentially an artefact of the metabolism study and results from the incorporation of the radio-label into natural compounds. This incorporation is obviously dependent on the positioning of the radio-label but usually results from the extensive metabolism of the xenobiotic. The products can be widely distributed throughout the biological system and result in a significant challenge for the metabolism chemist. In the metabolism of chlorethoxyfos in the goat (Ryan, 1993), total degradation of the xenobiotic was evident with the formation, and expiration, of $^{14}CO_2$. Although the majority of the $^{14}CO_2$ will be expired, either an intermediate metabolite or the CO_2 itself becomes involved in the carbon pool and is used in the biosynthesis. In the case of chlorethoxyfos, the major metabolites in the excreta were glycine and oxalic acid while radioactivity in the milk consisted of ^{14}C-lactose and milk proteins. In the tissues, the radio-atom was associated with proteins, mainly as ^{14}C-glycine and serine. Incorporation of metabolites into protein was also reported for pesticides of different chemical structures, e.g. benomyl/carbendazim (FAO, 1999k), dicloran (FAO, 1999l), dimethoate (FAO, 1999m), ferbam (FAO, 1997i); and thiram (FAO, 1997j). Huhtanen (1997) described the incorporation of ^{14}C into lipids, lactose and protein following dosing of hens and goats with ^{14}C-acephate. In this particular example, the position of the radio-label resulted in different patterns of natural incorporation: ^{14}C-carbonyl label provided incorporation into lipids, whereas ^{14}C-SCH$_3$ resulted in greater incorporation into carbohydrates and proteins.

In plant metabolism studies, the radio-label is frequently found incorporated into starch. This is particularly evident when some of the xenobiotic reaches the soil and where the compound is extensively mineralized. The $^{14}CO_2$ is absorbed by the plants and assimilated during photosynthesis. In the metabolism of azoxystrobin (Joseph, 1999) significant proportions of the total radioactive residue (TRR) were incorporated into naturally occurring compounds, in grapes as fruit sugars, in wheat as starch and in peanuts as fatty acids, sugars and amino acids.

TEST SUBSTANCES USED IN METABOLISM STUDIES

In order to detect and identify the many possible metabolites in the biological matrix, the pesticide is routinely radio-labelled. Radioisotopes are relatively cheap to incorporate and allow rapid and automated detection of residues in liquid and solid samples or fractions and recent technological advances provide a highly sensitive means of detection and quantification following chromatographic separation, e.g. liquid scintillation counting and phosphorimaging. Although advances in mass spectrometry (MS) and nuclear magnetic resonance (NMR) spectroscopy have provided a sensitive and selective means for detecting and identifying metabolites, these cannot replace the use of radio-labelling since the techniques are unable to determine the degree of extraction of residues. They are, however, complementary and the application of fluorine NMR spectroscopy in conjunction with the use of a radio-label has been utilized to assist in the identification of biotransformation pathways (Serre et al., 1997; Aubert and Pallett, 2000; Ratcliffe and Roschen, 1998).

The two major issues facing the metabolism chemist in obtaining the optimal test substance are the type of isotope to incorporate and the position of the label. The degree of difficulty of the synthesis and the scale and cost are other issues which need to be factored into the decision but are outside the scope of this review. These issues are described in detail by Harthoorn *et al.* (1985). The radio-label used is obviously governed by the atoms in the structure of the pesticide and by the predicted metabolic reactions. In pesticides, the most common elements are carbon, hydrogen, oxygen, nitrogen, phosphorus, sulfur and chlorine. In regulatory metabolism studies, the lack of a reasonably long-lived radioisotope of nitrogen or oxygen means that the label of choice is usually carbon-14 or hydrogen-3 (FAO, 1999n; FAO, 1997k) but radioisotopes of phosphorus (FAO, 1998b; FAO, 1999d) and sulfur (FAO, 1997l) have also been used. Problems with the use of the latter three isotopes include short half-lives, health hazards, isotope effects and isotopic exchange with hydrogen. The advantages of carbon-14 are (i) relatively low-energy β-emission – less problems with the need for shielding, and (ii) long half-life, i.e. 5730 years – no problems with loss of detection within the time of the study. The half-lives and emission energies of some of the common radio isotopes are shown in Table 3.6.

The positioning of the radio-label is crucial and is strategically placed to provide the maximum amount of data relating to the transformation pathway of the pesticide and is thus positioned in a stable part of the molecule. In many cases, the use of one labelled form of the pesticide is insufficient to fully define its metabolic fate, in which case further labelled forms are synthesized and separate studies conducted.

In the examples shown below for pirimicarb, predictive metabolism would suggest that the initial metabolism would include dealkylation of the exocyclic nitrogen and the loss of the carbamate group. The most stable part of the molecule is therefore the pyrimidine ring. In the case of permethrin, a rapid cleavage of the ester is expected with the formation of the alcohol and acid moieties. The radio-label would then be placed in the cyclopropyl and benzyl positions (Figure 3.23).

To prepare the test or dose material, the radiolabelled pesticide will be isotopically diluted with unlabelled material to provide a suitable specific activity, i.e. labelled-to-unlabelled ratio, for the study. This also serves to decrease the

Table 3.6 Half-lives and β-emission energies for some radioisotopes used in pesticide metabolism studies

Isotope	Half-life	Energy (MeV)
H-3	12	0.02
C-14	5730	0.16
P-32	14	1.71
S-35	88 days	0.17
Cl-36	310 000 years	0.71

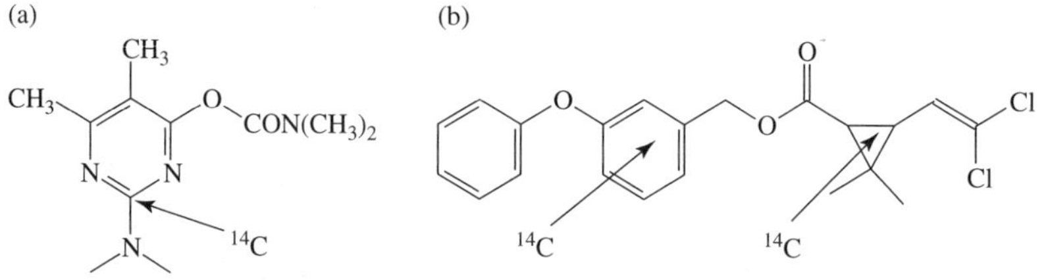

(a)   (b)

Figure 3.23 Radio-labelling positions for (a) pirimicarb, and (b) permethrin

instability of high-specific-activity samples through autoradiolysis where there
is direct or indirect interaction of the radiation with a molecule causing decom-
position. The latter effects are also reduced by dissolving the solids in solution,
storage at very low temperatures or the addition of scavengers to reduce secondary
interactions, e.g. the addition of a small amount of ethanol.

The use of the radio-label requires that studies are carried out in controlled
areas; for plants this can be either in small protected field plots or in pots housed
in suitable growing environments. These controls allow the radiochemical to be
applied in a controlled manner, thus limiting any exposure to the operator and
the environment. This also facilitates 'clean-up' procedures following the 'in-life'
phase of the study.

Many atoms routinely found in pesticides also exist as 'stable isotopes', i.e.
non-radioactive isotopes, which can prove particularly valuable in structure elu-
cidation (Table 3.7).

By admixing stable label and unlabelled material, a distinctive mass spectral ion cluster can be achieved and in a similar way to the positioning of the
radio-atom with the judicious placement of the stable label the ion cluster can
provide extensive metabolite identification. The use of ^{13}C and ^{15}N also provides
a useful means of obtaining nuclear magnetic resonance spectral data. Hendley
(1982) described an experiment to investigate the use of stable isotope labelling
in pesticide metabolism. In this investigation, a primary metabolite of pirimicarb

Table 3.7 Stable isotopes of atoms routinely found in pesticides

Most abundant isotope	'Stable isotope', natural abundance (%)
H-1	H-2 (0.02)[a]
C-12	C-13 (1.11)[a]
N-14	N-15 (0.36)[a]
O-16	O-17 (0.04)[a]
	O-18 (0.20)
S-32	S-34 (4.22)
Cl-35	Cl-37 (24.23)[a]
Br-79	Br-81 (49.31)[a]

[a]Isotope with a nuclear magnetic moment.

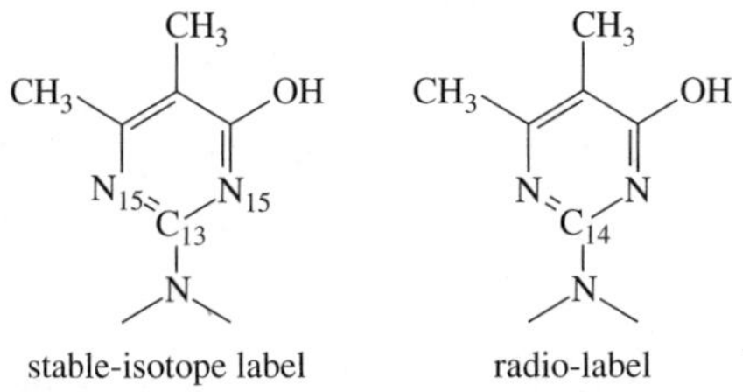

Figure 3.24 Positions of stable-isotope labels and radio-labelling for pirimicarb metabolites (Hendley, 1982)

was labelled with ^{13}C and ^{15}N (Figure 3.24). This stable labelled compound was admixed with unlabelled material in the ratio of 5:2 and approximately 2% of ^{14}C-labelled material was also added as a tracer. This cocktail provided an isotopic cluster 3 mass units apart (2 ^{15}N atoms and 1 ^{13}C atom). The investigation proceeded to use mass spectrometry and nuclear magnetic resonance spectroscopy to identify metabolites. The author concluded that the use of stable isotopes provided a complement to radiolabelling in the identification of metabolites.

In another example, the metabolism of methomyl in goat and chicken was facilitated by the inclusion of ^{13}C methomyl (Reiser et al., 1997).

Further examples for using multiple labelling or a mixture of compounds labelled at different positions include the following: fenbuconazole, phenyl and triazole ring (FAO, 1998g); ferpropinorph, morpholine and phenyl ring (FAO, 1996b); carbofuran, one of the geminal dimethyl group with ^{13}C and in other molecules with ^{14}C phenyl group (FAO, 1997l); folpet, trichloromethyl moiety and benzene ring (FAO, 1999o); kresoxim-methyl, 3-positions with ^{14}C and ^{13}C (FAO, 1999p); tebufenozide, 3-positions (FAO, 1997m) (Figure 3.25).

ACHIEVING REALITY IN METABOLISM STUDIES

If the biotransformation pathway determined in metabolism studies is to reflect the nature of the residues in 'normal-use' conditions, then the test substance must be applied in a manner which simulates actual practice, e.g. in plants the pesticide should be formulated and applied at realistic times and harvested at the appropriate pre-harvest intervals. In plant studies, the formulation of the pesticide fulfils two basic requirements, i.e. to provide a convenient and safe product which will not deteriorate over time in a range of environments, and to maximize the inherent biological activity by improving absorption. This latter point will be an important consideration for metabolism studies. Types of formulation include soluble concentrates, emulsifiable concentrates, water-dispersible powders or granules, suspension concentrates, emulsions and microencapsulated

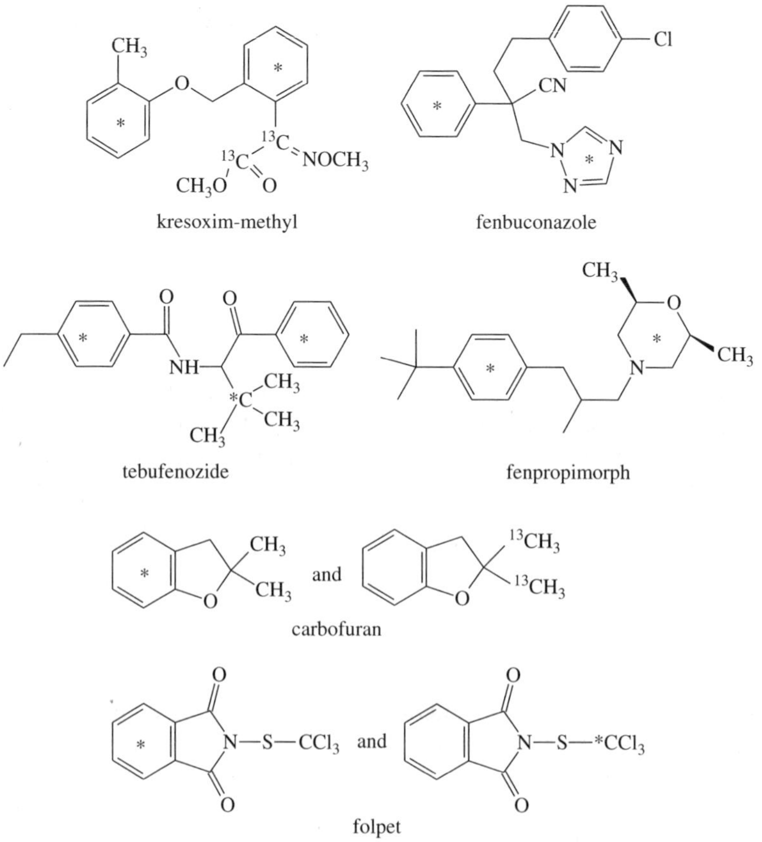

Figure 3.25 Radio-labelling positions used for a range of pesticides

suspensions. The necessity to approximate the formulation for metabolism studies is that the formulation can increase the uptake of the pesticide into the crops and thus the quantitative metabolism. Two of the more common formulations are emulsifiable concentrates (ECs) and suspension concentrates (SCs).

EC formulations essentially consist of the active ingredient, emulsifiers and solvents which are diluted with water to give an emulsion. In an SC formulation, a finely ground suspension of the solid pesticide is formed in water with the addition of dispersing agents. The preparation of an SC is difficult on a small-scale and requires that the active ingredient is milled to the specified particle size. The production of small-scale formulations can prove problematical and as formulation technologies develop, e.g. the use of encapsulated products, the formulation of radiolabelled products on a small scale will need to be considered. The use of the radio-label greatly facilitates the assurance that the formulation is homogenous prior to application.

In the cases of soil-applied chemicals, where the major factors for uptake will be adsorption and degradation in the soil, the effects of the formulation are likely to be minimal. Hence, for confined crop rotation studies the use of formulation may not be necessary.

In livestock studies, the route of exposure is generally well defined, i.e. either oral or dermal. In the case of oral dosing, the purpose is to reflect the likely exposure and is mostly carried out by using the parent compound. However, it is clear from previous sections of this chapter that residues on feed items may comprise a complex mixture of compounds. Where plant metabolites are also animal metabolites, then it can be assumed that by feeding the parent compound exposure to the metabolite will have also been tested and no additional studies will be required. If, however, a 'plant-unique' metabolite is generated, then additional livestock feeding may be required.

DESIGN AND CONDUCT OF CORE REGULATORY STUDIES

The overall objectives of metabolism studies are three-fold, as follows:

(i) To define the major components of the residues and their distribution within the system.
(ii) To indicate which components of the residue should comprise the residue definition and require method development and residue analysis.
(iii) To demonstrate the efficiency of the extraction procedures and thus validate residue methods.

PLANT METABOLISM

The regulatory position in terms of guidelines for plant metabolism studies is generally well established and to a large extent harmonized.

Plant metabolism studies are generally conducted on the target crops for the pesticide; however, some extrapolation within crop groupings is acceptable, e.g. the US Environment Protection Agency (EPA) states that a study in beans would be representative for all legumes but would not be extrapolated to root crops (Anon, 1996). Where multiple-use patterns for the chemical are planned, then it is accepted that where the nature of the residues are the same, for the same use pattern, in representative crops from three different crop groupings, then no further studies are required. Crops must, however, be representative for the categories for the intended use.

Application should be carried out by using formulated material to the whole plant at the recommended growth stages, with samples taken to approximate agricultural practices, e.g. immature harvests of cereal are taken to simulate the practice of grazing by livestock. Where possible, the studies should be conducted in the field to allow the influence of climatic conditions; however, in some countries specific radiochemical safety regulations exclude this practice and in some cases it may not be possible to use field studies for practical reasons. In these cases, it is acceptable to use glasshouse or climatic chambers.

Where residues in commodities are expected to be low at the highest label rate, then it is recommended that the studies are carried out at exaggerated rates, phytotoxicity permitting.

The use of *in vitro*-type studies using cell tissue culture, plant parts or enzyme systems can provide data to define and understand mechanisms and to provide larger quantities of metabolites for identification.

CONFINED CROP ROTATION

Confined crop rotation studies provide data on the nature and amount of pesticide residues which are taken up by the following or rotated crops. The regulatory position for this type of study is not so well defined as that for plant metabolism but, unlike primary plant metabolism studies, the data are used to establish crop rotation restrictions or to indicate the need for limited field trials. In terms of conduct and analysis, these studies represent a significant commitment in terms of resource and cost.

In the case of the EPA, these studies are now regulated by the Office of Prevention, Pesticides and Toxic Substances (OPPTS) 860 guidelines. In the latest guidance, the following recommendations are given for the conduct of the study:

(i) Sandy loam soil treated at the highest maximum field rate and in a manner consistent with practice, i.e. multiple application. The aging period, in this instance, begins at the last application.
(ii) Crops are sown after appropriate soil aging periods, e.g. 30, 120 and 365 d.
(iii) Crops should be representative of those expected in agricultural practice and where possible should include crops from three crop groupings (specified as root, small grain and leafy vegetable). Soybean may be substituted for a leafy vegetable.
(iv) Crops should be harvested at the appropriate intervals.

In the case of the EU, a proposal has been made for a tiered approach which recognizes the overall cost and relevance of these studies. In the proposed first stage, use is made of existing data, relating to the persistence of the pesticide in soil and the theoretical calculation of the expected residues in crops based on a relative transition factor. Where further testing is required, a model test is recommended based on a 'worst-case' scenario, e.g. using the soil with the slowest rate of degradation and 30 d planting interval. Where residues exceed the official maximum residue limit (MRL), a full test would be required.

LIVESTOCK METABOLISM

The purpose of livestock metabolism studies is essentially to define the nature of the residues present in tissues, milk and eggs intended for human consumption. Additionally, the studies define the distribution of the residues in tissues and the validation of the efficiency of the extraction techniques used for residue analytical testing. Studies are carried out with typical livestock species used in agricultural commerce and are required whenever a pesticide is applied directly to animals or when treated plant commodities are used for animal feed. Typically, the studies are carried out in ruminants and poultry, although if the use pattern of the pesticide specifically targets other species then extra studies may be required. Metabolism studies are carried out in representative species from these groups, usually lactating goats or cows and laying hens. In the case of the former, the goat is usually the species of choice in the interests of economy of scale, the use of less radio-label, and the amount of excreta produced. If the metabolism in the ruminant, poultry and rat are the same, then the metabolism is considered to be the same in all species and further studies are not required. If the converse is true, a pig study may be required to define the nature of the residues in a mono-gastric species, i.e. the pig.

It is important that the animals are obtained in advance of the study and allowed several days to acclimatize to the surroundings and to minimize the stress experienced from the transportation and also to adjust to any changes in diet and for any medication, e.g. worming. Changes in diet or the previous use of antibiotics are particularly important in ruminants where these can invoke changes in rumen populations.

Treatment is carried out to closely approximate expected exposure, as follows:

(a) *For ingested residues* – oral dosing is usually carried out over a period of several days to allow the residues in tissues, milk, and eggs to reach a steady state. Where the primary use of the study is to identify the nature of the residues and to develop methodologies, then the dosing need only be carried out for a period of 3 d. If, however, the intent is to use the study to demonstrate that the residues in tissues, milk and eggs will be negligible at the expected dietary burden, then the study should be continued for extended periods, i.e. 4–7 d for the ruminant and 10–14 d for hens. The dosing (test) material should reflect the major component of the terminal residue in treated crops. The dose rate should be at least equal to the theoretical dietary burden, although in most cases this will be very low, less than 0.1 mg/kg feed, and the identification of the residues will be commensurately low and difficult. It is therefore usual for these studies to be carried out using a dose rate equivalent to a feed level of 10 mg/kg to facilitate the detection, isolation and characterization of metabolites. The regulatory requirement for livestock studies depends on the authority; the USEPA requires studies for the pesticide whenever the pesticide is applied to the crop or crop parts are used in feed. EU guidance only requires studies when the theoretical dietary burden calculated on an as-received basis is > 0.1 mg/kg feed. In order to accurately define the amount of radioactivity administered, the test material is dosed either in a gelatine capsule or through oral gavage. Ruminants can conveniently be dosed twice a day at milking and poultry once a day. This minimizes the degree of handling, thus reducing stress and hence the chances of reduced lactation and egg yield.

(b) *For dermal applications* – the radiolabelled chemical is formulated and applied in a way that reflects the proposed use pattern. Where this occurs, the animal is allowed to groom itself, thus allowing some of the chemical to be ingested. In this case, the results of the oral study may well be sufficient to define residues resulting from the dermal treatment.

To minimize the potential for radioactive contamination and ensure that a balance of radioactivity can be achieved, animals under test are housed in stalls and metabolism cages. In the case of cows, the animal is fitted with a harness which leads to a separating device which allows for the separation of faeces and urine. The use of a catheter for the collection of urine is less desirable because of the chance of infection and the need for antibiotics which may affect the gut bacteria, hence leading to possible changes in the metabolism in the rumen. Goats are usually housed in metabolism cages which allow separation of urine and faeces through wire mesh floors. Conventional battery cages are appropriate for hens where eggs can roll to the front of the cage and excreta is collected under the mesh floor. To acclimatize the animals, they are placed in the containers for several days before the first dose. It is usual to use a single cow or goat and up to five hens. In order to avoid enzyme induction, animals should not be pre-conditioned with the chemical.

Samples of milk, eggs and excreta are taken throughout the dosing period. The animals are usually killed within 24 h of the final dose and tissues are taken *post mortem*. For ruminants, meat, a mixture of hind and fore quarter, fat renal and subcutaneous, liver and kidneys are taken, while in the hen, meat, fat and liver are taken. Analysis of the excreta provides a material balance and a source of metabolites likely to be present in tissues, milk and eggs.

Where the material balance is low, e.g. below 60 %, then further studies may be required to measure the production of volatile products.

PROCESSING (SEE ALSO CHAPTER 4)

Although not routinely considered as a core metabolism study, investigations into the nature of the residues following unit processes in the preparation of food items are key in understanding the residues present in food items. For this reason, these studies will be included. The main objectives of the processing studies have been defined by the 1999 FAO/WHO Joint Meeting on Pesticide Residues (FAO, 2000a) as follows:

- to obtain information about the breakdown or reaction products which require separate risk assessment;
- to determine quantitative distribution of residues in various processed products, thus allowing the estimation of processing factors for products which may be consumed;
- to allow more realistic estimates to be made for the chronic and acute dietary intake of pesticides.

Processing studies should simulate commercial or household practices as closely as possible. Such studies should be carried out with materials containing aged or incurred residues following the application of the pesticide with maximum dosage and the shortest pre-harvest interval according to the recommended agriculture practice. As such, these studies are more routinely carried out by using commodities which are treated as part of normal field trials. If the residue levels in the processed fractions are expected to be low, exaggerated dose rates may be applied to facilitate the identification and quantitation of the residues. Following processing, the balance of the residue of concern before and after the process is evaluated. In cases where a loss of residue is found, then it may be necessary to conduct further studies using a radiolabelled pesticide to identify the nature of the loss. In the European Guidelines (Anon, 1997a,b), the stability of the pesticide is investigated in a range of conditions which are similar to unit processes, i.e. cooking, juicing, oil extraction, and preserve preparation. The guidance argues that the most likely reaction occurring during these processes is hydrolysis since enzyme activity would be inactivated during such processes. Representative hydrolytic conditions include temperatures of 90 °C, and a pH 7 for 20 min to represent pasteurization, 100 °C and a pH of 5 for 60 min to represent baking and boiling, and 120 °C and a pH of 6 for up to 60 min to represent sterilization. Only individual products, which would represent a greater content than 0.05 mg/kg in the final processed commodity, need to be identified.

EVALUATION OF METABOLISM DATA

Metabolism studies are complex investigations leading to the definition of a biotransformation pathway and an understanding of the nature of the residues in food and feed items. These data form the foundation of the risk assessment by identifying the metabolites whose relevance and significance must be assessed. In this instance, the relevance is defined as the inherent hazard and the significance of the levels of likely exposure.

DETERMINATION OF THE BIOTRANSFORMATION PATHWAY AND THE RESIDUE OF CONCERN

Within the guidelines for the conduct of regulatory studies, trigger values are provided as guidance for when a component needs to be identified or characterized. In general, components exceeding 0.05 mg/kg or 10 % total radioactive residue (TRR) in food items require identification wherever possible and need to be characterized, while a component between 0.01 and 0.05 mg/kg requires its chemical behaviour to be characterized. Components at levels below 0.01 mg/kg are considered to be of no concern unless the parent compound has particular toxicology concerns.

Residues identified as arising from the incorporation of the radio-atom into natural components can be considered to be of having no significance and are therefore of no concern. Similarly, metabolites identified as simple molecules, e.g. simple aliphatic structures, may be considered to be of no relevance.

Where radioactive residues are characterized as being bound into natural macromolecules, the exocon may itself not have been identified and even where identification has been achieved the question of the relevance of the residue needs to be addressed in terms of the bioavailability. A discussion of the methods by which bioavailability can be assessed has been described by Skidmore *et al.* (1998).

Metabolites resulting from phase I and II metabolism can be assessed directly, although the significance of conjugated metabolites may need to be considered more in terms of their bioavailability. The process of determining the relevance of the metabolite is by comparison with the biotransformation in the mammalian species used for the toxicology assessment. Where the metabolite is also found in mammalian metabolism, then it is considered that the toxicology has been assessed in the main toxicology package and in many cases the metabolite is considered to have no relevance. There is some question remaining as to the levels at which a metabolite must be present in a mammalian metabolism study to be determined as non-relevant but this is addressed on a case-by-case basis. In instances where a unique plant metabolite is present, then its relevance must be assessed by using a 'package' of appropriate mammalian toxicology studies and structure–activity assessments. Special attention is also required for the assessment of the toxicological significance of metabolites formed by photolysis. These metabolites are unlikely to be formed by the metabolic processes in animals and may require specific mammalian toxicology studies. For instance, a photo-isomer of fipronil, designated as fipronil desulfinyl (Figure 3.26) showed higher toxicity than the parent compound, and separate toxicological tests had to be carried out with that degradation product (WHO, 1998). An example where a photolysis product forms part of the residue definition is abamectin: 8,9-Z-avermectin B_{1a} is a photoproduct of abamectin B_{1a} and is included in the residue definition (FAO, 1998j).

In terms of their significance, the levels of metabolites found in the metabolism studies are also used to define the residue definition, and hence the need for the studies to be carried out in a manner that simulates reality.

In cases where a very complex biotransformation pathway is defined, then it may be required to measure part or all of the residue by using a common moiety approach, e.g. alachlor and acetochlor residues in corn are measured, for USA regulatory purposes, following strong-base hydrolysis where the metabolites are hydrolysed to substituted anilines.

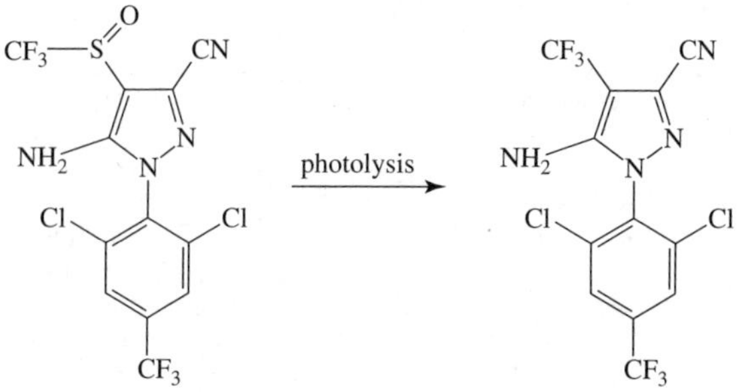

Figure 3.26   Chemical structures of fipronil and its major photoproduct

DEFINITION OF RESIDUES FOR REGULATORY AND RISK ASSESSMENT PURPOSES

From the above process, the definition of the residue for monitoring, the maximum residue limit (MRL – *the maximum amount of a substance that can legally be present in food and feed items*), and for dietary safety assessment purposes, can be defined. These may not be the same, since for dietary exposure calculations it may be desirable to include parent and metabolites of toxicological concern while it may be more appropriate to use an indicator residue for monitoring compliance with the MRL.

The definition of residues for monitoring purposes should be suitable for checking the compliance with the registered and recommended use patterns, i.e. good agriculture practice (GAP), and at the same time be as simple as possible to facilitate the analysis of a large number of samples. The 1998 FAO/WHO Joint Meeting on Pesticide Residues (JMPR) has therefore recommended separate residue definitions for monitoring purposes and for chronic and acute risk assessment purposes where appropriate.

Some examples of these are given in Table 3.8.

DEVELOPMENT OF ANALYTICAL METHODS

The development of analytical methods is facilitated by using samples from the metabolism studies to optimize the efficiency of the proposed extraction procedure. These samples represent 'aged residues' and by comparing the extraction efficacy of the analyte obtained from the residue extraction method with the exhaustive extractions used in the metabolism study can fully validate the residue analytical method procedures.

Unfortunately, the comparison of the efficiency of extraction applied in multiresidue methods is rarely compared with the exhaustive extraction used in metabolism studies, and consequently the extraction efficiency of the regulatory method cannot be confirmed. The 1998 FAO/WHO Joint Meeting on Pesticide Residues recommended the following (FAO, 1999t):

Table 3.8 Definition of residues for enforcement and dietary exposure assessment (FAO, 1998h, 1999q, 1999q,r,s, 1999r, 1999s, 2000b)

Pesticide (commodity)	Checking compliance with MRL[a]	Dietary exposure assessment
Bitertanol (plant)	Bitertanol	Bitertanol
Bitertanol (animal)	Bitertanol	Sum of bitertanol, p-hydroxybitertanol and acid-hydrolysable conjugates of p-hydroxybitertanol
Carbofuran (plant)	Sum of carbofuran and 3-hydroxy carbofuran expressed as carbofuran	Sum of carbofuran and 3-hydroxy carbofuran, free and conjugated, expressed as carbofuran
Chlorothalonil (plant)	Chlorothalonil	Chlorothalonil
Chlorothalonil (animal)	Chlorothalonil	Sum of chlorothalonil and 4-hydroxy-2,5,6-trichloroisophthalonitrile, expressed as chlorothalonil
Fenpropimorph (plant)	Fenpropimorph	Fenpropimorph
Fenpropimorph (animal)	2-Methyl-2-{4-[2-methyl-3-(cis-2,6-dimethylmorpholin-4-yl)propyl]} propionic acid expressed as fenpropimorph	
Glyphosate (plant)	Glyphosate	Sum of glyphosate and aminomethylphosphonic acid expressed as Glyphosate
Kresoxim-methyl (plant)	Kresoxim-methyl	Kresoxim-methyl
Kresoxim-methyl (animal)	α-(p-Hydroxy-o-tolyloxy)-o-tolyl(methoxyimino) acetic acid expressed as kresoxim-methyl	
Quintozene (plant)	Quintozene	Sum of quintozene, pentachloroaniline and methyl pentachlorophenyl sulfide expressed as quintozene
Quintozene (animal)	Sum of quintozene, pentachloroaniline and methyl pentachlorophenyl sulfide expressed as quintozene	–
Thiabendazole (plant)	Thiabendazole	Thiabendazole
Thiabendazole (animal)	Sum of thiabendazole and 5-hydroxy thiabendazole	Sum of thiabendazole, 5-hydroxy thiabendazole and 5-hydroxy thiabendazole sulfate

[a]MRL, maximum reside limit.

Comparative extraction efficiency studies including the frequently used extraction solvents, such as acetone/water, ethyl acetate and acetonitrile/water should be carried out on samples from metabolism studies for the compounds which are expected to be included in the residue definition(s).

These results should be made publicly available.

REFERENCES

Akhtar, M., Humayoun Mahadevan, S. and Paquet, A. (1994). Comparative metabolism of deltamethrin and 3-phenoxybenzoic acid in chickens, *J. Environ. Sci. Health, Part B*, **29**, 369–394.

Anderson, J. J., Shalaby, L. M. and Berg, D. S. (1999). Metabolism of ^{14}C-flusilazole in the goat, *J. Agric. Food Chem.*, **47**, 2439–2446.

Anon (1996). *Nature of the Residues in Plants and Livestock Guidelines*, United States Environmental Protection Agency Pesticide Residue Chemistry Test Guidelines, OPPTS 860 1300, United States Environmental Protection Agency, Washington DC, USA (August, 1996).

Anon (1997a). *Commission of the European Communities Guidelines for the Generation of Data Concerning Residues (as provided in Annex II, part A and Annex III, part A)*, Directive 91/414/EEC, Commission of the European Communities, Brussels, Belgium (January, 1997).

Anon (1997b). *Commission of the European Communities (Guideline Appendix E, Processing Studies)*, 7035/VI/95 Rev. 5, Commission of the European Communities, Brussels, Belgium (July, 1997).

Aubert, S. and Pallett, K. E. (2000). Combined use of ^{13}C- and ^{19}F-NMR to analyse the mode of action and metabolism of the herbicide isoxaflutole, *Plant Physiol. Biochem.*, **38**, 517–523.

Bounds, S. V. J. and Hutson, D. H. (2000). The comparative metabolism of agrochemicals in plants and mammals, in *Metabolism of Agrochemicals in Plants*, Roberts, T. (Ed.), Wiley, Chichester, UK, pp. 179–209.

Bray, L. D., Heard, N. E., Overman, M. C., Vargo, J. D., King, D. L., Lawrence, L. J. and Phelps, A. W. (1997). Hydrolysis of prosulfuron at pH 5: evidence for a resonance-stabilised triazine cleavage product, *Pesticide Sci.*, **51**, 56–64.

Breithaupt, D., Zorn, H. and Schwack, W. (1998). Photoinduced additions of the fungicide anilazine and chlorothalonil to plant cuticles, in *Proceedings of the 9th International Congress on Pesticide Chemistry*, London, UK, August 1998, Abstract 5A-027.

Briggs, G. G., Bromilow, R. H. and Evans, A. A. (1982). Relationship between lipophilicity and root uptake and translocation of non-ionised chemicals by barley, *Pesticide Sci.*, **13**, 495–504.

Bromilow, R. H., Chamberlain, K. and Evans, A. A. (1990). Physiochemical aspects of phloem translocation of herbicides, *Weed Sci.*, **38**, 305–314.

Caldwell, J. and Marsh, M. V. (1983). Interrelationships between xenobiotic metabolism and lipid biosynthesis, *Biochem. Pharmacol.*, **32**, 1667–1672.

Capps, T. M., Barringer, M. V., Eberle, W. J., Brown, D. R. and Sanson, D. R. (1996). Identification of a unique glucosylsulfate conjugate metabolite of profenofos in cotton, *J. Agric. Food Chem.*, **44**, 2408–2411.

Chukwudebe, A. C., Wislocki, P. G., Sanson, D. R., Halls, T. D. J. and Vanden-Heuvel, W. J. A. (1994). Metabolism of thiabendazole in laying hen and lactating goats, *J. Agric. Food Chem.*, **42**, 2964–2969.

Church, D. C. (1993). *The Ruminant Animal: Digestive Physiology and Nutrition*, Waveland Press, Il, USA.

Clark, T., Clifford, D. R., Deas, A. H. B., Gendl, P. and Watkins, D. A. M. (1978). Photolysis, metabolism and other factors influencing the performance of triadimefon as a powdery mildew fungicide, *Pesticide Sci.*, **9**, 497–506.

Cole, D. J. (1994). Detoxification and bioactivation of agrochemicals in plants, *Pesticide Sci.*, **42**, 209–222.

Dean, G. M., Kirkpatrick, D., Riseborough, J., Biggs, S. R. and Hawkins, D. (1988). Identification of metabolites of vinclozolin in hen liver, in *Proceedings of the Brighton Crop Protection Conference – Pests and Diseases*, Brighton, UK, 1988, pp. 693–698.

Dorough, H. W. (1980). Classification of radioactive pesticide residues in food producing animals, *J. Environ. Pathol. Toxicol.*, **3**, 11–19.

Dubelman, A. M., Solsten, T. R., Fujiwara, H. and Mehrsheikh, A. (1997). Metabolism of halosulfuron-methyl by corn and wheat, *J. Agric. Food Chem.*, **45**, 2314–2321.

FAO (1995). *Pesticide Residues in Food – 1995 Evaluations Part I, Residues*, FAO Plant Production and Protection Paper 137, Food and Agriculture Organization of the United Nations, Rome, p. 632.

FAO (1996a). *Pesticide Residues in Food – 1995 Evaluations Part I, Residues*, FAO Plant Production and Protection Paper 137, Food and Agriculture Organization of the United Nations, Rome, p. 11.

FAO (1996b). *Pesticide Residues in Food – 1995 Evaluations Part I, Residues*, FAO Plant Production and Protection Paper 137, Food and Agriculture Organization of the United Nations, Rome, p. 190.

FAO (1997a). *Pesticide Residues in Food – 1996 Evaluations Part I, Residues*, FAO Plant Production and Protection Paper 142, Food and Agriculture Organization of the United Nations, Rome, p. 349.

FAO (1997b). *Pesticide Residues in Food – 1996 Evaluations Part I, Residues*, FAO Plant Production and Protection Paper 142, Food and Agriculture Organization of the United Nations, Rome, pp. 444–445.

FAO (1997c). *Pesticide Residues in Food – 1996 Evaluations Part I, Residues*, FAO Plant Production and Protection Paper 142, Food and Agriculture Organization of the United Nations, Rome, p. 447.

FAO (1997d). *Pesticide Residues in Food – 1996 Evaluations Part I, Residues*, FAO Plant Production and Protection Paper 142, Food and Agriculture Organization of the United Nations, Rome, pp. 254–258.

FAO (1997e). *Pesticide Residues in Food – 1996 Evaluations Part I, Residues*, FAO Plant Production and Protection Paper 142, Food and Agriculture Organization of the United Nations, Rome, p. 532.

FAO (1997f). *Pesticide Residues in Food – 1996 Evaluations Part I, Residues*, FAO Plant Production and Protection Paper 142, Food and Agriculture Organization of the United Nations, Rome, pp. 253–254.

FAO (1997g). *Pesticide Residues in Food – 1996 Evaluations Part I, Residues*, FAO Plant Production and Protection Paper 142, Food and Agriculture Organization of the United Nations, Rome, p. 291.

FAO (1997h). *Pesticide Residues in Food – 1996 Evaluations Part I, Residues*, FAO Plant Production and Protection Paper 142, Food and Agriculture Organization of the United Nations, Rome, pp. 354–357.

FAO (1997i). *Pesticide Residues in Food – 1996 Evaluations Part I, Residues*, FAO Plant Production and Protection Paper 142, Food and Agriculture Organization of the United Nations, Rome, p. 243.

FAO (1997j). *Pesticide Residues in Food – 1996 Evaluations Part I, Residues*, FAO Plant Production and Protection Paper 142, Food and Agriculture Organization of the United Nations, Rome, p. 540.

FAO (1997k). *Pesticide Residues in Food – 1996 Evaluations Part I, Residues*, FAO Plant Production and Protection Paper 142, Food and Agriculture Organization of the United Nations, Rome, p. 244.

FAO (1997l). *Pesticide Residues in Food – 1996 Evaluations Part I, Residues*, FAO Plant Production and Protection Paper 142, Food and Agriculture Organization of the United Nations, Rome, p. 96.

FAO (1997m). *Pesticide Residues in Food – 1996 Evaluations Part I, Residues*, FAO Plant Production and Protection Paper 142, Food and Agriculture Organization of the United Nations, Rome, p. 348.

FAO (1997n). *Pesticide Residues in Food – 1996 Evaluations Part I, Residues*, FAO Plant Production and Protection Paper 142, Food and Agriculture Organization of the United Nations, Rome, pp. 533–543.

FAO (1998a). *Pesticide Residues in Food – 1997 Evaluations Part I, Residues*, FAO Plant Production and Protection Paper 146, Food and Agriculture Organization of the United Nations, Rome, p. 370.

FAO (1998b). *Pesticide Residues in Food – 1997 Evaluations Part I, Residues*, FAO Plant Production and Protection Paper 146, Food and Agriculture Organization of the United Nations, Rome, p. 588.

FAO (1998c). *Pesticide Residues in Food – 1997 Evaluations Part I, Residues*, FAO Plant Production and Protection Paper 146, Food and Agriculture Organization of the United Nations, Rome, p. 365.

FAO (1998d). *Pesticide Residues in Food – 1997 Evaluations Part I, Residues*, FAO Plant Production and Protection Paper 146, Food and Agriculture Organization of the United Nations, Rome, pp. 358–360.

FAO (1998e). *Pesticide Residues in Food – 1997 Evaluations Part I, Residues*, 364. FAO Plant Production and Protection Paper 146, Food and Agriculture Organization of the United Nations, Rome, pp. 357–359.

FAO (1998f). *Pesticide Residues in Food – 1997 Evaluations Part I, Residues*, FAO Plant Production and Protection Paper 146, Food and Agriculture Organization of the United Nations, Rome, pp. 96–98.

FAO (1998g). *Pesticide Residues in Food – 1997 Evaluations Part I, Residues*, FAO Plant Production and Protection Paper 146, Food and Agriculture Organization of the United Nations, Rome, p. 351.

FAO (1998h). *Pesticide Residues in Food – 1997, Report 1997*, FAO Plant Production and Protection Paper 145, Food and Agriculture Organization of the United Nations, Rome, pp. 221–231.

FAO (1998i). *Pesticide Residues in Food – 1997 Evaluations Part I, Residues*, FAO Plant Production and Protection Paper 146, Food and Agriculture Organization of the United Nations, Rome, pp. 529–531.

FAO (1998j). *Pesticide Residues in Food – 1997, Report 1997*, FAO Plant Production and Protection Paper 145, Food and Agriculture Organization of the United Nations, Rome, p. 4.

FAO (1999a). *Pesticide Residues in Food – 1998 Evaluations Part I/2, Residues*, FAO Plant Production and Protection Paper 152/2, Food and Agriculture Organization of the United Nations, Rome, pp. 696–697.

FAO (1999b). *Pesticide Residues in Food – 1998 Evaluations Part I/2, Residues*, FAO Plant Production and Protection Paper 152/2, Food and Agriculture Organization of the United Nations, Rome, pp. 818–819.

FAO (1999c). *Pesticide Residues in Food – 1998 Evaluations Part I/1, Residues*, FAO Plant Production and Protection Paper 152/1, Food and Agriculture Organization of the United Nations, Rome, p. 336.

FAO (1999d). *Pesticide Residues in Food – 1998 Evaluations Part I/2, Residues*, FAO Plant Production and Protection Paper 152/2, Food and Agriculture Organization of the United Nations, Rome, p. 392.

FAO (1999e). *Pesticide Residues in Food – 1998 Evaluations Part I/1, Residues*, FAO Plant Production and Protection Paper 152/1, Food and Agriculture Organization of the United Nations, Rome, pp. 108–111.

FAO (1999f). *Pesticide Residues in Food – 1998 Evaluations Part I/1, Residues*, FAO Plant Production and Protection Paper 152/1, Food and Agriculture Organization of the United Nations, Rome, pp. 317–319.

FAO (1999g). *Pesticide Residues in Food – 1998 Evaluations Part I/1, Residues*, FAO Plant Production and Protection Paper 152/1, Food and Agriculture Organization of the United Nations, Rome, p. 7.

FAO (1999h). *Pesticide Residues in Food – 1998 Evaluations Part I/1, Residues*, FAO Plant Production and Protection Paper 152/1, Food and Agriculture Organization of the United Nations, Rome, p. 74.

FAO (1999i). *Pesticide Residues in Food – 1998 Evaluations Part I/1, Residues*, FAO Plant Production and Protection Paper 152/1, Food and Agriculture Organization of the United Nations, Rome, p. 821.

FAO (1999j). *Pesticide Residues in Food – 1998 Evaluations Part I/2, Residues*, FAO Plant Production and Protection Paper 152/2, Food and Agriculture Organization of the United Nations, Rome, p. 921.

FAO (1999k). *Pesticide Residues in Food – 1998 Evaluations Part I/1, Residues*, FAO Plant Production and Protection Paper 152/1, Food and Agriculture Organization of the United Nations, Rome, p. 109.

FAO (1999l). *Pesticide Residues in Food – 1998 Evaluations Part I/1, Residues*, FAO Plant Production and Protection Paper 152/1, Food and Agriculture Organization of the United Nations, Rome, p. 336.

FAO (1999m). *Pesticide Residues in Food – 1998 Evaluations Part I/1, Residues*, FAO Plant Production and Protection Paper 152/1, Food and Agriculture Organization of the United Nations, Rome, p. 389.

FAO (1999n). *Pesticide Residues in Food – 1998 Evaluations Part I/1, Residues*, FAO Plant Production and Protection Paper 152/1, Food and Agriculture Organization of the United Nations, Rome, p. 3.

FAO (1999o). *Pesticide Residues in Food – 1998 Evaluations Part I/1, Residues*, FAO Plant Production and Protection Paper 152/1, Food and Agriculture Organization of the United Nations, Rome, p. 640.

FAO (1999p). *Pesticide Residues in Food – 1998 Evaluations Part I/2, Residues*, FAO Plant Production and Protection Paper 152/2, Food and Agriculture Organization of the United Nations, Rome, p. 817.

FAO (1999q). *Pesticide Residues in Food – 1998, Report 1998*, FAO Plant Production and Protection Paper 148, Food and Agriculture Organization of the United Nations, Rome, pp. 215–225.

FAO (1999r). *Pesticide Residues in Food – 1998 Evaluations Part I/2, Residues*, FAO Plant Production and Protection Paper 152/2, Food and Agriculture Organization of the United Nations, Rome, p. 910.

FAO (1999s). *Pesticide Residues in Food – 1998 Evaluations Part I/2, Residues*, FAO Plant Production and Protection Paper 152/2, Food and Agriculture Organization of the United Nations, Rome, p. 1129.

FAO (1999t). *Pesticide Residues in Food – 1998, Report 1998*, FAO Plant Production and Protection Paper 148, Food and Agriculture Organization of the United Nations, Rome, p. 11.

FAO (2000a). *Pesticide Residues in Food – 1999, Report 1999*, FAO Plant Production and Protection Paper 153, Food and Agriculture Organization of the United Nations, Rome, p. 12.

FAO (2000b). *Pesticide Residues in Food – 1999, Report 1999*, FAO Plant Production and Protection Paper 153, Food and Agriculture Organization of the United Nations, Rome, pp. 219–223.

Feung, C., Hamilton, R. H. and Mumma, R. O. (1973). Metabolism of 2,4-dichlorophenoxyacetic acid. V. Identification of metabolites in soybean callus tissue cultures, *J. Agric. Food Chem.*, **21**, 637–640.

Goldman, P. (1982). *Metabolic Basis for Detoxication. Metabolism of Functional Groups*, Jackoby, W. B. R., and Caldwell, J. (Eds), Academic Press.

Gonzalez, F. J. (1992). *International Encyclopedia of Pharmacology and Therapeutics – Pharmacogenetics of Drug Metabolism*, Kalow, W. (Ed.), Pergamon Press, New York.

Gross, D., Laanio, T., Dupuis, G. and Esser, H. O. (1979). The metabolic behaviour of chlorotoluron in wheat and soil, *Pesticide Biochem. Physiol.*, **10**, 49–59.

Guy, R. H., Hadgraft, J. and Bucks, D. A. (1987). Transdermal drug delivery and cutaneous metabolism, *Xenobiotica*, **17**, 325–343.

Harthoorn, P. A., Gillam, M. A. and Wright, A. N. (1985). The synthesis of radiolabelled pesticides and related compounds, in *Progress in Pesticide Biochemistry and Toxicology*, Vol. 4, Hutson, D. H. and Roberts, T. R. (Eds), Wiley, Chichester, UK, pp. 261–355.

Hatton, P. J., Cole, D. J. and Edwards, R. (1996). Influence of plant age on glutathione levels and glutathione transferases involved in herbicide detoxication in corn (*Zea mays*) and giant foxtail (*Setaria faberi*), *Pesticide Biochem. Physiol.*, **54**, 199–209.

Hendley, P. (1982). The potential of stable isotopes in pesticide metabolism studies, in *Progress in Pesticide Biochemistry*, Vol. 2, Hutson, D. H. and Roberts, T. R. (Eds), Wiley, Chichester, UK, pp. 35–70.

Holland, P. (1996). Glossary of terms relating to pesticides, *Pure Appl. Chem.*, **68**, 1167–1193.

Hotchkiss, S. A. and Caldwell, J. (1989). The absorption and disposition of xenobiotics in skin, in *Intermediary Xenobiotic Metabolism in Animals*, Hutson, D. H., Caldwell, J. and Paulson, G. D. (Eds), Taylor and Francis, London, pp. 65–79.

Huber, R., Hamm, R., Ohnsorge, U. and Turk, W. (1988). The metabolism of cycloxydim in soybeans, in *Proceedings of the Brighton Crop Protection Conference – Pests and Diseases*, Brighton, UK, 1988.

Huhtanen, K. L. (1997). Characterisation of ^{14}C-acephate residues in laying hens and lactating goats, in *Proceedings of the 213th Meeting of the American Chemical Society*, San Francisco, CA, USA, Part 1, Abstract 085.

Hungate, R. E. (1988). Introduction to the ruminant and the rumen, in *The Rumen Microbial Ecosystem*, Hobson, P. N. (Ed.), Elsevier Applied Science, London, UK, pp. 1–19.

Ishihara, K., Shiotani, H., Soeda, Y. and Ono, S. (1988). Fate of sethoxydim in sugar beet, *J. Pesticide Sci.*, **13**, 231–237.

Joseph, R. S. I. (1999). Metabolism of azoxystrobin in plants and animals, in *Pesticide Chemistry and Bioscience – The Food Environment Challenge*, Brooks, G. T. and Roberts, T. R. (Eds), The Royal Society of Chemistry, Cambridge, UK, pp. 265–278.

Kaneko, H., Matsuo, M. and Miyamoto, J. (1986). Differential metabolism of fenvalerate and granuloma formation. Identification of a cholesterol ester derived from a specific chiral isomer of fenvalerate, *Toxicol. Appl. Pharmacol.*, **83**, 148–156.

Kreuz, K. and Martinoia, E. (1999). Herbicide metabolism in plants. Integrated pathways of detoxification, in *Pesticide Chemistry and Bioscience – The Food Environment Challenge*, Brooks, G. T. and Roberts, T. R. (Eds), The Royal Society of Chemistry, Cambridge, UK, pp. 279–287.

Laignelet, L., Riviere, J. L. and Lhuguenot, J. C. (1992). Metabolism of an imidazole fungicide prochloraz in the rat after oral administration, *Food Chem. Toxicol.*, **30**, 575–583.

Langebartels, C. and Harms, H. (1985). Analysis for nonextractable (bound) residues of pentachlorophenol in plant cells using a cell wall fractionation procedure, *Ecotox. Environ. Safety*, **10**, 268–279.

Levine, W. G. (1983). *Biological Basis of Detoxication*, Caldwell, J. and Jackoby, W. B. (Eds), Academic Press, London.

Li, Y., Barefoot, A. C., Reiser, R. W., Fogiel, A. J. and Sabourin, P. J. (1997). Identification of thiocyanate as the principle metabolite of oxamyl in lactating goats, *J. Agric. Food Chem.*, **45**, 962–966.

Machin, A. F., Rogers, H., Cross, A. J., Quick, M. P., Howells, L. C. and Janes, N. F. (1975). Metabolic aspects of the toxicology of diazinon. I. Hepatic metabolism in the sheep, cow, pig, guinea pig, rat, turkey, chicken and duck, *Pesticide Sci.*, **6**, 461–473.

McCall, S. N., Jurgens, P. and Ivanetich, K. M. (1983). Hepatic microsomal metabolism of the dichloroethanes, *Biochem. Pharmacol.*, **32**, 207–214.

Metcalf, R. L., Fukuto, T. R. and March, R. B. (1957). Plant metabolism of dithio-systox and thimet, *J. Econ. Entomol.*, **50**, 338–345.

Mikami, N., Wakabayashi, N., Yamada, H. and Miyamoto, J. (1984). New conjugated metabolites of 3-PBA in plants, *Pesticide Sci.*, **15**, 531–542.

Mikami, N., Wakabayashi, N., Yamada, H. and Miyamoto, J. (1985a). The metabolism of fenvalerate in plants; the conjugation of the acid moiety, *Pesticide Sci.*, **16**, 46–58.

Mikami, N., Baba, Y., Katagi, T. and Miyamoto, J. (1985b). Metabolism of the synthetic pyrethroid fenpropathrin in plants, *J. Agric. Food Chem.*, **33**, 980–987.

Monson, K. D. (1991). Release and characterisation of bound benomyl and carbendazim metabolites in animal tissues via Raney nickel desulfurisation and acid dehydration, *J. Agric. Food Chem.*, **39**, 1808–1811.

Ohori, Y. and Aizawa, H. (1983). *In vitro* metabolism of fluoroimide in the cell suspension of apple leaves, *J. Pesticide Sci.*, **8**, 223–227.

Owen, W. J. (1987). Herbicide detoxification and selectivity, in *Proceedings of the Brighton Crop Protection Conference – Weeds*, Vol. 1, Brighton, UK, pp. 309–318.

PSD (1995). *Evaluation on Fenbuconazole*, Report No. 128, Pesticide Safety Directorate, York, UK.

PSD (1996). *Evaluation on Bromuconazole*, Report No. 160, Pesticide Safety Directorate, York, UK.

Quistad, G. B., Staiger, L. E. and Schooley, D. A. (1978). Environmental degradation of the miticide cycloprate (hexadecyl cyclopropanecarboxylate). Bovine metabolism, *J. Agric. Food Chem.*, **26**, 71–75.

Ratcliffe, R. G. and Roscher, A. (1998). Prospects for *in vivo* NMR methods in xenobiotic research in plants, *Biodegradation*, **9**, 411–422.

Reiser, R. W., Barefoot, A. C., Dietrich, R. F., Fogiel, A. J., Johnson, W. R. and Scott, M. T. (1991). Application of microcolumn liquid chromatography–continuous-flow fast atom bombardment mass spectrometry in environmental studies of sulfonylurea herbicides, *J. Chromatog., Part A*, **554**, 91–101.

Reiser, R. W., Dietrich, R. F., Djanegara, T. K. S., Fogeil, A. J., Payne, W. G., Ryan, D. L. and Zimmerman, W. T. (1997). Identification of a novel animal metabolite of methomyl insecticide, *J. Agric. Food Chem.*, **45**, 2309–2313.

Roberts, T. (1998). *Metabolic Pathways of Agrochemicals*, Part 1, *Herbicides and Plant Growth Regulators*, The Royal Society of Chemistry, Cambridge, UK.

Roberts, T. and Hutson, D. (1999). *Metabolic Pathways of Agrochemicals*, Part 2, *Insecticides and Fungicides*, The Royal Society of Chemistry, Cambridge, UK.

Ruzo, L. O. and Casida, J. E. (1979). Degradation of decamethrin on cotton plants, *J. Agric. Food Chem.*, **27**, 572–575.

Ryan, D. L. (1993). Metabolism of chlorethoxyfos in lactating goat, in *Proceedings of the 205th Meeting of the American Chemical Society*, Denver, USA, Part 1, Abstract Agro 075.

Sandermann, H. (1994). Higher plant metabolism; the 'green liver' concept, *Pharmacogenetics*, **4**, 225–241.

Sandermann, H., Scheel, D. vd Trenck, K. T. (1983). Metabolism of environmental chemicals by plants-copolymerization into lignin, *J. Appl. Polym. Sci.*, **37**, 407–420.

Sandermann, H., Arjmand, M., Gennity, I., Winkler, R. and Struble, C. B. (1990). Animal bioavailability of defined xenobiotic lignin metabolites, *J. Agric. Food Chem.*, **38**, 1877–1880.

Scheline, R. R. (1978). *Mammalian Metabolism of Plant Xenobiotics*, Academic Press, London.

Schlagbauer, B. G. L. and Schlagbauer, A. W. J. (1972). The metabolism of carbamate pesticides – a literature analysis: Part 1, *Residue Rev.*, **42**, 1–90.

Schmidt, B., Thiede, B. and Rivero, C. (1994). Metabolism of pesticide metabolites 4-nitrophenol and 3,4-dichloroaniline in carrot (*Daucus carota*) cell suspension cultures, *Pesticide Sci.*, **40**, 231–238.

Schroeder, P. (1998). Glutathione conjugation of xenobiotics, the metabolism of glutathione conjugates in plants, *J. Exp. Bot.*, **49** (Supplement) P3.09, p. 22.

Serre, A. M., Roby, C., Roscher, A., Nurit, F., Euvrard, M. and Tissut, M. (1997). Comparative detection of fluorinated xenobiotics and their metabolites through ^{19}F NMR or ^{14}C label in plant cells, *J. Agric. Food Chem.*, **45**, 242–248.

Sidelmann, U. G., Gavaghan, C., Carless, H. A. J., Spraul, M., Hofmann, M., Lindon, J. C., Wilson, I. D. and Nicolson, J. K. (1995). 750 MHz directly coupled HPLC–NMR: Application to the sequential characterisation of the positional isomers and anomers of 2-, 3- and 4-fluorobenzoic acid glucuronides in equilibrium mixtures, *Anal. Chem.*, **67**, 4441–4445.

Skidmore, M. W., Worthington, P. A., Hand, L. and Tseriotis, G. (1990). Metabolism studies of hexaconazole in temperate cereals and the synthesis of the metabolites, in *Proceedings of the Brighton Crop Protection Conference – Pests and Diseases*, Brighton, UK, 1998, pp. 1035–1040.

Skidmore, M. W., Paulson, G., Kuiper, H. A., Ohlin, B. and Reynolds, S. (1998). Bound xenobiotic residues in food commodities of plant and animal origin, *Pure Appl. Chem.*, **70**, 1423–1447.

Smith, L. W. and Chang, F. Y. (1973). Aminotriazole metabolism in *Cirsium Arvense (L.) scop.* and *Pisum Sativum L.*, *Weed Res.*, **13**, 339–350.

Sonobe, H., Kemps, Laverne R., Mazzola, E. P. and Roach, J. A. G. (1981). Isolation and identification of a new conjugated carbofuran metabolite in carrots. Angelic acid ester of 3-hydroxycarbofuran, *J. Agric. Food Chem.*, **29**, 1125–1129.

Sterling, T. M. (1994). Mechanisms of herbicide adsorption across plant membranes and accumulating plant cells, *Weed Sci.*, **42**, 263–276.

Stevens, J. L. and Jones, D. P. (1989). The mercapturic acid pathway: biosynthesis, intermediary metabolism and physiological disposition, in *Coenzymes and Cofactors*, Part B, *Glutathione*, Dolphin, D., Poulson, R. and Avramovic, O. (Eds), Wiley, New York, pp. 45–84.

Still, G. C. (1967). Metabolism of 3,4-dichloropropionanilide in plants. Metabolic fate of 3,4-dichloroaniline moiety, *Science*, **159**, 992–993.

Subba-Rao, R. V., Onisko, B. C., Nguyen, E., Ortiz, D. and Wei, Y. (1997). Metabolism of fonofos in peanuts, in *Proceedings of the 213th Meeting of the American Chemical Society*, San Francisco, CA, USA, Abstract Agro 035.

Villedieu, J. C., Calmon, M. and Calmon, J. P. (1994). Mechanisms of dicarboximide ring opening in aqueous media: procymidone, vinclozolin and chlozolinate, *Pesticide Sci.*, **41**, 105–115.

Walker, L. M., Hatzios, K. K. and Wilson, H. P. (1994). Absorption, translocation and metabolism of ^{14}C-thifensulfuron in soybean (*Glycine max.*), spurred anoda (*Anoda cristata*) and velvetleaf (*Abutilon theophrasti*), *J. Plant Growth Reg.*, **13**, 27–32.

Wolf, A. E., Dietz, K. J. and Schroeder, P. (1996). Degradation of glutathione-*S*-conjugate by carboxypeptidase in the plant vacuole, *FEBS Lett.*, **384**, 31–34.

WHO (1998). *Pesticide Residues in Food – 1997 Evaluations Part II: Toxicological and Environmental*, WHO/PCS/98.6, World Health Organization, Geneva, Switzerland.

4 Effects of Food Preparation and Processing on Pesticide Residues in Commodities of Plant Origin

GABRIELE TIMME and BIRGITT WALZ-TYLLA
Bayer CropScience AG, Leverkusen, Germany

INTRODUCTION

When pesticides are used in or on plants or plant products, residues frequently occur on the raw agricultural commodities (RACs). The highest residue levels to be expected on the crops are found from supervised field trials performed according to Good Agricultural Practice (GAP) under maximum allowed conditions. Data resulting from these trials form the basis for the establishment of national and international maximum residue limits (MRLs) in food and feed. Whereas MRLs are a useful tool to control the proper application of a pesticide, they are only of limited value for a realistic assessment of the potential risk through the

Pesticide Residues in Food and Drinking Water: Human Exposure and Risks.   Edited by Denis Hamilton and Stephen Crossley
© 2004 John Wiley & Sons, Ltd   ISBN: 0-471-48991-3

dietary intake of pesticide residues. Monitoring studies for estimating residues in human food ('Total diet studies') have consistently shown that using MRLs as a basis for calculating human dietary consumption of pesticide residues overestimate actual intakes considerably. In an evaluation of data from total diet studies from eight countries, residues of 70 pesticides used up less than 1 % each of the respective acceptable daily intakes (ADIs) in 84 % of 243 cases; even the highest values found did not approach half of the ADIs (Frehse, 1992). A supplementary evaluation of more recent studies supported these findings (Frehse, 1997).

The World Health Organization (WHO) recommended a tiered approach for the dietary risk assessment (WHO, 1997). Among other factors, this concept takes into account the fact that many crops are prepared or processed either commercially or in the household prior to consumption. Residues on the RACs normally decline during storage, transport, preparation, and household and commercial processing of food.

Two types of processing studies have to be distinguished: investigations to determine the effect on either the nature or the magnitude of the residues. The nature of the residues may be addressed by performing processing studies with radio-labelled compounds using laboratory small-scale procedures, or by using a concept of model hydrolysis studies under representative conditions. To investigate the effect of processing on the magnitude of the residues, field studies should be conducted according to GAP with subsequent processing of the harvested crops. In addition, these studies deliver information on the transfer of residues into commodities which may be used as animal feed.

In certain situations, regulatory authorities (e.g. US Environmental Protection Agency (EPA), EU Commission) require studies on the effect of food preparation and processing on pesticide residues before pesticide products are authorized in the respective country or countries. The same is true for international bodies (e.g. Food and Agriculture Organization of the United Nations (FAO), WHO) when they evaluate data for the establishment of international MRLs (e.g. Codex Alimentarius Limits (CXLs)). The trigger for the requirement of processing studies is different, however. The US EPA (Regulation OPPTS 860.1520) requires processing data whenever there is a possibility of residue levels in processed foods or feeds exceeding the level present in the RAC. In those cases where residues concentrate in the processed product, MRLs are set for this commodity. In Europe (EU Commission Directive 91/414), the prerequisites for deciding whether the need for processing studies has been triggered are as follows:

- significant residues in the RAC (generally defined as > 0.1 mg/kg, unless the compound has a high acute or chronic toxicity);
- importance of the RAC and the processed product in the human and animal diet;
- the first step in the dietary risk assessment shows that the Theoretical Maximum Daily Intake (TMDI) uses up more than 10 % of the Acceptable Daily Intake (ADI).

Farm animal metabolism and transfer studies are generally necessary before a pesticide is authorized when the pesticide is used on animal feed and significant residues remain on these commodities. However, the transfer of crop residues to commodities of animal origin is usually rather low, when feed, containing pesticide residues, is consumed by farm animals. Therefore, regulatory authorities do not routinely require processing studies for animal products.

Several reviews and many papers on specific studies with plant material have been published during the last 20 years. The effect of processing on pesticide residues was summarized in a comprehensive review by Holland *et al.* (1994).

This present chapter describes the general concepts for investigating the impact of processing on the nature and the magnitude of residues in and on food and feed. For illustration, relevant examples are quoted from the paper by Holland *et al.*, along with other information from the literature which has been published since this review. Furthermore, the designs of representative procedures for the most widely used processes in industry and the home are presented for the following:

- food preparation (washing, peeling and trimming, and hulling)
- cooking
- juicing
- brewing and vinification
- canning
- milling and baking
- oil production
- drying

The results of the processing studies are used as follows:

- to recognize reductions and concentrations of residues in food and feed items
- to interpret the reasons for changes in the residue concentrations (loss versus redistribution versus change in moisture content)
- to calculate transfer factors
- to perform a more realistic dietary risk assessment

The evaluation of the literature compiled in the following confirm the conclusions drawn by several authors, namely that washing and cleaning, which are the initial steps in most processing procedures, frequently reduce residue levels, particularly of non-systemic compounds. Many other types of processing (e.g. milling of cereals and polishing of rice) result in a significant lowering of residue levels. In some cases, however, residues may be concentrated in processed fractions, thus resulting in higher levels than those present in the RACs. There are basically two types of processes where residues can typically concentrate. In the first type, the concentration is based on the loss of water during processing, e.g. in the preparation of tomato paste or dried apple pomace. In the second type of process, the RAC is separated into different components, one of which may contain the bulk of the residues (e.g. oil from oil seeds, or bran from grain). Some pesticides decompose during processing procedures, such as boiling and heating, but degradation products of toxicological significance may be formed only in exceptional cases.

EFFECTS OF FOOD PROCESSING ON THE NATURE OF PESTICIDE RESIDUES

Depending upon the type of process involved and upon the chemical nature of the residue in the RAC, differences in the nature of the residue in the processed commodities and the RAC may have to be determined. Such investigations are best performed by using radio-labelled chemicals. However, the difficulty then is to conduct such a study in a manner representative of the conditions prevailing in normal practice. Moreover, it is difficult to produce and to handle sufficient amounts of treated crops for processing, and to remove contamination of the processing equipment caused by radioactive compounds.

REPRESENTATIVE HYDROLYTIC CONDITIONS

The major factors influencing the stability of residues in typical processes are temperature, pH, water content and chemical nature of the residue. Hydrolysis is most likely to affect the nature of the residue during most processing operations, because processes such as heating would generally inactivate enzymes present in the substrate, leaving simple hydrolysis as a degradation mechanism. Since the substrate itself is not likely to have a major effect (apart from governing the pH level in some situations), Buys *et al.* (1993) developed a concept based on a range of hydrolytic conditions that usually prevail in food processing operations. These operations typically involve higher temperatures for much shorter periods of time and, in some cases, more extreme pH values than those employed in hydrolysis studies intended to investigate the environmental fate of a plant protection product (25 °C, pH 5, 7 and 9 for one month).

Table 4.1 summarizes typical conditions prevailing in several types of processing. From the details described in this table, three representative sets of conditions are defined in Table 4.2. They should be used, according to Buys *et al.* (1993), for an appropriate simulation of the respective processing operations. The extreme conditions, which would be required to mimic the refinement of oil, were not included in this set of representative conditions. In the processing of oil seeds, the major interest is directed toward the possible concentration or reduction of residues during pressing or the extraction of oil. In most cases, residues will be very low at this stage, and only in exceptional circumstances will further studies on the nature of the residues be necessary. In contrast, hydrolytic conditions during the preparation of wine are very mild compared with those in other processes and are, therefore, also not included in Table 4.2.

Table 4.1 Significant parameters during processing operations

Type of processed products	Critical operation	Temperature (°C)[a]	Time (min)	pH
Cooked vegetables	Boiling	100	15–50	4.5–7.0
Preserves				
fruits	Pasteurization	90–95	1–20	3.0–4.5
vegetables	Sterilization	118–125	5–20	4.5–7.0
Fruit juice	Pasteurization	82–90	1–2	3.0–4.5
Oil	Refinement	190–270	20–360	6.0–7.0
Beer	Brewing	100	60–120	4.1–4.7
Red wine[b]	Heating of mash	60	2[c]	2.8–3.8
Bread	Baking	100–120	20–40	4.0–6.0

[a] Temperatures in or on the different commodities during processing.
[b] The mash of white grapes is not heated.
[c] Subsequently either chilled quickly or allowed to cool slowly (overnight).

Table 4.2 Representative hydrolytic conditions simulating processing conditions

Processes represented	Temperature (°C)	Time (min)	pH
Pasteurization	90	20	4
Baking, brewing and boiling	100	60	5
Sterilization	120	20	6

CONDUCT OF STUDIES

Depending upon the potential range of agricultural uses of a particular plant protection product, one or more of the representative hydrolysis situations given in Table 4.2 should be investigated (an autoclave will be needed for temperatures above 100 °C). These studies should generally be conducted with a radio-labelled form of the active ingredient to maximize the chances of identifying the residue components produced during hydrolysis.

In order to avoid contamination of the equipment by radioactivity, special care has to be taken. The laboratory personnel have to wear protective clothes to prevent contamination of the skin. Furthermore, the used, wasted and lost radioactivity has to be balanced.

In cases where the residues in the RAC consist primarily of a metabolite of the active ingredient, the need to conduct hydrolysis studies with that metabolite should be considered on a case-by-case basis. For example, comparison of its structure with that of the parent compound and with the hydrolysis products of the parent compound may suggest that additional studies are not necessary.

INTERPRETATION OF RESULTS

An individual hydrolysis product need not be identified, if it is clear by calculation that its concentration in the final processed commodities will be less than 10 % of the total radioactive residue applied on the RAC, or less than 0.05 mg/kg (whichever is the greater). Such evaluation should take into account the magnitude of the product found in the hydrolysis study (as a proportion of the starting material), dilution or concentration during the processing, and the initial residue levels in the raw commodity. If the hydrolysis products are identical with the transformation products already identified as the residue of toxicological significance in the RAC, processed commodities can be analysed according to the same residue methodology principles as are employed in the analyses of the RAC. Products formed in the hydrolysis studies not already identified as metabolites in plants may require a separate dietary or toxicological risk assessment and different analytical approaches.

EXAMPLES

The ethylene bisdithiocarbamate (EBDC) fungicides, e.g. mancozeb, are often used as examples to illustrate the formation of toxicologically relevant metabolites during processing procedures. Ethylenethiourea (ETU) is an EBDC degradation product of putative carcinogenicity which is also found in plant metabolism studies. However, under certain processing conditions, the formation of ETU is accelerated. The conversion of EBDCs to ETU (or, similarly, the formation of propylenethiourea (PTU) from propylene bisdithiocarbamates, e.g. propineb) is particularly favoured by high pH, heat and an intensive contact of surface residues on harvested crops with a liquid (e.g. during the production of red wine). On the other hand, Vogeler *et al.* (1977) was able to show that there was hardly any difference between grapes and wine concerning the level of degradation products (including PTU). An excellent overview of processing studies conducted with RACs treated with dithiocarbamates is given by Holland *et al.* (1994).

Although no examples were reported of pesticides where food processing has resulted in the production of metabolites other than those known from metabolism studies on plants or animals, the proportions of various metabolites formed during processing may differ from those found in field or laboratory studies on plants. As metabolites are generally more polar than active ingredients, changes in proportions of individual metabolites between processing fractions can also be expected. Alary *et al.* (1995) studied the degradation of radio-labelled captan residues during the processing of apples to sterilized puree using laboratory small-scale processing (125 °C for 20 min and at pH 4.0). The results were compared with those obtained in buffer medium mimicking the same process in the absence of the food matrix. Extensive degradation of this fungicide occurred, leading to its complete disappearance and genesis of derived compounds, gases and adducts to the apple matrix. The same pattern of degradation was observed in both cases.

These data support the postulate that the degradation conditions of a molecule in a complex biological medium may be simulated by specifically designed hydrolysis studies. Reactions *in vitro*, however, do not take into account the formation of adducts with highly reactive molecules in plants. When more study results according to the guideline developed by Buys *et al.* (1993) become available, this approach will be validated using different compounds belonging to a variety of chemical classes.

EFFECTS OF FOOD PROCESSING ON THE MAGNITUDE OF PESTICIDE RESIDUES

After the nature of the residues formed during processing has been clarified and the appropriate compounds (active ingredient and relevant metabolites) to be analysed have been identified, processing studies are conducted with RACs that normally undergo processing in the home or under commercial conditions. The process may be physical (e.g. peeling) or may involve the use of heat or chemicals.

These types of processing studies are intended:

- to provide information on the transfer of residues from the RAC to the processed products, in order to calculate reduction or concentration factors;
- to enable a more realistic estimate to be made of the dietary intake of pesticide residues;
- to establish MRLs for residues in processed products where necessary, according to requirements of national regulatory authorities or international standards.

HOUSEHOLD AND COMMERCIAL PROCEDURES

The technology to be used in these types of processing studies should always correspond as closely as possible to the actual conditions that are normally used in practice. Thus, processed products that are prepared in the household, e.g. cooked vegetables, should be produced using the equipment and preparation techniques that are normally used in the home. On the other hand, for the preparation of industrially produced food, commercial practices should be followed as closely as possible.

Studies on the effect of household preparation and commercial processing can be substituted, one for the other, where the recipes used are similar. To cover the most extreme cases occurring in practice, those representative processes should be used that most likely lead to higher residues in important processed products. Nevertheless, careful consideration needs to be given to the most plausible scenario for processing which may affect particular residues in the food, so that the primary question on the likely residue intake by consumers can be economically answered.

REPRESENTATIVE PROCEDURES

Since any processing study has to be conducted according to Good Laboratory Practice (GLP) principles, simulation of commercial processing in the laboratory is usually preferable to conducting studies in pilot plants or industrial premises.

If this is impossible, such that technical scale (pilot plant) operation is necessary, care must be taken to extend GLP principles as far as possible into that industrial processing environment.

A Standard Operating Procedure (SOP) or a study protocol must describe in detail how to proceed for each process, indicating at least:

- the size of the field sample (see also following paragraph)
- apparatus and ingredients
- individual steps of the procedure
- technical parameters of the study (time, temperature, etc.).

A one-page flow sheet describing the conduct of the procedure should be prepared. Preferably, RAC samples used in processing studies should contain field-treated quantifiable residues as close as possible to the MRL, so that measurable residues are obtained, and transfer factors for the various processed commodities can be determined.

A transfer factor gives the ratio of the residue concentration in the processed commodity to that in the RAC. For example, if the residue concentration is 0.5 mg/kg in olives and 0.2 mg/kg in olive oil, the transfer factor is $0.2/0.5 = 0.4$. A factor < 1 (= reduction factor) indicates a reduction of the residue in the processed commodity, whereas a factor > 1 (=concentration factor) indicates a concentration effect of the processing procedures. Enhancing the residues either by increasing the application rates, shortening the pre-harvest interval (PHI, which is the time between the last application of the pesticide and harvest of the RAC) or spiking the RAC with the active ingredient and its metabolites *in vitro* is not, as a rule, desirable. Spiking is only acceptable if the RAC residues can be shown to consist only of surface residues. However, in some cases, especially where residues in the RAC are close to the analytical limit of determination, field treatment at exaggerated rates or shortened PHIs is advisable to obtain sufficient residue levels for the processing studies.

Food Preparation

The first step in household or commercial food processing is the preparation of food using various mechanical processes, such as removing damaged or spoiled items or parts of crops, washing, peeling, trimming or hulling. This often leads to significant declines in the amount of pesticide residues in the remaining edible portions (Petersen *et al.*, 1996; Schattenberg *et al.*, 1996; Çelik *et al.*, 1995).

Household washing procedures are normally carried out with running or standing water at moderate temperatures. Commercial washing processes are performed:

- in wash baths with or without movement
- on transportation belts by spraying with water
- in special washing machines, e.g. by spraying air into the washing baths.

Detergents, chlorine or ozone can be added to the wash water to improve the effectiveness of the washing procedure (Ong *et al.*, 1996). If necessary, several washing steps can be conducted consecutively.

Table 4.3 shows examples of the effects of washing on the residue levels of different pesticides applied to fruits and vegetables. The effects depend on the physicochemical properties of the pesticides, such as water solubility, hydrolytic rate constant, volatility, and octanol–water partition coefficient (P_{OW}), in conjunction with the actual physical location of the residues; washing processes lead to reductions of hydrophilic residues which are located on the surface of crops. In addition, the temperature of the washing water and the type of washing have an influence on the residue level. As pointed out by Holland *et al.* (1994), hot washing and the addition of detergents are more effective than cold water washing. Washing coupled with gentle rubbing by hand under tap water for 1 min dislodges pesticide residues significantly (Barooah and Yein, 1996). Systemic and lipophilic pesticide residues are not removed significantly by washing.

The outer leaves of vegetables often contain residues of pesticides applied during the growing season. Therefore, peeling or trimming procedures reduce the residue levels in leafy vegetables. Peeling of root, tuber and bulb vegetables with a knife is common household practice. Commercial peeling of fruits and vegetables is conducted according to three different procedures (Heiss, 1990):

(a) *Steam peeling.* In special vessels, the crops are heated with water steam for about 1 min. Through sudden easing of pressure, the peels are removed or loosened. Relevant crops for this procedure are root and bulb vegetables and sensitive and soft fruit such as peaches.

(b) *Lye peeling.* The crops are washed in a lye bath containing 0.5–20 % sodium hydroxide for a few minutes at increasing temperatures. Subsequently, neutralization is achieved with citric acid. The procedure can be used for the peeling of root and bulb vegetables and tomatoes.

(c) *Mechanical peeling.* The peel is removed in drums containing rough surfaces on the inner walls or with peeling waltzes. In addition, machines with peeling knives or peeling knife drums are available. The mechanical procedure can be used for root and bulb vegetables.

In some cases, a combination of steam and mechanical peeling is employed.

Table 4.4 shows examples of the effects of peeling and trimming on the residue levels of different pesticides applied to fruits and vegetables. These examples show that most of the residue concentration is located in or on the peel. Peeling of the RACs may remove more than 50 % of the pesticide residues present in the commodity. Thus, removal of the peel achieves almost complete removal of residues, so leaving little in the edible portions. This is especially important for fruits which are not eaten with their peels, such as bananas or citrus fruits.

Table 4.3 Effect of washing on pesticide residue levels

Crop[a]	Pesticide	Processing procedure	Transfer factor	Reference
Apple	Azinphos-methyl (organophosphate)	Washed several times with distilled water	0.9	Çelik *et al.* (1995)
	Azinphos-methyl (organophosphate)	Ozone wash at pH 4.5–10.7 (30 min), 21 °C	0.2–0.6	Ong *et al.* (1996)
	Captan (phthalimide)	Ozone wash at pH 4.5–10.7 (6 min), 21 °C	0	Çelik *et al.* (1995)
	Diazinon (organophosphate)	Washed several times with distilled water	0.9	Çelik *et al.* (1995)
	Ethion (organophosphate)	Washed several times with distilled water	0.7	Çelik *et al.* (1995)
	Methidathion (organophosphate)	Washed several times with distilled water	0.8	Çelik *et al.* (1995)
	Phosalone (organophosphate)	Washed several times with distilled water	0.7	Çelik *et al.* (1995)
	Pirimicarb (carbamate)	Washed several times with distilled water	0.7	Çelik *et al.* (1995)
Broad bean	Pirimiphos-methyl (organophosphate)	Washing	0.3	Kamil *et al.* (1996)
	Malathion (organophosphate)	Washing	0.3	Kamil *et al.* (1996)
Brinjal fruit (egg plant)	Quinalphos (organophosphate)	Washing under tap water	0.2	Barooah and Yein (1996)
Grape	Methidathion (organophosphate)	Washed several times with distilled water	0.8	Çelik *et al.* (1995)
	Phosalone (organophosphate)	Washed several times with distilled water	0.6	Çelik *et al.* (1995)
Orange	2-Phenylphenol (phenol)	Washing at 20 °C	1	Reynolds, (1996)
		Washing at 50 °C	1	
	Imazalil (imidazole)	Washing at 20 °C	0.8	Reynolds, (1996)
		Washing at 50 °C	0.6	
	Thiabendazole (benzimidazole)	Washing at 20 °C	0.9	Reynolds, (1996)
		Washing at 50 °C	0.7	
Peppers	Phosalone (organophosphate)	Washed several times with distilled water	0.6	Çelik *et al.* (1995)
	Pirimicarb (carbamate)	Washed several times with distilled water	0.7	Çelik *et al.* (1995)
Tomato	Diazinon (organophosphate)	Washed several times with distilled water	0.9	Çelik *et al.* (1995)
	Pirimicarb (carbamate)	Washed several times with distilled water	0.6	Çelik *et al.* (1995)

[a]RAC, raw agricultural commodity.

Table 4.4 Effect of peeling and trimming on pesticide residue levels

Crop[a]	Pesticide	Processed commodity, procedure	Transfer factor	Reference
Apple	Azinphos-methyl (organophosphate)	Apple, after washing and peeling	0.3	Çelik *et al.* (1995)
	Diazinon (organophosphate)	Apple, after washing and peeling	0.3	Çelik *et al.* (1995)
	Ethion (organophosphate)	Apple, after washing and peeling	0.3	Çelik *et al.* (1995)
	Methidathion (organophosphate)	Apple, after washing and peeling	0.4	Çelik *et al.* (1995)
	Phosalone (organophosphate)	Apple, after washing and peeling	0.2	Çelik *et al.* (1995)
	Pirimicarb (carbamate)	Apple, after washing and peeling	0.2	Çelik *et al.* (1995)
Carrot	Chlorfenvinphos (organophosphate)	Peeled and trimmed carrots	0.2	Reynolds (1996)
	Pirimiphos-methyl (organophosphate)	Peeled and trimmed carrots	0.2	Reynolds (1996)
	Quinalphos (organophosphate)	Peeled and trimmed carrots	0.2	Reynolds (1996)
	Triazophos (organophosphate)	Peeled and trimmed carrots	0.2	Reynolds (1996)
Banana	Aldicarb (carbamate)	Peel	1–2	FAO (1996)
		Pulp	0.6–1	
	Fenarimol (pyrimidinyl carbinol)	Pulp	0.3–1	FAO (1994a)
	Tebuconazole (azole)	Peel	1–2	FAO (1994b)
		Pulp	0.8–1	
Grapefruit	Profenofos (organophosphate)	Peel	3	FAO (1994b)
		Pulp	0.1	
Lemon	Phosalone (organophosphate)	Peel	1–4	FAO (1994b)
		Pulp	< 0.1	
	Profenofos (organophosphate)	Peel	3	FAO (1994b)
		Pulp	0.1	
Mandarin	Acephate (organophosphate)	Peel	0.4	FAO (1994a)
		Pulp	1	
	Fenpyroximate (pyrazole)	Peel	5	FAO (1995)
		Pulp	0.1–0.2	
Orange	Acephate (organophosphate)	Peel	1–3	FAO (1994a)
		Pulp	0.2–1	
	Methamidophos (organophosphate)	Peel	2–3	FAO (1994b)
		Pulp	0.1–0.5	
	Profenofos (organophosphate)	Peel	3	FAO (1994b)
		Pulp	< 0.1	

[a]RAC, raw agricultural commodity.

However, the peel from commercial peeling processes can be used as animal feed or for the production of essential oils (citrus) or pectin (citrus, apple, etc.). For such industrial processes, it is important to realize that especially non-systemic surface residues are often concentrated in the peel. For systemic pesticides, peeling may not be as effective, as shown by Sheikhorgan *et al.* (1994). After application of thiometon on cucumbers, no reduction of residue levels could be detected in the peeled cucumbers.

Under the Codex Alimentarius, as in other international standards, MRLs refer to the whole fruit, which is appropriate for assessing compliance with GAP. These MRLs are of limited significance, however, in assessing dietary exposure to pesticides from fresh fruits, which are peeled (Holland *et al.*, 1994).

Commercial hulling processes, e.g. of beans, grains or oil seeds, are conducted with peeling waltzes. Hulling of cereal grains and oil seeds often leads to reductions of the residue levels in the remaining part of the grains or seeds, because most of the pesticide residues are located in or on the hulls. Table 4.11 below shows that residues can concentrate in the hulls (bran) of cereals in the milling process, while Table 4.12 below shows residue transfer factors for hulls of different oil seeds. Hulling of broad beans resulted in a considerable reduction of organophosphate residues in the hulled beans (Table 4.5).

Table 4.5   Effect of hulling on pesticide residue levels in beans (Kamil *et al.*, 1996)

Crop[a]	Pesticide	Processed commodity	Transfer factor
Broad bean seed	Malathion (organophosphate)	De-hulled beans	0.1
	Pirimiphos-methyl (organophosphate)	De-hulled beans	0.1

[a] RAC, raw agricultural commodity.

Cooking

The following processes can be conducted on a household or on a commercial scale, but each would have different equipment:

- cooking in water at 100 °C (e.g. potatoes and other vegetables);
- cooking under pressure at > 100 °C (e.g. fast cooking of small portions);
- cooking under low pressure at < 100 °C (e.g. cooking and concentrating of fruits and vegetables in the production of paste, jams, ketchup, etc.);
- steaming under normal or higher pressure (e.g. vegetables);
- steaming with a small amount of water and addition of fat at normal temperature for vegetables and fruits (without fat);
- frying in fat (e.g. potatoes);
- baking at 130–180 °C (e.g. bread or other bakery products);
- roasting with hot air or hot contact plates (e.g. coffee, cocoa, nuts and biscuits);
- microwave cooking.

Because of the large variability of conditions, relevant information regarding the processes used must always be recorded, particularly in cooking experiments. Several studies were reported on the dissipation of pesticides in crops during cooking.

Cooking procedures at different temperatures, the duration of the process, the amount of water or food additives, and the type of system (open or closed) may have an impact on the residue level. Normally, residues are reduced during the cooking process by volatilization in open systems or by hydrolysis in closed systems. In any case, adding cooking liquid dilutes the residues. Table 4.6 shows examples of the effects of cooking on the residue levels of different pesticides applied to fruits, vegetables and cereals.

In addition to the studies summarized in Table 4.6, the behaviour of the organophosphorus pesticides chlorfenvinphos, fenitrothion, isoxathion, methidathion and prothiophos during cooking was examined by Nagayama (1996) with green tea leaves, spinach and fruits. These pesticides decreased during the cooking process corresponding to the boiling time. According to their water solubility, some pesticides were translocated from the raw materials into the cooking water. On the other hand, the pesticides remained in the processed food according to their octanol–water partition coefficient, which is an indicator of the hydrophilic or lipophilic properties of the compound.

In exceptional cases, cooking processes may cause pesticide degradation, yielding a reaction product of toxicological significance. For example, daminozide is degraded to UDMH (1,1-dimethylhydrazine), which is much more potent than the parent compound (Leparulo-Loftus *et al.*, 1992). Another example is the formation of ETU from EBDCs during heating processes (Petersen *et al.*, 1996), as discussed previously.

Juicing

Fruit and vegetable consumption has become more popular during the past decade, as consumers learn more about health benefits attributed to fruit and vegetable constituents by current nutrition studies. Likewise, consumption of fruit and vegetable juices has increased, too. World-wide, 80–90 % of fruit juices are produced from apples and oranges. The European market for vegetable juices makes up 0.5–3 % of the total European juice market only, with about 90 % of the vegetable juices being produced from tomatoes. The remaining 10 % of the vegetable juices are produced mainly from spinach, carrots, celery or cabbage fermented with lactic acid (Schobinger, 1987). Processed juices are marketed as fruit or vegetable juices or fruit or vegetable nectars, refreshment beverages or lemonades. Processed juices or food can also be prepared from intermediate products such as juice concentrates or fruit/vegetable pastes.

Juicing combines the following main steps: preparation, extraction, pressing, clarification, filtration, concentration and preservation (heating or freezing). After washing processes, special kinds of preparation may be needed for the different

Table 4.6 Effect of cooking on pesticide residue levels

Crop[a]	Pesticide	Processed commodity, procedure	Transfer factor	Reference
Broad bean	Malathion (organophosphate)	Cooked beans, under pressure	< 0.1[b]	Kamil et al. (1996)
		Cooked beans, common method	< 0.1[b]	
	Pirimiphos-methyl (organophosphate)	Cooked beans, under pressure	0.1	Kamil et al. (1996)
		Cooked beans, common method	0.1	
Buckwheat	Chlorpyrifos-methyl (organophosphate)	Noodles	0.6	Tsumura et al. (1994)
	Fenitrothion (organophosphate)	Noodles	0.4	Tsumura et al. (1994)
	Malathion (organophosphate)	Noodles	0.4	Tsumura et al. (1994)
	Methyl bromide (methane)	Noodles	0.2	Tsumura et al. (1994)
Cowpea	Endosulfan (cyclodiene, benzodioxathiepin oxide)	Cooked cowpeas	0.1–0.6	Kumari et al. (1996)
	Lindane (gamma-HCH)	Cooked cowpeas	0.1–0.2	Kumari et al. (1996)
Orange	2-Phenylphenol (fungicide)	Marmalade, microwave cooked	0.9	Reynolds (1996)
		Marmalade, conventionally cooked	0.5	
	Imazalil (imidazole)	Marmalade, microwave cooked	1.6	Reynolds (1996)
		Marmalade, conventionally cooked	1.5	
	Thiabendazole (benzimidazole)	Marmalade, microwave cooked	1	Reynolds (1996)
		Marmalade, conventionally cooked	0.8	

[a] RAC, raw agricultural commodity.
[b] Residues in cooked beans were below the limit of detection.

fruits and vegetables. Most pome fruit and small stone fruits can be used for direct juice extraction, and no peeling is needed. Small stone fruits such as apricots and plums must be pitted. Cherries may be pressed with the pit intact. In grape juice processing, a stemmer and crusher removes residual stems and leaves from the grapes and performs the initial crushing of the fruits (Somogyi *et al.*, 1996). In order to maximize juice yield and colour and flavour extraction, a hot break process or enzyme treatment is often used. Vegetables can be cooked or fermented with lactic acid before pressing or straining.

Generally, the same processing equipment can be used in the commercial production of both vegetable and fruit juices. Different pressing equipment is used, however, whereby the hydraulic rack and frame presses are very common systems. Pressing procedures normally yield juice and wet pomace. The latter is dried to a water content of < 10 %. The juice is subsequently clarified in most cases. The clarification step removes solids from the juice, but this often requires multiple steps and possible pre-treatments. Clarification can be achieved enzymatically, using pectinase enzymes, and non-enzymatically, e.g. by the addition of gelatin, casein, or tannic acid–protein combinations, or by using decanters and finishers (Kilara and Van Buren, 1989). The subsequent juice filtration can be conducted with different equipment, e.g. diatomaceous earth filtration, pressure filtration, rotary vacuum filtration or membrane filtration. Fruit juice concentration offers the advantage of reducing the bulk of juice, the storage volume and the transportation costs. The juice is concentrated by evaporation of water, including, as a main step, stripping volatile substances from which aromas can be recovered. This is handled by partial evaporation or by a steam stripping process. For preservation, the juices are pasteurized, sterilized or frozen (in the case of concentrate). In special cases, the addition of preservatives, e.g. sorbic acid, is allowed.

Depending upon the solubility and penetration properties of pesticides, the residues distribute either in the juice or in by-products like pomace. As citrus and apple pomace are initial products for pectin production, investigating the pesticide residue level in pomace as well as in juice is important. Furthermore, fruit pomace may be fed to farm animals.

Table 4.7 shows examples of the effect of juicing on the residue levels of different pesticides applied to fruits and tomatoes. These examples demonstrate that the pesticide residue levels in juice are reduced by the juicing process. The highest increase was observed in dry pomace, which can be attributed to the fact that the moisture level in dry pomace is lower than in wet pomace. In the studies conducted with citrus fruits, most of the residues were located on the peel.

Brewing and Vinification

Most beers are brewed from barley malt, a portion of additional carbohydrate sources such as corn or rice, and hops, yeast and water. The ground barley malt and adjunct grains are mixed with water and mashed. In a lauter tun, this mash

Table 4.7 Effect of juicing on pesticide residue levels

Crop[a]	Pesticide	Processed commodity	Transfer factor	Reference
Apple	Captan (phthalimide)	Juice	< 0.1–0.3	FAO (1994a, b)
		Wet pomace	3–4	
		Dry pomace	2–4	
	Fenarimol (pyrimidinyl carbinol)	Juice	< 0.1[b]	FAO (1995)
		Wet pomace	4–5	
		Dry pomace	5–18	
	Fenpyroximate (pyrazole)	Juice	< 0.1–0.8[b]	FAO (1995)
		Wet pomace	2–3	
	Fenthion (organophosphate)	Juice	0.8	FAO (1995)
		Dry pomace	4	
	Metiram (dithiocarbamate)	Juice	< 0.1[c]	FAO (1995)
	Tebuconazole (azole)	Juice	0.1	FAO (1994b)
		Pomace	18	
	Tebufenozide (diacylhydrazine)	Juice	0.2	FAO (1996)
		Wet pomace	3	
	Teflubenzuron (benzoylurea)	Juice	< 0.5[b]	FAO (1996)
		Wet pomace	2–10	
		Dry pomace	22–120[b]	
	Thiram (dithiocarbamate)	Juice	0.3	FAO (1996)
		Wet pomace	1	
		Dry pomace	4	
	Ziram (dithiocarbamate)	Juice	0.1	FAO (1996)
		Wet pomace	1	
		Dry pomace	2	
Citrus	Aldicarb (oxime carbamate)	Juice	0.3–0.9	FAO (1994a)
		Wet peel	1–2	
		Dry peel	0.6–2	
	Chlorpyrifos (organophosphate)	Juice	< 0.1[b]	FAO (1995)
		Dry peel	2–4	
	Fenthion (organophosphate)	Juice	< 0.1[b]	FAO (1995)
		Peel	4	
Grape	Tebuconazole (azole)	Juice	0.4	FAO (1994b)
		Wet pomace	1.8	
		Dry pomace	6	
Peach	Methamidophos (organophosphate)	Juice	0.2	FAO (1996)
		Juice	0.5	
Tomato	Acephate (organophosphate)	Juice	1	FAO (1996)
		Wet pomace	0.6	
		Dry pomace	1	
	Buprofezin (thiadiazinone)	Juice	0.1	FAO (1995)
		Wet pomace	23	
		Dry pomace	34	

[a] RAC, raw agricultural commodity.
[b] Residues in RAC or juice were below the limit of determination.
[c] Results obtained from 42 residue studies.

is then separated into wort and non-soluble parts of the grains (brewer's grain). The wort is subsequently boiled in the brew kettle together with hops, whereby the bitter resins and oils are extracted from the hop cones. The leafy hop cones are removed after the boiling by a strainer as the wort leaves the brew kettle. The wort is then clarified, cooled and pitched with yeast, and after several weeks of fermentation, lagering and final clarification, the wort becomes beer.

Table 4.8 shows examples of the effect of brewing on the residue levels of different pesticides applied to barley and hops. Fenarimol residues were higher in the dried hop cones because of the loss of water during the drying of the fresh cones. On the other hand, there was a significant decrease of residues in spent hops, which can be used in animal feed. Also in spent yeast, which is of minor importance for the human diet, the residues were reduced. The residue concentrations of all pesticides were significantly lower in the beer, which could be expected due to the high dilution during the brewing process. Depending on the type of beer, 22 kg of barley and 400 g of hops per 1 hl (hectolitre) of beer can be used. This leads to dilution factors of 4.5 for barley and 250 for hops. In Europe, processing studies on hops are only required when residues are higher than 5 mg/kg of dried cones.

For vinification, grapes are stemmed (especially red grapes) and crushed. The resulting mash is promptly pressed, yielding must and pomace. This process removes the stems with their bitter tannins and simultaneously breaks the skin of the grapes. In the case of red grapes, the mash is generally fermented before pressing in order to extract the red colour from the grape skin. After pressing, yeast is added to the must and the alcoholic fermentation starts. As the fermentation comes to an end, requiring three to fourteen days for red wines and ten days

Table 4.8 Effect of brewing on pesticide residue levels

Crop[a]	Pesticide	Processed commodity	Transfer factor	Reference
Barley (grain)	Chlormequat (quaternary ammonium)	Malt Malt sprouts Beer	0.7 0.3 < 0.1	FAO (1994a)
Hops (fresh cones)	Fenarimol (pyrimidinyl carbinol)	Dried hops Spent hops Spent yeast Beer	1–3 0.2 < 0.1 < 0.1[b]	FAO (1995)
Hops (dried cones)	Fenpyroximate (pyrazole)	Spent hops Spent yeast Beer	0.1–0.2 < 0.1 < 0.1[b]	FAO (1995)
	Iprodione (dicarboximide)	Beer	< 0.1	FAO (1994a)

[a] RAC, raw agricultural commodity.
[b] Residues in beer were below the limit of determination.

Table 4.9   Effect of vinification on pesticide residue levels

Pesticide	Processed commodity, procedure	Transfer factor	Reference
Aldicarb (oxime carbamate)	Fresh juice	0.5–0.8	FAO (1994a)
	Pomace	0.9–3	
	New wine	0.3–0.7	
	Aged wine	0.2–0.6	
Fenpyroximate (pyrazole)	Wine	< 0.1[a]	FAO (1994a)
Metiram (dithiocarbamate)	Must	0.2–1	FAO (1995)
	Wine	< 0.1[a]	
Penconazole (azole)	Must	< 0.1[a]–0.4	FAO (1995)
	Wet pomace	2–3	
	Dry pomace	4–8	
	Wine	< 0.1[a]–0.4	
Tebuconazole (azole)	Must	0.4–2	FAO (1994b)
	Wine	< 0.1–0.9	
Tebufenozide (diacylhydrazine)	Must, mash fermentation with skin	1	FAO (1996)
	Must, mash fermentation without skin	< 0.1–0.4	
	Pomace	2–4	
	Wine	< 0.1–0.7	

[a]Residues in must/wine were below the limit of determination.

to six weeks for white wines, the wine is clarified. Clarification is often begun by racking (the process of allowing the wine to settle) and then transferring the supernatant wine to a second barrel or tank. Several clarification steps are usually employed before the wine is bottled.

Table 4.9 shows examples of the effect of vinification on the residue levels of different pesticides applied to grapes. Processing of grapes to must and wine significantly reduces the residues in wine. However, fermentation of mash still containing the skin can result in higher residues in must due to an intensive extraction of the skin during fermentation. This can be attributed to the fact that the residue concentration on the grape skin (pomace) is normally higher than within the grape berries. Must obtained from mash not containing skins shows reduced residue concentrations. These effects of processing grapes are independent of the different grape varieties and climatic conditions.

Canning

Canning is used to preserve a large variety of fruit products, including berries, whole fruits and fruit sections, or vegetables. The canning process combines a number of operations, such as separation of non-edible from edible parts, cleaning, size reduction, blanching or pre-cooking, filling, closing, thermal processing, cooling and storing. The temperature used in the thermal processing depends on the specific micro-organism flora in the RAC which is influenced by the acid content of the crop: pasteurization up to 100 °C with pH ≤ 4.5, and sterilization above 100 °C with pH > 4.5.

Table 4.10 shows examples of the effect of canning on the residue levels of different pesticides applied to fruit and fruiting vegetables. The residue concentration in canned food is governed by dilution (e.g. canned fruits or vegetables) or concentration (e.g. tomato paste) and the thermal process. Due to the higher proportion of dry matter in tomato paste (for two-fold concentrated paste, 28–30 %), the residue concentration in the paste is often higher than in the puree (dry matter, 8–12 %).

Table 4.10   Effect of canning on pesticide residue levels

Crop[a]	Pesticide	Processed commodity	Transfer factor	Reference
Apple	Fenpyroximate (pyrazole)	Apple sauce	0.3–0.8[b]	FAO (1995)
	Fenthion (organophosphate)	Apple sauce	0.5	FAO (1995)
	Tebuconazole (azole)	Apple sauce	0.5	FAO (1994)
	Teflubenzuron (benzoylurea)	Apple sauce	0.3	FAO (1996)
Cherry	Teflubenzuron (benzoylurea)	Preserve	0.7	FAO (1996)
Peach	Methamidophos (organophosphate)	Canned peaches	0.5–0.8	FAO (1996)
Tomato	Acephate (organophosphate)	Canned tomato	0.5	FAO (1996)
		Puree	2	
		Paste	4	
	Benomyl (benzimidazole)	Puree	0.7	FAO (1994a)
	Buprofezin (thiadiazinone)	Puree	0.6	FAO (1995)
		Paste	1	
	Ethephon (organophosphonate)	Puree	0.6	FAO (1994a)
		Paste	0.4	

[a]RAC, raw agricultural commodity.
[b]Residues in apple sauce were below the limit of determination (< 0.05 mg/kg).

Milling and Baking

The object of milling grain is the separation of the endosperm from hulls (bran) and germ. First, the grains are cleaned by removing impurities, e.g. stones, metals, foreign seeds, etc. Then the grains are conditioned to optimum water content, for easier separation of the bran from the endosperm during milling. From the same kind of cereal, e.g. a wheat mix, it is possible, by stream selection, to manufacture several grades of flour that differ to a degree in granulation, colour, protein content, chemical composition and physical properties. Wheat flours are produced from hard, soft and durum wheat depending upon their intended usage (Wolff, 1982).

The production of baker's ware includes several steps, beginning with the preparation of the sponge dough in which, greatly simplified, flour, water, yeast and yeast food are combined and mixed. The resulting mixture is called a 'sponge'. The latter is then fermented for several hours and afterwards mixed with more water, flour and other ingredients. After a rest period, the sponge is separated into pieces and allowed to relax for a few minutes in order to become pliable. Then the pieces or loaves are baked, which is followed by cooling of the loaves, slicing and wrapping.

Table 4.11 shows examples of the effect of milling and baking on the residue levels of different pesticides applied to rye, sorghum and wheat. By milling and baking, residues were mainly reduced in flour and bread and mostly in wholemeal flour and wholemeal bread. A higher concentration of residues was observed only in the bran because most of the residues are located in the exosperm of the grains.

Oil Production

Studying the behaviour of pesticide residues during the production of commercial vegetable oils from oil seeds or fruits is important especially with regard to lipophilic residues, which are more likely to be retained in the oil fraction than are hydrophilic residues. The concentration of pesticide residues in the oil is affected by different procedures.

Different crude oils (e.g. cotton seed oil, sunflower seed oil and rapeseed oil) are produced in several steps, which may affect the pesticide residue levels:

- cleaning – for all kinds of oil seeds
- hulling and peeling – for cotton seed or sunflower seed
- flaking with waltzes – for rapeseed
- crushing – for cotton seed or sunflower seed
- conditioning – for all kinds of oil seeds

The preparation of vegetable seed oil can be performed by one of three procedures, as follows:

- direct screw pressing
- direct solvent extraction
- pre-press-solvent extraction (combined procedure)

The most common and probably the most economical process is the pre-press–solvent extraction, as it utilizes a combination of two processes (Wolff, 1983). Solvent extraction is most advantageous for seeds with a low oil content ($< 20\%$, e.g. cotton seed), but this extraction can also be used for direct extraction of rapeseed (40–45 % oil) in high-capacity processing plants. Solvent extraction may take place in either percolation-type extraction, where the solvent is allowed

Table 4.11 Effect of milling and baking on pesticide residue levels

Crop[a]	Pesticide	Processed commodity	Transfer factor	Reference
Rye	Chlormequat (quaternary ammonium)	Bran	4	FAO (1994a)
		Flour	1	
		Bread	0.2	
		Wholemeal flour	2	
		Wholemeal bread	1	
Sorghum	Aldicarb (oxime carbamate)	Bran	1–5	FAO (1994a)
		Flour	0.3, 0.5[b]	
Wheat	Bifenthrin (pyrethroid)	Bran	3–4	FAO (1996)
		Flour	0.3	
		Bread	0.1	
		Wholemeal flour	0.8–0.9	
		Wholemeal bread	0.2–0.3	
	Chlormequat (quaternary ammonium)	Bran	3	FAO (1994a)
		Flour	0.2	
		Bread	0.1	
		Wholemeal flour	0.8	
		Wholemeal bread	0.4	
	Glufosinate-ammonium (phosphinico amino acid)	Coarse bran	2[c]	FAO (1994a)
		Fine bran	0.7[c]	
		Flour	0.1[c]	
		Wholemeal bread	0.2[c]	
	Malathion (organophosphate)	Bran	0.9	FAO (1996)
		Flour	0.2	
		Bread	< 0.1	
		Wholemeal flour	0.5	
		Wholemeal bread	0.1–0.2	
	Tebuconazole (azole)	Bran	1	FAO (1994b)
		Flour	< 0.1	

[a] RAC, raw agricultural commodity.
[b] Residues in flour were below the limit of determination.
[c] Average value.

to percolate through the seedbed, or in filtration-type extraction, where the solvent and the seed mass are slurried and filtered. After a solvent–oil extraction, the solvent is removed from the miscella (solvent plus oil) by distillation. The solvent is removed from the meal in a desolventizer toaster in which the meal is subjected to a temperature gradient from 80 to about 110 °C. The extracted meal can be used as animal feed, and the crude oil can be used for the manufacture of soap, glycerine, lecithin or fatty acid manufacturing.

Crude oils contain various natural impurities, which give an unpleasant flavour and colour to the oil. The content of free fatty acids can cause spoilage, which prevents storage or further processing steps. Therefore, the crude oils must be refined before consumption. Refining includes the following four main steps:

- cleaning with acid (e.g. phosphoric acid) and water at about 90 °C
- neutralizing with sodium hydroxide at about 90 °C
- bleaching with Fuller's earth
- deodorizing with water steam at high temperatures (190–270 °C).

Regarding these stages, deodorizing is the most effective step in reducing pesticide residues in vegetable oils. However, further modifications (e.g. hydrogenation or winterization) of fats and oils can also reduce residues.

The loss of pesticide residues during the processing of crude and refined cotton seed oil can be exemplified by a study with aldicarb (Tunçbilek et al., 1997). In the latter, ^{14}C-labelled aldicarb was applied to cotton plants grown in Turkey. The crude oil was produced by hulling the delinted seed and crushing the seeds with a hammer. Then, the cotton seed was extracted with hexane in a Soxhlet apparatus for 12 h. Subsequently, the solvent was removed from the extract by evaporation. The crude oil contained 27 % of the total radioactivity from the harvested seeds. Most of the residue remained in the cake. After the refining process, the refined oil contained only 35 % of the original radioactivity in the crude oil, equivalent to 9.5 % of the radioactivity in the harvested seeds.

Table 4.12 shows further examples of the effect of oil preparation on the residue levels of different pesticides applied to oil seeds. In most cases, especially for hydrophilic pesticides, a reduction of residues in crude and refined oil was observed, whereas more lipophilic pesticides, such as parathion-methyl, can concentrate in the oil. A concentration of residues in the hulls and presscake is explainable by residues being located mainly on the surface of the oil seeds.

Fruits from oil palms and olive trees can be used for the production of crude and refined oil; olive oil is the more important vegetable oil for the European market.

The best olive oil quality for consumption is produced by pressing procedures, without solvent extraction of the olive press cake or refining the crude oil. In Europe, native olive oil is classified into nine different quality grades depending on the physicochemical and organoleptic properties of the oil (Anon, 1987): native olive oil extra, native olive oil, normal native olive oil, lampant oil, refined olive oil, olive oil, crude olive oil, refined olive-residue oil and olive-residue oil.

Table 4.12 Effect of oil preparation on pesticide residue levels in oil seeds

Crop[a]	Pesticide	Processed commodity, procedure	Transfer factor	Reference
Cotton seed	Aldicarb (oxime carbamate)	Hulls	3–5	FAO (1994a)
		Meal, extracted	0.1–0.5	
		Crude oil	$< 0.1^b$	
		Refined oil	$< 0.1^b$	
	Methamidophos (organophosphate)	Hulls	27	FAO (1994b)
		Crude oil	$< 0.1^b$	
		Refined oil	$< 0.1^b$	
Maize	Iprodione (dicarboximide)	Meal	0.5–0.6	FAO (1994a)
		Crude oil	0.3–0.6	
		Refined oil	$< 0.1^b$	
Peanut, nut dry	Aldicarb (oxime carbamate)	Hulls	2–3	FAO (1994a)
		Kernels	0.1–0.7	
		Meal	$< 0.1^b$–1.2	
		Crude oil	$< 0.1^b$	
Peanut, nut meat	Iprodione (dicarboximide)	Hulls	11–16	FAO (1994a)
		Crude oil, solvent extracted	3	
		Crude oil, screw pressed	$< 0.1^b$	
		Refined oil	$< 0.1^b$	
Rapeseed	Chlormequat (quaternary ammonium)	Oil	$< 0.1^b$	FAO (1994a)
Soya bean	Clethodim (cyclohexanedione oxime)	Hulls	1	FAO (1994a)
		Meal	1	
		Crude oil	0.1	
		Refined oil	$< 0.1^b$	
		Crude lecithin	2	
		Soapstock	1	
	Diquat (bipyridylium)	Hulls	2	FAO (1994a)
		Meal	0.7	
		Crude oil	$< 0.1^b$	
		Refined oil	$< 0.1^b$	
		Soapstock	0.1	
	Parathion-methyl (organophosphate)	Hulls	20	FAO (1994b)
		Crude oil	5	
		Refined oil	4	
Sunflower seed	Clethodim (cyclohexanedione oxime)	Hulls	1	FAO (1994a)
		Presscake, solvent extracted	2	
		Presscake, screw pressed	2	
		Crude oil, solvent extracted	$< 0.1^b$	
		Crude oil, screw pressed	$< 0.1^b$	
		Refined oil	$< 0.1^b$	

[a] RAC, raw agricultural commodity.
[b] Residues in crude and/or refined oil were below the limit of determination.

The physicochemical parameters include the content of free fatty acids, peroxide, and waxes or cholesterol.

The processing of native olive oil starts with removing damaged fruits, washing the fruits and subsequently crushing the olives in either a rolling or discs mill to produce olive paste. The paste is kneaded in mixing or kneading devices, with the addition of salt in order to disintegrate the oil cells. Afterwards, the paste is pressed in open straining presses or in continuously operating screw presses. Then the oil–water emulsion is cleared by centrifugation in order to separate the native oil from the water and the impurities. The olive presscake, which may be used for animal feed, is dried until approximately 10 % moisture remains.

Table 4.13 shows examples of the effect of oil preparation on the residue levels of different pesticides applied to olives. Most of the residues were concentrated in olive oil as a result of their lipophilic nature, whereas the hydrophilic dimethoate was reduced by the processing procedure.

A monitoring study by Lentza-Rizos (1994) on olive fruits and olive oil including eight organophosphorus insecticides and one metabolite showed that the residues in commercially packed oil (native and refined) either contained no determinable residues of the insecticides or low residue concentrations in the case of fenthion, fenthion sulfoxide and chlorpyrifos. A relatively high concentration of fenthion in oil was to be expected because fenthion was used for the control of *Dacus oleae* (olive fruit fly), with some treatments close to harvest. Furthermore, fenthion is lipophilic, indicated by a log (octanol–water partition coefficient) greater than 4 (log $P_{OW} = 4.84$).

Processing studies with olives treated with ethephon (ethylene generator) indicated very low residue levels (< 0.01–0.012 mg/kg) in the olive oil when the olives were harvested six to seven days after treatment (FAO, 1994a).

Table 4.13   Effect of the preparation of olive oil (processed commodity) on pesticide residue levels in olive fruits (Cabras *et al.*, 1997)

Pesticide[a]	Transfer factor
Azinphos-methyl	3
Diazinon	3
Dimethoate	0.2
Methidathion	3
Parathion-methyl	5
Quinalphos	3

[a] Organophosphates.

Drying

The aim of food preservation by drying is the reduction of the water activity in the crop to a level where microbial growth will not occur. Fruits and vegetables can be dried by several procedures, e.g. by the sun (only fruits) or mechanically by dehydrators. The more widely dried fruits include raisins, prunes, dates, apples, figs, apricots and peaches. Many are sun-dried to yield a product with a water content of less than 25 %. A heat process reducing the final water content to 1–5 % must be employed in order to obtain dehydrated fruits, which are utilized for reprocessing. It is common to treat fruits with sulfur dioxide (1–4.9 mg/l aqueous solutions) prior to drying to inhibit browning (Somogyi and Luh, 1986). Common dried vegetables include dehydrated potatoes, dried beans, powdered onions, garlic and bell peppers.

Table 4.14 shows examples of the effect of drying on the residue levels of different pesticides applied to fruits. In most cases, an increased concentration of residues was observed in the raisins and dried apples, due to the fact that the relative amount of dry matter is higher in the dried fruits and, consequently, also the residue concentration, expressed as mg/kg, is increased.

Table 4.14 Effect of drying on pesticide residue levels

Crop[a]	Pesticide	Processed commodity, procedure	Transfer factor	Reference
Apple	Captan (phthalimide)	Dried apples	0.8–2	FAO (1994a)
	Tebuconazole (azole)	Dried apples	0.5	FAO (1994b)
Grape	Aldicarb (oxime carbamate)	Raisins	0.4–2	FAO (1994a)
	Captan (phthalimide)	Raisins	1	FAO (1994a)
		Raisin waste	3–26	
	Carbendazim (benzimidazole)	Raisins	1–2	FAO (1994a)
		Raisin waste	4–5	
	Ethion (organophosphate)	Raisins	2–3	FAO (1994a)
		Raisin waste	6–10	
	Fenarimol (pyrimidinyl carbinol)	Raisins	0.2–2	FAO (1995)
	Penconazole (azole)	Raisins	2–6[b]	FAO (1995)
		Raisin waste	4–9[b]	
	Tebuconazole (azole)	Raisins, sun-dried	1	FAO (1994b)
		Raisins, oven-dried	2	
		Raisin waste, sun-dried	4	
		Raisin waste, oven-dried	3	
Plum	Captan (phthalimide)	Prunes	0.1	FAO (1994a)

[a] RAC, raw agricultural commodity.
[b] Total residues as DCBA.

EVALUATION OF RESULTS

Based on the results of processing studies, it will be possible to achieve the following:

• recognize reductions and concentrations of residues in food and feed items;

- interpret the reasons for changes in the residue concentrations (loss versus redistribution versus change in moisture content);
- calculate transfer factors;
- perform a more realistic dietary risk assessment.

If more than one processing study is available for a particular pesticide on a given RAC, the mean transfer factor should be used when deriving an MRL for the processed food or in assessing the chronic dietary risk.

The following calculations may be performed with the aid of the transfer factor:

(a) to set an MRL for a processed commodity, the MRL of the RAC is multiplied by the mean transfer factor;
(b) to estimate the contribution of processed food to the chronic dietary intake of pesticide residues, the median residue found in the RAC (median from a set of field trials) is multiplied by the mean transfer factor;
(c) to estimate the contribution of processed food to the acute dietary intake of pesticide residues, the highest residue found in the RAC is multiplied by the highest transfer factor.

The principles and procedures of dietary risk assessments are described later in Chapters 7 and 8.

REFERENCES

Alary, J., Bescos, D., Monge, M. C., Debrauwer, L. and Bories, G. F. (1995). Laboratory simulation of captan residues degradation during apple processing, *Food Chem.*, **54**, 205–211.

Anon (1987). *Leitsätze für Speisefette und Speiseöle*, Lebensmittelrecht/Text EL 43, Anhang 2/62, C. H. Beck'sche Verlagsbuchhandlung, München, Germany.

Barooah, A. K. and Yein, B. R. (1996). Residues of quinalphos in/on brinjal (*Solanum Melongena*) fruits, *Pesticide Res. J.*, **8**, 43–48.

Buys, M., Timme, G., Tournayre, J.-C., Walz-Tylla, B. and Whiteoak, R. (1993). *Guidelines for Processing Studies*, European Crop Protection Association, Brussels.

Cabras, P., Angioni, A., Garau, V. L., Melis, M., Pirisi, F. M., Karim, M. and Minelli, E. V. (1997). Persistence of insecticide residues in olives and olive oil, *J. Agric. Food Chem.*, **45**, 2244–2247.

Çelik, S., Kunç, S. and Asan, T. (1995). Degradation of some pesticides in the field and effect of processing, *Analyst*, **120**, 1739–1743.

FAO (1994a). *Pesticide Residues in Food – 1994 Evaluations Part I, Residues*, FAO Plant Production and Protection Paper 131/1, Food and Agriculture Organization of the United Nations, Rome.

FAO (1994b). *Pesticide Residues in Food – 1994 Evaluations Part II, Residues*, FAO Plant Production and Protection Paper 131/2, Food and Agriculture Organization of the United Nations, Rome.

FAO (1995). *Pesticide Residues in Food – 1995 Evaluations Part I, Residues*, FAO Plant Production and Protection Paper 137, Food and Agriculture Organization of the United Nations, Rome.

FAO (1996). *Pesticide Residues in Food – 1996 Evaluations Part I, Residues*, FAO Plant Production and Protection Paper 142, Food and Agriculture Organization of the United Nations, Rome.

Frehse, H. (1992). *Pesticide Residues in Food*, GIFAP Position Paper, Brussels Dépôt Légal D/1993/2537/4.

Frehse, H. (1997). Pesticide residues in food, paper presented at the IBC UK Conference *Pesticide Residues in Food and Dietary Risk Assessment*, London, 27 June 1997, ECPA 6 Av.E.Van Nieuwenhuyse, Brussels, Belgium.

Heiss, R. (1990). *Lebensmitteltechnologie – Biotechnologische chemische mechanische und thermische Verfahren der Lebensmittelverarbeitung* Springer-Verlag, Berlin, pp. 187–188.

Holland, P. T., Hamilton, D., Ohlin, B. and Skidmore, M. W. (1994). Effects of storage and processing on pesticide residues in plant products, *Pure Appl. Chem.*, **66**, 335–356.

Kamil, M. E.-D. M., Abou-Zahw, M. M. and Hegazy, N. A. (1996). Efficiency of some technological processes on reducing the residues of malathion and pirimiphos-methyl in mature broad bean seeds, *Nahrung*, **40**, 277–281.

Kilara, A. and Van Buren, J. (1989). Clarification of apples juice, in *Processed Apple Products*, Downing, D. L. (Ed.), Van Nostrand Reinhold, New York, Ch. 4, pp. 83–96.

Kumari, B., Kumar, R., Malik, M. S., Naresh, J. S. and Kathpal, T. S. (1996). Dissipation of endosulfan and lindane on sunflower seeds and cowpea pods, *Pesticide Res. J.*, **8**, 49–55.

Lentza-Rizos, C. (1994). Monitoring pesticide residues in olive products: organophosphorus insecticides in olives and oil, *J. Assoc. Off. Anal. Chem.*, **77**, 1097–1100.

Leparulo-Loftus, M., Petersen, B. J., Chaisson, C. F. and Tomerlin, J. R. (1992). Dietary exposure assessment in the analysis of risk from pesticides in foods, in *Food Safety Assessment*, Finley, J. W., Robinson, S. F. and Armstrong, D. J. (Eds), American Chemical Society, Washington, DC, ch. 21, pp. 214–229.

Nagayama, T. (1996). Behavior of residual organophosphorus pesticides in foodstuffs during leaching or cooking, *J. Agric. Food Chem.*, **44**, 2388–2392.

Ong, K. C., Cash, J. N., Zabik, M. J., Siddiq, M. and Jones, A. L. (1996). Chlorine and ozone washes for pesticide removal from apples and processed apple sauce, *Food Chem.*, **55**, 153–160.

Petersen, B., Tomerlin, R. J. and Barraj, L. (1996). Pesticide degradation: exceptions to the rule, *Food Technol.*, **50**, 221–223.

Reynolds, S. (1996). Effects of Processing on Pesticide Residues, SCI Lecture Papers Series, Paper No 0081, Society of Chemical Industry, London.

Schattenberg, H. J. III, Geno, P. W., Hsu, J. P., Fry, W. G. and Parker, R. P. (1996). Effect of household preparation on levels of pesticide residues in produce, *J. Assoc. Off. Anal. Chem.*, **79**, 1447–1453.

Schobinger, U., Askar, A., Brunner, H. R., Crandall, P. G., Daepp, H. U., Dittrich, H. H., Feric, M., Glunk, U., Herrmann, K., Korth, A., Lüthi, H. R., Schumann, G., Šulc, D., Tanner, H. and Weiss, J. (1987). Frucht- und Gemüsesäfte – Technologie, Chemie, Mikrobiologie, Analytik, Bedeutung, Recht, Eugen Ulmer GmbH & Co., Stuttgart-Hohenheim, Germany, pp. 24 and 263.

Sheikhorgan, A., Talebi, K., Zolfagharieh, H. R. and Mashayekhi, S. (1994). *Thiometon Residues in Cucumber*, IAEA-SM-343/27, International Atomic Energy Agency, Vienna, Austria.

Somogyi, L. P. and Luh, B. S. (1986). Dehydration of fruits, in *Commercial Fruit Processing*, Woodruff, J. G. and Luh, B. S. (Eds), AVI Publishing Company, Westport, CT, USA, pp. 353–405.

Somogyi, L. P., Ramaswamy, H. S. and Hui, Y. H. (1996). Biology, Principles and Applications, in *Processing Fruits: Science and Technology*, Vol. 1, pp. 67–94.

Tsumura, Y., Hasegawa, S., Sekiguchi, Y., Nakamura, Y., Tonogai, Y. and Ito, Y. (1994). Residues of post-harvest application pesticides in buckwheat after storage and processing into noodles, *J. Food Hygienic Soc. Jpn*, **35**, 1–7.

Tunçbilek, A., Aysal, P. and Halitligil, M. B. (1997). Loss of label during processing cotton oil grown with ^{14}C-aldicarb, *Bull. Environ. Contam. Toxicol.*, **58**, 213–218.

Vogeler, K., Dreze, Ph., Rapp, A., Steffan, H. and Ullemeyer, H. (1977). Distribution and metabolism of propineb in apples and grapes, and of its degradation products propylenethiourea and ethylenethiourea in apples, *Bayer AG, Pflanzenschutz-Nachrichten*, **30**(1), 72–97.

WHO (1997). *Guidelines for Predicting Dietary Intake of Pesticide Residues* (revised), Programme of Food Safety and Food Aid, WHO/FSF/FOS/97.7, World Health Organization, Geneva, Switzerland.

Wolff, I. A. (1982). *CRC Handbook of Processing and Utilization in Agriculture*, Vol. II (Part 1), *Plant Products*, CRC Press, Boca Raton FL, USA, pp. 91–121.

Wolff, I. A. (1983). CRC Handbook of Processing and Utilization in Agriculture, Vol. II (Part 2), *Plant Products*, CRC Press, Boca Raton, FL, USA, pp. 145–297.

5 Toxicological Assessment of Agricultural Pesticides

MIKE WATSON
Ricerca Inc., Painesville, OH, USA

INTRODUCTION

Toxicology is that branch of science that deals with the potentially harmful effects of chemicals. That part of toxicology described in this present chapter is associated with determining the potential for adverse effects of agricultural pesticides in order to perform risk assessments as part of the regulatory approval process. It is important to remember that these experiments are designed to help answer a quite specific question – Is the proposed use of the agricultural pesticide acceptably safe? It therefore stands to reason that different proposed uses of agricultural pesticides will require different toxicology testing programs. The toxicology testing scheme described in this chapter is based on the assumption that the proposed pesticide use might result in exposure of operators applying the pesticide in the field and that residues of a pesticide-related chemical might be detectable in a food commodity. This situation generally requires the most extensive toxicology testing. If potential human exposure is more restricted, then a more restricted toxicology testing program may be appropriate.

Pesticide Residues in Food and Drinking Water: Human Exposure and Risks.   Edited by Denis Hamilton and Stephen Crossley
© 2004 John Wiley & Sons, Ltd   ISBN: 0-471-48991-3

A fundamental concept of toxicology is that of the dose–response relationship. This is linked to the biological concept of variability. Individual members of a species display different characteristics. Therefore, it may be expected that different members of a species react to a greater or lesser extent when exposed to any given chemical. For each member of the species there will be some level of exposure that causes no discernible reaction. In toxicology this is referred to as the 'no-adverse-effect level'. In practice, groups of animals are used in toxicology studies, with each group receiving a different dose of the test material, and observations are compared to an untreated 'control' group. In this manner, it is possible to estimate the no-adverse-effect level for the population being studied. By studying reactions in different animal species, toxicologists can extrapolate between species and predict what might happen if humans were exposed to the material being tested. In general terms, the objective of toxicology testing is to define a safe exposure level for humans.

GUIDELINES

The results of toxicity testing of agricultural pesticides are subject to detailed scrutiny by regulatory authorities as part of a process of approval prior to sale and use of the product. In order to try to ensure that regulatory authorities receive the type and quality of data they want to see in this review process, these same regulatory authorities have developed guidelines, which list the toxicity tests to be conducted and detail how to perform the tests. In some cases, regulatory authorities act through supra-national organizations, such as the Organization for Economic Co-operation and Development (OECD).

The OECD has published the most comprehensive set of guidelines detailing how to perform toxicology studies as *The OECD Guidelines for Testing Chemicals*. This volume is composed of four sections:

- Section 1 – Physical-chemical properties (blue pages)
- Section 2 – Effects on biotic systems (green pages)
- Section 3 – Degradation and accumulation (yellow pages)
- Section 4 – Health effects (pink pages)

The toxicology guidelines relevant to this chapter are contained in Section 4. These OECD guidelines are in a process of continual development. Chemical manufacturers and trade organizations work with government authorities in individual countries to update old guidelines and propose new guidelines to the OECD, which organizes a review process leading to acceptance and publication of such guidelines. Studies conducted in accordance with the OECD Guidelines will be acceptable to regulatory authorities in all member countries of the OECD (although there are sometimes contentious issues with older OECD Guidelines that a regulatory authority may argue are no longer scientifically valid). It is not the objective of this chapter simply to re-publish the OECD Guidelines, but rather to refer the reader to documents such as these guidelines for details of study design and to present here some background information which gives reasons for conducting studies, points to consider while performing studies and some commentary on interpretation of results and their use in risk assessment.

The OECD Guidelines are restricted to how to perform studies and do not give any details on when to perform the different studies. Such information is available as guidance from the different regulatory authorities that will be reviewing the data package as part of a marketing submission. As mentioned earlier, the required extent of testing will be determined by the potential for human exposure resulting from the proposed use of the agricultural pesticide. In general, testing will be required in a number of different areas, as follows:

- Acute toxicity testing
- Short-term repeat dose toxicity testing
- Long-term repeat dose toxicity/carcinogenicity testing
- Reproductive toxicity testing
- Genotoxicity
- Specialized studies

Each of these topics will be considered in more detail in subsequent sections of this chapter.

It is not possible to end a section dealing with guidelines without a mention of Good Laboratory Practice (GLP) Regulations. These guidelines were first promulgated in response to fabrication of data submitted to a regulatory authority. The GLP regulations do not have anything to do with the scientific aspects of study conduct, but do impact study quality – by regulating aspects such as record keeping and ensuring that any study can be easily 'reconstructed' from the raw data records of the study. Virtually all toxicology studies submitted to any regulatory authority must be conducted in compliance with the GLP regulations. Since their introduction in the 1970s, the GLP regulations have placed a large cost burden on laboratories conducting regulatory toxicology studies, but all reputable laboratories now have the required infrastructure in place. The GLP regulations also placed a cost burden on governments, by setting up inspection and approval units to ensure compliance with the regulations. Although the cost has been high, the introduction of the GLP regulations has resulted in better data being available to regulatory decision makers and is therefore well justified.

ACUTE TOXICITY TESTING

There are several definitions of 'acute toxicity'. In general, all agree that acute toxicity is concerned with adverse effects occurring within a short time of administration of a single dose of a test material. In a regulatory framework, acute toxicity testing is conducted in order to satisfy the classification, packaging and labelling requirements associated with transport of chemicals. It follows that testing must be performed with the active ingredient of the agrochemical and with the formulated product.

Acute toxicity testing required by regulatory authorities is often referred to as a "six pack" in view of the number of tests required:

- Acute oral toxicity
- Acute dermal toxicity
- Acute inhalation toxicity
- Dermal irritation
- Ocular irritation
- Dermal sensitization

The acute oral toxicity test is generally performed only in rodents. The standard methodology is described in OECD Guideline 401. This test was formerly known as the LD_{50}. Determination of the lethal dose that kills 50 % of a population of animals was used as a classification basis and to compare one chemical with another. In the assessment and evaluation of the toxic characteristics of a substance, acute toxicity testing is generally performed by the probable route of exposure in order to provide information on health hazards likely to arise from short-term exposure by that route. However, the acute oral toxicity test provides much more useful data than just an indication of the dose required to kill. Data obtained from an acute study may serve as a basis for hazard categorization, labelling, or child-resistant packaging and may also serve to designate pesticides that may be applied only by certified applicators. It is also an initial step in establishing a dosage regimen in sub-chronic and other studies and may provide information on absorption and the mode of toxic action of a substance. Acute toxicity testing has been the subject of much criticism in view of the number of animals used. New methods have been developed to allow generation of reliable data by using fewer animals than were used in the past. Use of these alternative test protocols when available is generally encouraged. Thus, for example, acute oral toxicity testing may be performed using the Fixed Dose Method (OECD Guideline 420), or the Acute Toxic Class Method (OECD Guideline 423), or the Up-and-Down Method (OECD Guideline 425). Some regulatory authorities are now insisting that acute toxicity data should be generated by using these newer methods, to such an extent that data generated in accordance with the older guidelines such as OECD 401 will be considered unacceptable. Alternative methods, using fewer animals, are not yet available through OECD for acute toxicity testing by routes of exposure other than oral.

The acute dermal toxicity test was historically performed by using rabbits. Animal usage and cost considerations have led to regulatory acceptance of this test being performed by using rats. Currently acceptable methodology is described in OECD Guideline 402. Acute inhalation toxicity is generally performed by using nose-only exposure in rats, with the currently acceptable methodology being described in OECD Guideline 403. Generation of the exposure atmosphere and monitoring of exposure, including attention to such details as particle size in the exposure atmosphere, require special skills.

Several factors should be considered in determining the corrosion and irritation potential of chemicals before testing is undertaken. Existing human experience and data and animal observations and data should be the first line of analysis, as this gives information directly referable to effects on the skin. In some cases, enough information may be available from structurally related compounds to make classification decisions. Likewise, pH extremes (pH < 2 or > 11.5) may indicate the potential for dermal effects. Generally, such agents are expected to produce significant effects on the skin. It also stands to reason that if a chemical is extremely toxic by the dermal route, a dermal irritation/corrosion study may not be needed. Likewise, if there is a lack of any dermal reaction at the limit dose (2000 mg/kg) in an acute toxicity study (for which observations of dermal reactions were made), a dermal irritation/corrosion study again may not be needed. It should be noted, however, that acute dermal toxicity and dermal irritation/corrosion testing might be performed in different species that may differ in sensitivity. *In vitro* alternatives that have been validated and accepted may also be used to help make classification decisions, although there are currently no OECD guidelines for these *in vitro* tests. The currently acceptable animal test methodology is described in OECD Guidelines 404 and 405.

The dermal sensitization study determines whether a product is capable of causing an allergic reaction. The guinea pig has been the animal of choice for predictive sensitization tests for several decades. Two types of tests have been developed: adjuvant tests in which sensitization is potentiated by the injection of Freunds Complete Adjuvant (FCA), and non-adjuvant tests. In OECD Guideline 406, the Guinea Pig Maximization Test (GPMT) of Magnusson and Kligman which uses adjuvant and the non-adjuvant Buehler Test are given preference over other methods and the procedures are presented in detail. The choice of which method to use for sensitization testing in agrochemical risk assessment is a very contentious issue. Results of an adjuvant test indicate whether the test material has any potential at all to induce a sensitization reaction. Non-adjuvant tests attempt to mimic the dermal exposure that occurs in use of the agrochemical and thereby predict whether a sensitization reaction would occur. Experience in dealing with specific regulatory authorities will dictate which testing methodology is required in any given situation. The greatest difficulty with these test methods is the subjective nature of the test end-point – an assessment of a dermal reaction and comparison with a dermal reaction in a different animal. New methods for sensitization testing have been developed which make the end-point more objective. One such method is the mouse lymph node assay, which uses measurement of incorporation of tritiated thymidine as the end-point. Validation work is underway to incorporate this method into the internationally accepted regulatory framework.

SHORT-TERM REPEAT-DOSE TOXICITY TESTING

In the assessment and evaluation of the toxic characteristics of a chemical, the determination of oral toxicity using repeated doses may be carried out after initial information on toxicity has been obtained by acute testing. These studies provide information on the possible health hazards likely to arise from repeated exposure over a period of time. The duration of exposure in the first repeat-dose study is generally 28 days. Currently acceptable test methodology is described in OECD Guideline 407. Toxicity studies of increasing duration will be required, as the potential for duration of human exposure increases. For an agricultural pesticide that has the potential to leave residues in food crops, the toxicology testing requirement will include studies covering long-term administration to animals. In practice, it is difficult to start a long-term test without doing a short-term test first, because of the difficulties of dose selection. Thus, 28-day studies are conducted, leading to 90-day studies, 1 year studies and eventually, in rodents, life-span, 2-year studies. OECD Guidelines 408 and 409 give examples of the acceptable test methodology.

The principles of study design for repeat-dose toxicology tests are the same, regardless of the duration of the study. The test substance is orally administered daily in graduated doses to several groups of experimental animals, one dose level per group throughout the test period. During the period of administration, the animals are observed closely each day for signs of toxicity. Laboratory investigations, including hematology, blood chemistry and urinalysis, may be performed to look for effects caused by the test material. Animals that die or are killed during the test are necropsied and at the conclusion of the test surviving animals are killed and necropsied. Following a detailed post-mortem examination, which generally includes organ weight analysis, a wide range of tissues is preserved and samples of tissues subjected to histopathological examination.

It is important to remember that guidelines can only give general details of study design. There is no substitute for experience in the conduct of toxicology studies and knowledge of specific test materials. Certain classes of test material require specific tests to be incorporated into toxicology studies. For example, organophosphate insecticides work by inhibition of acetylcholine esterase, an enzyme crucial to nerve transmission. In toxicology studies with organophosphates it is essential to measure the activity of acetylcholine esterase in plasma, red blood cells and brain tissue. Such investigations would not be required with pyrethroid insecticides, but would be required for carbamate insecticides, which also work by inhibition of acetylcholine esterase. Measurement of acetylcholine esterase activity in animals treated with carbamate insecticides presents technical challenges in that the enzyme inhibition is rapidly reversible, which is not the case with the inhibition caused by an organophosphate. Sound science must lead both study design and experimental performance, rather than simple adherence to a published study guideline.

As mentioned earlier, the overall objective, in a regulatory setting, is to determine what level of exposure is safe. In a research setting, of course, toxicology testing serves other purposes. Development of new and better agricultural pesticides is aided by knowledge of which chemicals, or which parts of chemical structures, cause unacceptable toxicity.

LONG-TERM REPEAT-DOSE TOXICITY/CARCINOGENICITY TESTING

The objective of a long-term carcinogenicity study is to observe test animals for a major portion of their life span for the development of neoplastic lesions during or after exposure to various doses of a test substance by an appropriate route of administration. Such an assay requires careful planning and documentation of the experimental design, a high standard of pathology, and unbiased statistical analysis. These requirements are well known and have not undergone any significant changes during the past several years. The currently acceptable test methodology is described in OECD Guidelines 451, 452 and 453.

It is generally accepted that agrochemicals of unknown carcinogenic potential should be tested on two animal species. Although an unequivocal carcinogenic effect in one species is considered as a warning of carcinogenic potential in humans, only negative findings in all species tested (at least two) can be regarded as adequate negative evidence. Rats and mice have been preferred because of their relatively short life span, the limited cost of their maintenance, their widespread use in pharmacological and toxicological studies, their susceptibility to tumour induction, and the availability of inbred or sufficiently characterized strains. As a consequence of these characteristics, a large amount of information is available on their physiology and pathology. In selecting the species and strain, it is important to be aware that there are particular susceptibilities. For instance, in general it is easier to induce liver tumours in the mouse than in the rat, and conversely it is easier to induce subcutaneous tumours in the rat than in the mouse. There is no scientific rationale to recommend inbred, outbred or hybrid strains over any others. The important requirement is that the animals be from well-characterized and healthy colonies. Random-bred animals or animals bred with maximum avoidance of inbreeding of a well-characterized colony are acceptable. The use of inbred strains has the advantage of the availability of animals with known characteristics, such as an average life span, and a predictable spontaneous tumour rate. Hybrid mice of two inbred strains can be used because they are particularly robust and long-lived. A good knowledge of the tumour profile of the animal strain throughout the life span is highly desirable in order to evaluate the results of experiments in a proper way.

Long-term bioassays for carcinogenesis have been initiated most commonly in weanling or post-weanling animals. This procedure has allowed the greater part of the life span for tumour development to occur coincident with the exposure to the test substance. Interest in the possible increased susceptibility of the neonate arose with the evidence that established the influence of host age on viral carcinogenesis. More recently, pre-natal exposure has been the subject of considerable experimentation. It has been demonstrated that some tissues, particularly nervous tissues, are more susceptible during foetal life than later. At present, there is only limited evidence that pre-natal exposure may reveal the carcinogenic potential of a chemical that would not have been revealed had the treatment started at a later age. Although this is generally not a concern for agrochemicals, pre-natal exposure to test material for animals on carcinogenicity studies is often an important aspect of study design in other sectors (most notably with food additives regulated in the USA by the Food and Drug Administration (FDA)).

The most contentious area of design of carcinogenicity studies is concerned with selection of dose levels. There is general agreement that for risk assessment purposes, at least three dose levels should be used, in addition to the concurrent control group. The OECD Guideline 451 states that the highest dose level should be sufficiently high to elicit signs of minimal toxicity without substantially altering the normal life span due to effects other than tumours. Signs of toxicity are those that may be indicated by alterations in levels of activity of certain serum enzymes or slight depression of body weight gain (less than 10 %). Although this definition is receiving increasingly widespread support, some regulatory authorities prefer to see more overt toxicity at the high dose level.

The US Environmental Protection Agency (EPA) has in the past referred to the highest dose in a carcinogenicity study being the 'maximum tolerated dose'. Further discussion of dose level selection can be found in *Principles for the Selection of Doses in Chronic Rodent Bioassays* (International Life Sciences Institute, 1997). There are other points that receive more general acceptance. For a diet-mixture, the highest concentration should generally not exceed 5 %. The lowest dose should not interfere with normal growth, development and longevity of the animal, and it must not otherwise cause any indication of toxicity. The intermediate dose(s) should be established in a mid-range between the high and low doses, depending upon the toxicokinetic properties of the chemical, if known. The selection of these dose levels should be based on existing data, preferably on the results of sub-chronic studies. Frequency of exposure is normally daily but may vary according to the route chosen. If the chemical is administered in the drinking water or mixed in the diet, it should be continuously available.

The exact duration of carcinogenicity studies is a further point that deserves special consideration. It is necessary that the duration of a carcinogenicity test comprises the majority of the normal life span of the animals to be used. It has been suggested that the duration of the study should be for the entire lifetime of all animals. However, a few animals may greatly exceed the average lifetime, and the duration of the study may be unnecessarily extended and complicate the conduct and evaluation of the study. Rather, a finite period covering the majority of the expected life span of the strain is preferred since the probability is high that, for the great majority of chemicals, induced tumours will occur within such an observation period.

Generally, the termination of the study should be at 18 months for mice and hamsters and 24 months for rats; however, for certain strains of animals with greater longevity or low spontaneous tumour rate, termination should be at 24 months for mice and hamsters and at 30 months for rats. However, termination of the study is acceptable when the number of survivors of the lower doses or control group reaches 25 %. For the purpose of terminating the study in which there is an apparent sex difference in response, each sex should be considered a separate study. In the case where only the high-dose group dies prematurely for obvious reasons of toxicity, this should not trigger termination. In order for a negative test to be acceptable, no more than 10 % of any group should be lost due to autolysis, cannibalism or management problems, and survival of all groups should be no less than 50 % at 18 months for mice and at 24 months for rats.

Histopathological examination and statistical analysis of the pathology data are the crucial end-points of carcinogenicity studies. In simplistic terms, the tumour incidence in treated animals is compared to the tumour incidence in untreated, control animals. However, there are many other points to consider. Treatment might induce a very low incidence of a tumour type that is hardly ever seen in animals of the species and strain used in the study. Benign tumours (which are localized in one tissue) can sometimes progress to malignant types (which spread throughout the body to different tissues) and treatment may accelerate this process. In a life-span study, a certain tumour incidence is expected and treatment may have no effect on the overall tumour incidence, but the tumours may arise earlier in treated animals. As is the case in many fields of scientific endeavour, experience is the key requirement in the ability to perform carcinogenicity studies and interpret their results.

There is no doubt that carcinogenicity testing is expensive and time-consuming, which has led to research aimed at reducing the time and cost burdens of these tests. Such efforts have been led by the pharmaceutical industry, which is obviously under pressures similar to those of the agrochemical industry, but changes will clearly have relevance in the safety testing of agrochemicals.

Research has been sparked by the advent of transgenic research and the availability of animal models genetically altered to be more susceptible to the effects of carcinogens than normal animals. The basic principle of these new test designs is that if an animal is more susceptible to the effects of carcinogens then the test may be completed in a shorter time. There have been validation studies of alternative designs using different animal strains and in the area of pharmaceutical testing it has been accepted that a life-span study in rats can be supported by a shorter, transgenic mouse test. However, close consultation is needed between sponsors of such studies and the regulatory authorities who want to be involved in the design of the study. There is evidently a potential for cost saving, since a study of some 13 weeks duration is a lot less costly than a life-span study, but the overall length of time taken for the testing program is not reduced since it is still necessary to perform the two-year rat study.

REPRODUCTIVE TOXICITY TESTING

The aim of reproduction toxicity studies is to reveal any effect on mammalian reproduction. For this purpose, both the investigations and the interpretation of the results should be related to all other toxicological data available to determine whether potential reproductive risks to humans are greater than, lesser than, or equal to those posed by other toxicological manifestations. Furthermore, repeated-dose toxicity studies can provide important information regarding potential effects on reproduction, particularly male fertility. The combination of the studies selected should allow exposure of mature adults and all stages of development from conception to sexual maturity. To allow detection of immediate and latent effects of exposure, observations should be continued through one complete life cycle, i.e. from conception in one generation through conception in the following generation. For convenience of testing this integrated sequence can be sub-divided into the following stages:

A Pre-mating to conception (adult male and female reproductive functions, development and maturation of gametes, mating behaviour, and fertilization).
B Conception to implantation (adult female reproductive functions, pre-implantation development, and implantation).
C Implantation to closure of the hard palate (adult female reproductive functions, embryonic development, and major organ formation).
D Closure of the hard palate to the end of pregnancy (adult female reproductive functions, foetal development and growth, and organ development and growth).
E Birth to weaning (adult female reproductive functions, neonate adaptation to extrauterine life, and pre-weaning development and growth).
F Weaning to sexual maturity (post-weaning development and growth, adaptation to independent life, and attainment of full sexual function).

Reproduction toxicity testing of agricultural pesticides has generally been performed by conduct of two study types – multi-generation studies, looking at general reproductive performance, and teratology studies to look specifically for abnormalities that might arise during pregnancy. The currently acceptable methodology is listed in OECD guidelines 414, 415 and 416.

In a multi-generation study, which is generally carried out only in rats, the test substance is administered in graduated doses to several groups of males and females. Males of the parental generation should be dosed during growth and for at least one complete spermatogenic cycle (approximately 70 days in the rat) in order to elicit any adverse effects on spermatogenesis by the test substance.

Females of the parental generation should be dosed for at least two complete oestrous cycles in order to elicit any adverse effects on oestrous by the test substance. Males and females from the same treated groups are allowed to mate and reproduce normally. After weaning of the offspring, the administration of the substance is continued to the offspring during their growth into adulthood. The offspring are then allowed to mate and reproduce and their offspring are followed until weaning, when the study is terminated.

In teratology studies, which are generally performed in rats and rabbits, the test substance is administered in graduated doses, for at least that part of the pregnancy covering the period of organogenesis, to several groups of pregnant experimental animals, with one dose being used per group. Shortly before the expected date of delivery, the mother is sacrificed, the uterus removed, and the contents examined for embryonic or foetal deaths, and live foetuses. Recent changes in the accepted study design for teratology studies concern the exact duration of dosing. Older study designs refer to dosing during the period of organogenesis, which is generally taken to mean days six to fifteen of gestation in rats and days six to eighteen of gestation in rabbits. Newer study designs state that as a minimum, the test substance should be administered daily from implantation (generally accepted as days three to four of gestation) to the day before caesarean section on the day prior to the expected day of parturition. Alternatively, if preliminary studies do not indicate a high potential for pre-implantation loss, treatment may be extended to include the entire period of gestation, from fertilization to approximately one day prior to the expected day of termination.

Reproduction toxicology studies present technical challenges in scheduling the work to be performed. This is because study events are scheduled on specific days of gestation or days of age of newborn animals. Due to the biological variability in length of gestation, this means that events do not coincide on the same calendar day. Teratology studies involve dosing pregnant animals and laboratories performing these studies have to decide between in-house breeding programs to maintain a supply of pregnant animals or purchase of pregnant animals. The latter option presents difficulties if animals are to be dosed throughout the entire period of gestation. In-house breeding programs are simple to operate for rodents. In-house programs to supply pregnant rabbits are generally more complex, involving either artificial insemination (which requires hormone injections to induce ovulation) or natural mating. Teratology studies present further scheduling challenges because the total number of animals required for a study cannot be necropsied on one day in even the largest of facilities conducting these studies. Studies are therefore split into a number of different 'sub-sets' starting on different days – sometimes in different weeks, in order to minimize weekend working. In study design, careful attention must be paid to splitting the different treatment groups evenly across the sub-sets in order to avoid introduction of bias. The final report for these studies amalgamates the different sub-sets back into a single set of data for statistical analysis and interpretation.

GENOTOXICITY

Genotoxicity tests can be defined as *in vitro* and *in vivo* tests designed to detect compounds that induce genetic damage directly or indirectly by various mechanisms. These tests should enable hazard identification with respect to damage to DNA and its fixation. Fixation of damage to DNA in the form of gene mutations, larger-scale chromosomal damage, recombination and numerical chromosome changes is generally considered to be essential for heritable effects and in the multi-step process of malignancy, a complex process in which genetic changes may play only a part. Compounds that are positive in tests that detect such kinds of damage have the potential to be human carcinogens and/or mutagens, i.e. may induce cancer and/or heritable defects. Because the relationship between exposure to particular chemicals and carcinogenesis is established for man, while a similar relationship has been difficult to prove for heritable diseases, genotoxicity tests have been used mainly for the prediction of carcinogenicity. Nevertheless, because germ line mutations are clearly associated with human disease, the suspicion that a compound may induce heritable effects is considered to be just as serious as the suspicion that a compound may induce cancer. In addition, the outcome of such tests may be valuable for the interpretation of carcinogenicity studies.

No single test is capable of detecting all relevant genotoxic agents. Therefore, the usual approach should be to carry out a battery of *in vitro* and *in vivo* tests for genotoxicity. Such tests are complementary rather than representing different levels of hierarchy. A three- or four-test battery is generally accepted as covering the end-points of concern. The generally accepted standard test battery includes a test for gene mutation in bacteria (the 'Ames test'), a test for gene mutation in mammalian cells, an *in vitro* test for chromosomal damage, and an *in vivo* test for chromosomal damage using rodent haematopoietic cells. In some guidelines, the test for gene mutation in mammalian cells and the *in vitro* test for chromosomal damage can be performed as one test, since the mouse lymphoma tk assay, which is primarily a gene-mutation assay using mouse lymphoma cells, can also detect chromosomal damage. The OECD guidelines 471 to 486 provide examples of currently acceptable test methods.

SPECIALIZED STUDIES

Further special studies may be required on a case-by-case basis. The requirement may be driven by the nature of the test material, or by the results obtained in the basic set of toxicology studies.

NEUROTOXICITY

The best example of special toxicity testing being required based on the nature of the test material arises when agricultural insecticides have as their mode of action an association with the nervous system. Investigation of toxicity to the nervous system is called 'neurotoxicology'. Pyrethroids and organophosphates, and carbamates that inhibit acetylcholine esterase, fall into such a class. In these cases, testing needs to consider the potential of these substances to cause specific types of neurotoxicity that might not be detected in other toxicity studies.

The currently acceptable methodology is discussed in OECD guideline 424, which describes a neurotoxicity screening battery developed for use in rodents that consists of a functional observational battery, motor activity measurement and neuropathology. The functional observational battery (commonly called an FOB) consists of non-invasive procedures designed to detect gross functional deficits in animals and to better quantify behavioural or neurological effects detected in other studies. The functional observational battery includes assessment of signs of autonomic function, such as degree of lacrimation and salivation, recording the presence or absence of piloerection and exophthalmus, and a ranking or count of urination and defecation, including polyuria and diarrhoea. Pupillary function such as constriction of the pupil in response to light or a measure of pupil size and the degree of palpebral closure are also recorded, along with a description of the incidence, and severity of any convulsions, tremors or abnormal motor movements, both in the home cage and the open field. A ranking of the animal's reactivity to general stimuli such as removal from the cage or handling, and the general level of activity during observations are also recorded. Descriptions and incidence of any posture and gait abnormalities observed are also important. Fore-limb and hind-limb grip strength are measured by using an objective procedure along with a quantitative measure of landing foot splay.

Sensorimotor responses to stimuli of different types are used to detect gross sensory deficits (pain perception may be assessed by reaction to a tail-pinch and the response to a sudden sound, e.g., click or snap, may be used to assess hearing). A description is recorded of any unusual or abnormal behaviours, excessive or repetitive actions (stereotypies), emaciation, dehydration, hypotonia or hypertonia, altered fur appearance, red or crusty deposits around the eyes, nose or mouth, and any other observations are recorded that may facilitate interpretation of the data. The motor activity test uses an automated device that measures the level of activity of an individual animal. The neuropathological techniques are designed to provide data to detect and characterize histopathological changes in the central and peripheral nervous system.

This battery is designed to be used in conjunction with general toxicity studies and effects should be evaluated in the context of both functional neurological changes and neuropathological effects, and with respect to any other toxicological effects seen with the test material.

Validation of the procedures used to examine for potential neurotoxicity is important. Measurement of activity can be validated by treating animals with amphetamine and chlorpromazine, so that it is known that the procedures and equipment in use in a specific laboratory can detect increases and decreases in activity. Neuropathology can be validated by using toxicants with known effects on the central and peripheral nervous system, such as acrylamide, trimethyl tin and hexachlorophene.

DELAYED NEUROTOXICITY

Certain organophosphorus substances have been observed to cause delayed neurotoxicity in man. This phenomenon is called 'delayed neurotoxicity' since the neurotoxicity is separated by a period of some days, up to weeks from the exposure to the toxic agent. The OECD guidelines 418 and 419 describe special toxicity tests, which have been developed to investigate the potential for organophosphate chemicals to cause a delayed neurotoxic response. In a delayed neurotoxicity test, the test substance is administered orally, either as a single dose or repeatedly to domestic hens that have been protected from acute cholinergic effects. The animals are observed for 21 days for behavioural abnormalities, ataxia and paralysis. Biochemical measurements, in particular activity of neuropathy target esterase in blood and nervous tissue, are undertaken on hens randomly selected from each group (normally 24 and 48 h after dosing). Twenty-one days after a single exposure, or at the end of the repeat dose treatment period, the remainder of the hens are killed and histopathological examination of selected neural tissues is undertaken.

Validation of this study type is generally included in each individual study as a positive control group that is treated with a known delayed neurotoxicant. An example of a widely used neurotoxicant is tri-o-cresylphosphate (TOCP).

DEVELOPMENTAL NEUROTOXICITY

'Developmental neurotoxicity' is concerned with potential functional and morphological hazards to the nervous system that may arise in the offspring from exposure of the mother during pregnancy and lactation. Internationally accepted guidelines are still under active development but this is an area of special interest for the United States Environmental Protection Agency (US EPA), which has developed a test procedure in which the test substance is administered to several groups of pregnant animals during gestation and early lactation, with one dose level being used per group. The test material is also administered directly to the offspring, with the latter being randomly selected from within litters for neurotoxicity evaluation. The evaluation procedure includes observations to detect gross neurologic and behavioural abnormalities, determination of motor activity, response to auditory startle, assessment of learning and, at necropsy, brain weights and detailed neuropathological evaluation, including morphological measurements of gross changes in the size or shape of brain regions such as alterations in the size of the cerebral hemispheres or the normal pattern of foliation of the cerebellum.

DERMAL PENETRATION STUDIES

In risk assessment for application of pesticides, the extent of dermal penetration in operators needs to be considered. The extent of dermal penetration is often assumed to be in line with accepted default values (often 1 % or 10 %, dependent upon the concentration of the formulation). Dermal penetration can also be measured in *in vivo* or *in vitro* experiments. These studies are usually performed by using radio-labelled pesticide formulations, following the permeation of the radio-label through a skin membrane. Both *in vivo* and *in vitro* experiments have their own advantages and disadvantages. *In vivo* studies incorporate effects of blood flow and other effects that can only be seen in an intact animal. However, these studies are rarely conducted in man. *In vitro* studies, using skin membranes, cannot completely match the true scenario of living skin with a blood flow. However, human skin membranes can be used in *in vitro* studies. For these reasons, the generally accepted approach to measuring dermal penetration is to carry out an *in vivo* study in the first instance to refine any default assumptions of dermal penetration that have been made in the risk assessment. If this refinement does not adequately resolve any issues in the risk assessment, then an *in vitro* study may be conducted in addition to the *in vivo* study. The *in vitro* study should include human skin and a sample of skin from the animal species used in the *in vivo* study (normally the rat). This comparison of animal and human skin can then be used to derive an adjustment factor to be applied to the results of the *in vivo* study. In the vast majority of cases, human skin provides a better barrier to penetration than does skin in the rat.

INTERPRETATION OF RESULTS

Interpretation of results of toxicology studies in the process of development of agricultural pesticides is directed towards protection of consumers (who might eat food containing residues of pesticide products) and operators (who apply pesticides in the field). Consumer risk assessment is performed by comparison of predicted dietary intakes of the pesticide with an 'allowable' exposure – the acceptable daily intake (ADI). Operator exposure risk assessment is performed in exactly the same way, comparing predicted exposure with an 'allowable' limit, in this case, the acceptable operator exposure level (AOEL). These acceptable limits are often referred to as 'reference doses'.

The ADI and the AOEL are derived in exactly the same way. A selected no-effect level from a toxicology study is divided by a 'safety' or 'uncertainty' factor. This factor is intended to allow for uncertainty (in extrapolation from animals to man, variation within species or uncertainty about the actual no-effect level) and concern (a higher factor might be required if the no-effect level was based on a serious effect of great concern).

Traditionally, the 'default' safety factor has generally been assumed to be a factor of 100. This is composed of a factor of 10 for inter-species extrapolation combined with another factor of 10 for intra-species variation. Normally, both of these factors are needed for the derivation of reference doses. In some rare cases, there may be reliable human data that might make one of the factors of 10 unnecessary.

The difference between the ADI and AOEL is a reflection of the shorter length of time for which operators are exposed, compared to the potential lifetime exposure over which consumers may be exposed. Whereas the toxicological database to be considered in derivation of an ADI includes the totality of the available toxicology data, the database to be included in derivation of an AOEL includes general toxicology studies up to 13 weeks duration, teratology studies and (possibly) effects seen in the first generation of the reproduction study.

Since operator exposure can be a sum of the exposures from different routes (dermal, inhalation and oral), the AOEL is often referred to as a 'systemic' AOEL. This means that the different routes of operator exposure can be added and compared with a single AOEL during the operator exposure risk assessment. This has one implication for derivation of an AOEL from oral toxicity studies. If oral absorption is low, the observed 'systemic' toxicity is actually a result of a proportion of the administered oral dose. In these cases, the observed no-effect level will need to be 'corrected' to allow for the low absorption. The extent of oral absorption can be estimated from metabolism studies. Well-absorbed compounds ($> 75\%$) usually require no correction factor. More poorly absorbed compounds would need to incorporate a correction factor which would reduce the AOEL compared to that without any incorporation of a correction factor.

Operator exposure assessments comparing measured or predicted operator exposure with a systemic AOEL must also take into account the extent of dermal penetration in operators. The extent of dermal penetration is often assumed to be in line with accepted default values (often 1 % or 10 %, dependent upon the concentration of the formulation). As described previously, dermal penetration can also be measured in *in vivo* or *in vitro* experiments.

Dermal penetration can sometimes be estimated from comparison of the acute oral and dermal toxicity studies. However, in cases where the acute toxicity of the pesticide in question is relatively low, this comparison provides little useful information. In some cases, toxicology studies conducted using dermal administration may be useful in operator exposure risk assessment. However, there are a number of issues that need to be resolved before using a dermal toxicity study to derive a dermal AOEL. The first concerns the actual route or routes of operator exposure. Use of a 'systemic' AOEL allows different routes of operator exposure to be combined for comparison to a single AOEL. A dermal AOEL can only be compared with dermal operator exposure. Thus, the use of dermal toxicity studies in operator exposure risk assessment is only really valid when dermal operator exposure is predominant and the exposure resulting from other routes (e.g. inhalation) can virtually be ignored. The other main issue to be resolved concerns which type of a dermal toxicity needs to be conducted. It is pointless conducting, for example, a 28-day dermal toxicity study in rats, if the basic toxicology database suggests that the most crucial no-effect level for operator exposure risk assessment comes from a teratology study. In this case, a dermal teratology study would be needed. Similarly, in cases where, for example, a 28-day dermal toxicity study may be useful, the basic toxicology database must be carefully reviewed to ensure that all of the correct parameters are investigated in the dermal toxicity study.

In recent years, regulatory authorities have reviewed the established methods of risk assessment for consumer safety and have established new procedures to look specifically at short-term consumer exposure – the amount of a pesticide residue that a person might eat in one meal, for example. This has meant that toxicologists have to determine yet another reference dose – the acute reference dose. However, this reference dose is determined in exactly the same manner as an ADI or an AOEL. A selected no-effect level from a toxicology study is divided by a 'safety' or 'uncertainty' factor. As with the difference between the ADI and the AOEL, the difference with the acute reference dose lies in the selection of which toxicology studies to consider. In logical terms, only acute, single-dose, toxicology studies need to be considered in the derivation of an acute reference dose. However, it is important to ensure that all relevant end-points have been investigated in the studies being considered.

CONCLUSIONS

This chapter is intended to give some insight into the justification for toxicology studies performed during agrochemical risk assessment and an indication of how results are interpreted. As in any field of scientific endeavor, training and experience are required for high-quality work. However, no amount of training and experience alone will lead to success – teamwork is essential. A toxicology study involves technicians working with animals, formulation technicians, clinical pathologists, veterinarians, pathologists and support staff, all co-ordinated by a study director. Training and experience in all sectors of the work must be supported by teamwork and good, clear communications. In fact, the requirement for excellent communications does not stop at the boundaries of this team. Continuous interactions among all of the areas of research involved in agrochemical product development are necessary. Toxicology study requirements can be changed by, for example, the results of plant and soil metabolism studies or by a decision to change the field application technology used for the final product. Communications and teamwork are the foundations for success.

FURTHER READING

Organization for Economic Cooperation and Development (OECD): *OECD Guidelines for Testing Chemicals* – Published by the OECD and available from the OECD Publication Service, 2 Rue Andre Pascal, 75775 Paris Cedex 16, France. New draft guidelines and lists of published final guidelines are available on the Internet at: [http://www.oecd.org/ehs/test/testlist.htm].

United States Environmental Protection Agency (US EPA): *EPA Toxicology Testing Guidelines* (and other pesticide testing guidelines) are available on the Internet at: [http://www.epa.gov/docs/OPPTS_Harmonized/870_Health_Effects_Test_Guidelines/].

Ballantyne, Marrs and Syversen (Eds), *General and Applied Toxicology*, 2nd Edn, Macmillan References Ltd, NY, 2000.

Krieger, Hodgson, Gammon, Ecobichon and Doull (Eds), *Handbook of Pesticide Toxicology*, 2nd Edn, Academic Press, NY, 2001.

6 Diets and Dietary Modelling for Dietary Exposure Assessment

J. ROBERT TOMERLIN[1] and BARBARA J. PETERSEN[2]

[1] US Environmental Protection Agency, Washington, DC, USA [2] Exponent, Inc., Washington, DC, USA

Pesticide Residues in Food and Drinking Water: Human Exposure and Risks. Edited by Denis Hamilton and Stephen Crossley
© 2004 John Wiley & Sons, Ltd   ISBN: 0-471-48991-3

BACKGROUND

GOVERNMENT RESPONSIBILITY

As discussed elsewhere in this book, people are exposed to pesticide residues through several different routes. They are exposed to such residues in the water they drink, in the air they breathe, and on the objects they touch. To date, probably no route of exposure to pesticides has captured the attention of the public as has the dietary route. To many people, the prospect of pesticide residues on the food they eat seems to pose an unreasonable and unacceptable risk.

Although the public in general do not want unnecessary exposure to pesticide residues from their food, it is important to recognize that pesticides do have societal benefits. In the United States, for example, the Food Marketing Institute reported that the percentage of income required by Americans to feed themselves has dropped nearly 50 % since 1900 (Food Marketing Institute, 1994). Furthermore, the National Academy of Sciences reported that approximately 10 % of the disposable income of a typical American family is used to purchase food (NRC, 1991), lower than any other country (CAST, 1992; Korb and Cochrane, 1989). In its report on pesticides in the diets of young children, the National Academy of Sciences acknowledged that the increased life-span of Americans is attributable, in part, to a plentiful supply of fruits and vegetables made possible by crop protection chemicals – pesticides (NRC, 1993).

Given that pesticides are useful for providing a wide variety of fruits and vegetables as part of a nutritious and healthy diet, most governments require that their use be strictly controlled, but not banned outright. Furthermore, pesticides may not be used on crops intended for human consumption unless their use is shown to be safe. In order to ensure that pesticide residues in food are not high enough to threaten public health, many governments around the world have developed techniques for assessing dietary exposure to pesticides for the populace of the country. In many places of the world, dietary exposure is termed 'dietary intake'. The term 'intake' will be used throughout this chapter, but the terms 'intake' and 'exposure' designate the same thing. In addition, the World Health Organization has developed techniques to assess dietary intake to pesticides on an international basis.

The purpose of this present chapter is to describe how dietary intake assessments are conducted and the scientific issues that are important for interpreting the results of such assessments. This chapter will also examine some of the scientific issues currently being discussed in the field of dietary intake assessment.

The amount of pesticide residue that is taken into the body can be measured through the direct analysis of foods or computed as the product of the amount of food consumed and the residue concentration on that food. Risk, however, can only be estimated relative to some measure of toxicity. Pesticide toxicology is described elsewhere in this book. For our purposes, it is sufficient to consider that dietary intake estimates typically are compared to the Allowable (or Acceptable) Daily Intake (ADI). The comparison of the intake estimate with the toxicity forms the dietary risk assessment.

LEGISLATIVE MANDATES

The government's responsibility for evaluating public health is often mandated by national legislation, either directly or indirectly. In some instances, calculation of dietary intake is assumed to be part of the overall process of protecting public health. In other cases, dietary intake is specifically stipulated as a technique for evaluating the potential impact of pesticide use on public health.

Many countries around the world estimate dietary intake of pesticides for their population. In addition, the directives of the European Commission require an assessment of dietary intake of pesticides as part of the pesticide authorization process (EC, 1991).

Internationally, the Codex Alimentarius Commission is probably one of the most active food standard setting bodies; its activities include the establishment of Maximum Residue Limits (MRLs) for pesticides in food. An MRL defines the legal maximum residue concentration of a pesticide in or on foods. Regional diets have been devised, a major purpose of which is to allow the calculation of pesticide intake levels (WHO, 1998).

In the United States, regulations regarding the use of pesticides are based upon two pieces of parallel legislation, i.e. the Federal Insecticide, Fungicide and Rodenticide Act (FIFRA (US Congress, 1947)) and the Federal Food, Drug and Cosmetic Act (FFDCA (US Congress, 1938)). Thus, FIFRA and the FFDCA provide a two-pronged approach to pesticides and food safety. Under FIFRA, pesticide use may be registered and tolerances established. Under the FFDCA, the tolerances may be enforced. Because the registration of pesticides involves an assessment of their safety, the responsibility for conducting dietary intake analyses is the responsibility of the US Environmental Protection Agency (EPA). A tolerance is the maximum amount of the pesticide that may be in or on the food following the legal use of the pesticide. The same concept is expressed internationally as the maximum residue limit (MRL) (FAO/WHO, 1997).

International pesticide regulation is extremely complicated, and is not the topic of this present chapter (see Chapter 10 – 'International Standards'). The remainder of this chapter will discuss the techniques used to estimate exposure to pesticide residues in food and some of the important issues in the field. However, we should remember that the dietary risk assessment process takes place in a regulatory context. That context may differ from country to country. Nonetheless, the main objective of all such assessments is to determine whether or not the use of pesticides on crops is safe.

DIETARY INTAKE MODELS

Dietary intake is often considered in two forms, namely chronic (long-term intake) and acute (short-term) intake. The chronic intake analysis answers the question – 'What is the potential for experiencing some type of toxicological effect if a compound is ingested regularly over a long time period?'. The acute intake assessment, however, answers the question – 'What is the potential that the food being eaten right now contains residues at high enough levels to cause some toxicological effect?'.

Dietary intake analyses may be of two basic types, i.e. deterministic or distributional. A deterministic intake analysis generally requires few resources, is conducted using summary data, and is determined relatively quickly (Federal Register, 1998). Compared to deterministic intake analyses, complex distributional analyses – such as those using probabilistic techniques – require significant technical resources to execute and evaluate. A distributional intake assessment can be made on the basis of variation in the consumption data, assuming uniformly high residue levels in all foods. Consequently, distributional assessments generally require more comprehensive databases and are much more complex. Therefore, both execution and evaluation times are greater.

The basic model for dietary intake assessment is relatively simple:

$$\text{Dietary Intake} = \text{Amount of Food Consumed} \times \text{Residue Concentration} \quad (6.1)$$

This basic model is used, from the simplest deterministic model to the most complicated probabilistic model that applies Monte Carlo techniques.

POINT ESTIMATES

Deterministic models typically use point values for the model inputs (food consumption and residue concentration) and yield point value estimates of intake. Internationally, the Theoretical Maximum Daily Intake (TMDI) is a product of mean food consumption estimates and MRL level residues. Similarly, in the United States, the initial Theoretical Mean Residue Contribution (TMRC) is a product of mean food consumption estimates and tolerance level residues. The TMRC is conceptually similar to the Theoretical Maximum Daily Intake (TMDI) calculated internationally. The TMDI is calculated from average food consumption estimates and MRL level residues (FAO/WHO, 1997).

In spite of the procedural differences between the TMDI and TMRC calculations, both seek to calculate the same statistic – the long-term intake. Both statistics are calculated using mean consumption estimates and the legal maximum residue permitted on registered crops. In addition, both statistics assume that 100 % of registered crops will contain residues. The TMDI is calculated according to the following equation:

$$\text{TMDI} = \frac{\Sigma[\text{Consumption (mg/person/day)} \times \text{MRL (mg/kg)}]}{\text{Body weight (kg)}} \qquad (6.2)$$

The methodology is different in the UK, where the TMDI is calculated as an estimate of high-end dietary intake, as follows:

$$\text{UK-TMDI} = \Sigma[(\text{Two highest 97.5th percentile intakes})$$

$$+ (\text{Mean population intakes from all other foods})] \qquad (6.3)$$

The equation for the TMRC is similar to that for the TMDI (bw = body weight):

$$\text{TMRC} = \Sigma[\text{Consumption (g/kg bw/day)} \times \text{Tolerance } (\mu g/g)] \qquad (6.4)$$

In the UK-TMDI, all of the intakes are calculated as the product of the amount of food consumed and the residue on the food. The difference between the two methods has to do with the consumption data used in calculating the TMDI. In the UK method, the two highest 97.5th percentile intakes are calculated from 97.5th percentile food consumption, to which the mean intakes from all other registered foods are added. Outside of the UK, the TMDI is calculated with the mean food consumption values only. It is clear that the typical TMDI calculation is very similar to the TMRC calculation in the United States.

The TMDI is also calculated on an international basis in the standard manner described above. However, the food consumption information is from the five regional diets for the food commodity (FAO/WHO, 1995). Again, the purpose of the TMDI or TMRC is to estimate chronic, or long-term, pesticide intake. Although the data available may be slightly different in different countries or at the international level, the basic concepts are the same.

Examples of TMDI and TMRC calculations are provided in a case study in the 'Worked Examples' section later in this chapter.

In recent years, questions about potential acute dietary risk to pesticide residues have been raised as a public health issue. Acute dietary intake and risk have been the topic of international conferences and workshops, including the 'Geneva Consultation', a joint FAO/WHO expert consultation on food consumption and risk assessment of chemicals held in 1997 (WHO, 1997) and the workshop on residue variability and acute dietary exposure held in York in the UK in 1998 (Harris *et al.*, 2000). Refinements to dietary intake calculations have been steadily evolving, and these workshops have provided the fora for achieving consensus on the developing methodologies.

Estimates of acute intake can be calculated at the international and national level and have been named National Estimates of Short-Term Intakes (NESTIs) and International Estimates of Short-Term Intakes (IESTIs). The intent of acute intake assessment is to estimate the intake and corresponding risk over a relatively short period of time. Ideally, the acute intake assessment would take into account possible variability in consumption and residue levels for individual commodity units, such as individual apples, peaches and other commodities that may be consumed in their entirety.

NESTI and IESTI values can be calculated by probabilistic or non-probabilistic methods. Typically, non-probabilistic methods can be employed to obtain preliminary or screening estimates. The Geneva consultation (WHO, 1997) identified development of consumption data for acute intake assessment as a significant data need. The Joint FAO/WHO Meeting on Pesticide Residues (JMPR) is currently considering options for improving food consumption data and computational methods for acute dietary intake assessment.

The NESTI methodology has been extensively developed for the UK. The UK Pesticide Safety Directorate has identified three different cases to apply to commodities for the acute assessment, depending on the nature of the commodity. The approach actually selected depends on whether composite or individual sample residue data are available and how the commodity is consumed. Similar equations are used for both the IESTI and NESTI calculations. The main differences in the methods have to do with the data used for the consumption part of the equations. Information for IESTI calculations is not generally available because the international consumption data do not report such detailed information.

The types of data that may be used to modify the dietary intake estimates are illustrated and described in more detail in the 'Worked Examples' Section below.

Case 1

The available composite residue data reflect the residue levels in the food commodity as consumed. This is the case where either the commodity is well mixed during processing, e.g. cereals, or where the normal portion size reflects the consumption of many units of the commodity, e.g. cherries.

In this case, the NESTI is calculated by using the following equation:

$$\text{NESTI} = \frac{\text{F} \times (\text{HR-P})}{\text{Mean bodyweight}} \tag{6.5}$$

where F is the full portion consumption data for the commodity unit, and HR-P is the highest residue level detected incorporating processing or edible portion factors.

In calculating the NESTI, the analyst should use the highest residue levels in composite samples (i.e. field composite of 1–4 kg of commodity within a field trial or replicate) from residue trials corresponding to Good Agricultural Practice (GAP). The magnitude of the residue used for estimating acute dietary intake should be adjusted for residues in the edible portion or for the effects of processing when such information is available.

Case 2

The available composite residue data do not reflect the residue levels in the food commodity as consumed. Such a situation may occur where the normal consumption of the commodity during a single eating occasion is typically no more than four discrete commodity units (e.g. apples). This situation may also occur for large commodities for which a significant portion of the commodity could be eaten in a single sitting (e.g. melon). It is also the case where the commodity always is processed before consumption and the process leaves the commodity in the form of individual consumable units (e.g. baked potatoes).

For these commodities, there is a potential for variation in residues between individual units and this is accounted for by the introduction of a variability factor v. The precise value of v will depend upon the nature of the compound and its application method; however, in the absence of data to establish this variability factor, default values are available. These values are introduced into the equation shown below which represents the worst case. A similar calculation would be made for the IESTI.

$$\text{NESTI} = \frac{[\text{U} \times (\text{HR-P}) \times v] + [(\text{F} - \text{U}) \times (\text{HR-P})]}{\text{Mean bodyweight}} \qquad (6.6)$$

in the above equation:

- U is the weight of the first commodity unit, or if the full portion consumption data is less than one commodity unit then U is equal to the full portion consumption data and the second term of the equation drops out.
- F is the full portion consumption data. Where the full portion consumption weight is less than or equal to one commodity unit, then the second term of the equation drops out.
- v is the variability factor, and has the following default values:
  - 5 for large commodities (unit weight over 250 g);
  - 7[1] for medium commodities (unit weight 25 to 250 g);
  - v does not apply for small commodities (unit weight less than 25 g) since the composite residue data reflect the residue level in the food commodity as consumed.
- HR-P[2] is the highest residue level detected incorporating processing or edible portion factors.

---

[1] Originally, the v factor for medium commodities was 10. However, the residue workgroup of the 1998 Residue Variability Workshop recommended that the v factor be reduced to 7 for applications other than granular soil formulations.

[2] It should be noted that, originally, the residue level for second and subsequent units (the second term in Case 2) was the Supervised Trials Median Residue (STMR). However, the report of the 2000 Joint Meeting on Pesticide Residues recommended the change to the highest composite residue (HR).

Case 3

Residue data on an individual commodity unit basis are available which reflect the residue levels in the food commodity as consumed. This is the case where individual residue data are available and follows the same process as Case 2 with the exception that variability factors are not included as they are not relevant when residues on unit samples are available. (Note that the UK Case 3 is a different situation from the JMPR Case 3, which applies to processed commodities produced on a large scale and subject to bulking and blending.)

The NESTI is calculated by using the following equation:

$$\text{NESTI} = \frac{[U \times (\text{HR-P}_{\text{ind}})] + [(F - U) \times \text{HR}]}{\text{Mean bodyweight}} \tag{6.7}$$

in this equation:

- U is the weight of the first commodity unit. If the full portion consumption weight is less than one commodity unit, then U is equal to the full portion consumption data and the second term of the equation drops out.
- F is the full portion consumption data.
- HR-P$_{\text{ind}}$ is the highest residue detected in individual commodity units incorporating processing or edible portion factors

In effect, the NESTI method used by the UK Pesticide Safety Directorate is basically a point estimate approach (PSD, 1999a). The PSD methodology sums the high-intake TMDI for the two most highly consumed foods and the mean TMDI for the remaining foods. The high intake is based upon the 97.5th percentile food consumption values, while the mean TMDI is based upon the mean consumption values. Each consumption value is multiplied by the MRL value. Even for the refinement represented by the NESTI approach for acute intakes, the calculated statistic is still a point estimate.

As shown in the preceding paragraphs, a deterministic assessment of intake is expressed as a single value and is calculated from point estimate inputs. The estimate could represent an upper-bound scenario (for example, tolerance levels on foods) or a statistical tendency (for example, average values from appropriate field trial data). Simple to compute deterministic estimates do not produce a measure of the range of potential exposures. Distributional analyses require greater resources both to prepare and review, but they can represent high levels of complexity.

DISTRIBUTIONAL ANALYSES

In contrast to deterministic techniques, distributional risk assessments use the entire range of the available data. The simplest type of distributional analyses is represented by the EPA's Tier 1 acute dietary assessment (EPA, 1996). The

Tier 1 assessment utilizes the entire food consumption distribution to calculate acute dietary intake. However, a single residue value is assumed for each food. For example, if the tolerance (MRL) on apples is 2 ppm (2 mg/kg), then 2 ppm would be used for estimating intake for all of the apples in the food consumption database, regardless of the amount of apple consumed. In effect, the Tier 1 assessment uses a deterministic approach, even though the entire consumption distribution is used. Instead of a single estimate of dietary intake such as provided by the TMRC or TMDI calculations, the Tier 1 assessment yields an exposure distribution, where the variation in exposure results from variability in the amount of food consumed (see also the section below on 'United States Methodologies').

In practice, the data used for a complex, probabilistic analysis spans the entire spectrum of data, from point estimates to full data distributions. The output of a probabilistic analysis is a distribution of dietary intake – not a point estimate. Instead of individual input values, data distributions reflecting a range of potential values are used. A computer simulation then repeatedly selects individual values from each distribution to generate a range and frequency of potential intakes. The output is a probability distribution of estimated intakes, from which intakes at any given percentile can be determined. Consideration of probabilistic analyses will be limited to intake (and not toxicity) assessments at the current time. The PSD has described its method for conducting probabilistic dietary intakes using this Monte Carlo technique (PSD, 1999a). In the United States, the EPA uses the Dietary Exposure Evaluation Model (DEEM^{TM}) (Novigen, 1996) to conduct acute Monte Carlo distributional assessments of dietary exposure.

A common approach to Monte Carlo sampling basically provides a hybrid Monte Carlo assessment. A true Monte Carlo approach would randomly sample from both the consumption and residue databases. The common model, however, would use all of the food consumption data and samples only from the residue distributions according to the following procedure, which is illustrated in Figure 6.1:

1. The consumption of food B by individual 1 on day 1 of the survey period is multiplied by a *randomly selected* residue value from the residue distribution for food B.
2. Step 1 is repeated for all foods A and O that are consumed by individual 1 on day 1. The residue value for Food A is sampled from distribution A and the residue value from Food O is sampled from distribution O.
3. An estimate of the total exposure for person 1 on day 1 is obtained by summing the exposure estimates for all of the foods.
4. Steps 1 to 3 are repeated 1000 times, still using the consumption data for person 1 on day 1.
5. The 1000 exposure estimates for person 1 on day 1 are stored as frequencies in exposure intervals.
6. Steps 1 to 5 are repeated for person 1 on subsequent days of the survey period.

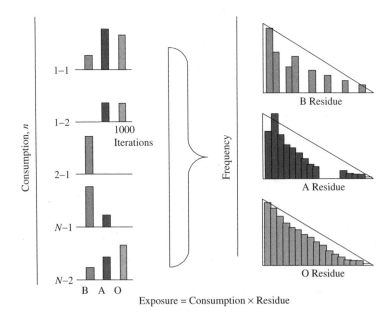

Figure 6.1 Illustration of the US Environmental Protection Agency Monte Carlo procedure for acute dietary exposure estimation (see text for further details) (Tomerlin., 2000). Reproduced from Tomerlin, J. R., *Food Additives and Contaminants*, **17**(7), 641–648 (2000), with permission of Taylor & Francis, Inc http://www.tandf.co.uk

7. Steps 1 to 6 are repeated for all individuals in the sub-population.
8. The frequency distribution of the exposure estimates for all individuals on all survey days is used to derive the percentile estimates.

Steps 1 to 3 constitute an iteration.

FOOD CONSUMPTION DATA

Food consumption data take many forms and may be collected by several different methods. The way in which the food consumption data are collected, or organized after collection, has a lot to do with the type of dietary intake estimate that may be calculated. Therefore, a discussion of the different types of food consumption surveys and the data they provide will lead to an understanding of the various kinds of intake estimates.

FOOD CONSUMPTION SURVEY METHODS

Some types of surveys provide information about the overall food supply. Probably the most common type of food supply survey is the food balance sheet. These sheets are typically compiled from national statistics about domestic food production, imported foods and food exports. Food balance sheets provide information about the amount of food that passes through the supply chain, but not about the amount of food that is actually consumed. In addition, food balance sheets do not provide information about the forms in which food is consumed. A significant deficiency of the food balance sheet is that information about food consumption by population sub-groups is not provided. Another weakness of the food balance sheet method is that sometimes it does not account for waste at either the household or individual levels. In some cases, information from household budget surveys allows scientists to adjust for errors in the food balance sheets.

Food intake surveys are of two general types, i.e. household surveys and surveys of individuals. Often, household surveys collect food consumption data as a secondary objective, with the primary objective being to collect economic information. Household surveys may be used to assess both gross and net food consumption on a household basis. A common application of the data from household surveys is to assess differences according to regional, urbanization, or socio-economic differences. As with the food balance sheet, the household survey does not account for waste, nor does it account for foods consumed outside of the home.

Surveys of individuals may be of several different types. Among the methods used for such surveys are dietary recall, dietary record or diary, and duplicate plate analysis.

Dietary recall methods are frequently used for food consumption surveys. Some recall surveys are conducted as face-to-face interviews between the survey participants and trained survey workers. The survey participant is asked to quantify the amount of food consumed over the previous 24 h period. In other surveys, the same type of information is obtained, but via a telephone interview. Frequently, the survey worker uses pictures, measuring equipment, and diagrams to aid the survey participants in quantifying their food consumption.

Data from dietary recall surveys may be used to estimate both chronic and acute dietary intake. All such surveys permit the calculation of mean consumption estimates, and therefore intake estimates such as the TMRC and TMDI. Depending on the design of the consumption survey and the detail of data collection, distributional dietary intake estimates may also be calculated. Generally, dietary recall surveys also record a great deal of information that permits the categorization of the population into various sub-groups.

In the diet history, survey respondents provide consumption information for a specified time period. One of the most common uses of diet history surveys is in evaluating various disease predictors. However, the design of such surveys sometimes makes them of limited use for intake estimation. Another weakness of the diet history surveys is that they focus on regular dietary patterns. Therefore, the data from diet histories are not suitable for assessing acute dietary intakes.

One of the least expensive types of food surveys to administer is the food-frequency survey. Most of these surveys address issues about usual consumption patterns. Such surveys have been used to evaluate consumption of fish and game or other specific foods that are consumed infrequently. Typically, the data from a food-frequency survey must be linked to a survey that provides information about quantity in order to estimate dietary intake of pesticides.

In contrast to dietary recall methods, surveys based upon the food diary method require survey participants to maintain a 'journal' for the foods they consume during the survey period. Survey participants may be asked to weigh or otherwise measure the amount of food that they consume. Food consumption information obtained from the food diary method is very detailed and provides the same general types of information as the dietary recall method. The data obtained from a food diary survey is suitable for estimating mean or median consumption, as well as the distribution of consumption and extreme values. As with the dietary recall method, extreme consumption values tend to be overestimated.

Duplicate plate analyses can provide researchers with direct measurements of pesticide residues in the food the subject eats, as well as information about the amount of food consumed. The survey participant provides the researcher with an exact duplicate of the foods he or she consumes. Unless data are collected for a given participant for several days, extreme consumption values tend to be overestimated. In addition, duplicate plate analyses are very expensive to administer, even if the costs for the chemical analyses are not included. Duplicate plate analyses are often used to address specific research issues and may not be suitable for general risk assessment.

Several countries conduct some form of total diet study, although specific methods may vary from country to country. The total diet study identifies a set of foods that represent the national diet and then sample them. After collection, the representative foods are analyzed for chemical residues, including pesticides. The total diet study provides information about the relative importance of the foods with respect to pesticide intake. Total diet studies are very useful for range finding. However, since the foods are not usually linked to any population group, it is not possible to estimate intake for different segments of the population.

SOURCES OF FOOD CONSUMPTION DATA

On an international basis, food balance sheets provide a comprehensive summary of a country's food supply over a defined time period. For each country, food balance sheets summarize the origin and utilization of raw agricultural commodities and some processed commodities. The food balance sheet 'balances' the amount of food produced in the country, the amount imported, and adjusts for any change in available supply that might have occurred since the beginning of the time period. Distinctions are made between foodstuffs that are exported, fed to livestock, used for seed, used in food processing, losses during storage and transportation, or used for human consumption. The food balance sheet compiles the total amount of food on a country-wide basis. The total balanced food supply is then divided by the population to yield a *per capita* consumption estimate.

A major advantage of food balance sheets is that they can be used for making comparisons between countries.

There is potential for error with food balance sheets, however, because the amount of food consumed may be lower than the quantity shown in the food balance sheet. The degree of error will depend on the extent to which losses of edible food during storage, preparation and cooking, wastage, or other losses are known and accounted for.

Since food balance sheets start with a relatively crude country-wide estimate of food quantity consumed, they do not give any indication of the dietary pattern differences among different sub-groups of the population. This is a serious limitation if exposure to children is the issue.

The Global Environment Monitoring System – Food Contamination Monitoring and Assessment Programme (GEMS/Food) (WHO, 1998) is another method for obtaining information about the dietary intake of contaminants on an international basis. The GEMS/Food program is implemented by the World Health Organization (WHO) via a network of WHO Collaborating Centres for Food Contamination Monitoring and participating institutions located in over 70 countries.

The GEMS/Food system maintains a database of food contamination monitoring data as well as preparing periodic assessment documents to provide a global overview of contaminants in food. These reports have addressed issues related to the levels of contaminants in various foods and in the total diet. GEMS/Food often calculates estimates of dietary intake of pesticides using the TMDI methodology. The program developed the GEMS/Food Regional Diets to permit calculation of TMDI estimates. The present regional diets provide mean consumption estimates for five regions around the world, namely Latin America, Europe, Middle-East Asia, Far-East Asia and Africa. Revised versions for 13 regions have been proposed (Barraj and Petersen, 1997).

In the UK, the Ministry of Agriculture, Fisheries and Food and the Department of Health have conducted three sets of food consumption surveys between 1983 and 1987. These three surveys comprise the National Diet and Nutrition Survey (NDNS) program. The 1986 'Infants Survey' recorded food consumption for 488 infants between the ages of six and twelve months. Food diaries were kept as a seven-day record of all of the food consumed. Food consumption was recorded by measurement devices and linked to serving size data to estimate consumption. Average body weights were used because individual body weights were not recorded (Mills and Tyler, 1992).

Food consumption data for school children were obtained in the 1983 'Schoolchildren Study.' This survey was a seven-day weighted dietary survey of 3367 school children of 10 to 11 and 14 to 15 years old. The target population was taken from Local Authority schools in which school meals were provided. As with the infants survey, individual body weights were not recorded, and so average body weights were used to express consumption relative to body weight. The survey 'over-sampled' Scottish school children and school children from poor families (Anon, undated).

A survey of children aged $1\frac{1}{2}$ to $4\frac{1}{2}$ years (the "toddlers survey") was conducted from 1992 to 1993 and provides information about food consumption patterns for young children. The survey sampled a total of 1675 individuals over a four-day sampling period. The survey was organized into four waves, with the waves corresponding to subject birth dates. Parents provided four-day weighed dietary intake records of all foods and beverages consumed by the child, both in and out of the home (Gregory et al, 1995).

The final survey in the series was the 1986–1987 'British Adults Survey', comprising 2197 adults between the ages of 16 and 65. The people sampled lived in private homes, i.e. not in group homes or other institutional settings. Pregnant women were excluded from the survey. Contrary to the infants and school children surveys, body weights were recorded in this survey (Gregory et al, 1990).

Food consumption data are also collected by agencies of other countries, notably in Germany, The Netherlands and France. However, at this time, data from the United States and the United Kingdom are readily available in electronic format, whereas data from other countries may require a considerable amount of work before being used for dietary intake calculations.

The Continuing Survey of Food Intakes by Individuals (CSFII) (USDA, 1992, 1993, 1994, 1995, 1996a, 1997a, 1997b) provides the food consumption data for use in dietary exposure calculations in the United States. The CSFII surveys replace the former vehicle for providing consumption information, i.e. the Nationwide Food Consumption Surveys. CSFII surveys were conducted in 1985 to 1986, 1989 to 1991 and 1994 to 1996. In each of the three years of the surveys, a nationally representative sample of individuals of all ages provided information on food intakes and socio-economic and health-related information. Information was obtained through in-person 24-h recall interviews, food intakes on three consecutive days (1989 to 1991 CSFII) or two non-consecutive days (1994 to 1996 CSFII). In an effort to increase the amount of information available for children's diets, the United States Department of Agriculture (USDA) also conducted the 1998 Supplemental Children's Survey (USDA, 2000), providing approximately 12 000 additional observations for children between the ages of one and nine years.

Food intake was recorded by time of day and by eating occasion (breakfast, brunch, lunch, dinner, supper and snack) as defined by the respondent. Separate entries were made in the survey databases for each food consumed. Quantities of foods and beverages consumed were recorded in household measures, weights, dimensions or common units (e.g. slice, piece, etc.). All quantities were converted to g/kg bw/day[3] by the USDA.

The surveys also recorded whether the food was consumed at home, taken from home and consumed away from home, or never brought into the home. When foods were obtained and eaten away from home, the location was specified

---

[3]bw, body weight.

as restaurant, cafeteria, school, day-care centre, community feeding programme, vending machine, store, or someone else's home. Foods obtained from fast-food carryout places were identified whether they were eaten at home or away from home.

In the 1989 to 1991 CSFII survey, approximately 10 400 individuals provided three days worth of data. Approximately 15 300 individuals provided data over two days in the 1994 to 1996 CSFII survey. Therefore, each of these CSFII surveys provided more than 30 000 food consumption observations.

The USDA has developed statistical sampling weights to compensate for over- and under-representation of certain population sub-groups in the unweighted sample due to the sample design (low-income households were 'over-sampled'), non-response and unequal interviewing across seasons and days of the week.

PESTICIDE RESIDUE DATA

Dietary intake, as we have already seen, is the product of the amount of food that is consumed and the magnitude of the residues on the food. The residue component of the dietary intake equation may take on a wide variety of values, although various calculations mandated by regulatory agencies may specify the particular type of residue value to be used. For example, in the United States, the theoretical mean residue concentration (TMRC) is supposed to use the tolerance value. The TMRC is comparable to the international theoretical maximum daily intake (TMDI) calculation, which uses the MRL as the residue value.

MAXIMUM RESIDUE LIMITS AND TOLERANCES

Most countries establish legal limits for the maximum residue concentration that is permitted on food. In most countries of the world, these limits are called maximum residue limits (MRLs). In the United States, such limits are known as 'tolerances'.

The MRL is calculated from studies conducted according to Good Agricultural Practice (GAP). Even so, the MRL represents the highest residue level expected under normal use conditions. US tolerances are based upon field trial studies that are conducted under conditions that maximize the potential for residues. They are conducted at the maximum application rate permitted on the pesticide label, with the maximum number of applications and the minimum pre-harvest interval (PHI), namely the minimum interval between the last application and harvest. Although different procedures are followed for establishing tolerances in the US and MRLs elsewhere, both statistics represent maximum residue concentrations. Such values may be used to calculate an initial estimate of dietary intake.

SUPERVISED TRIALS MEDIAN RESIDUES AND ANTICIPATED RESIDUES

Sometimes, dietary intake estimates calculated with tolerances or MRLs exceed acceptable toxicology limits. In such cases, revised dietary intake estimates may be calculated by using residues derived from field trials. In most countries of the world, supervised trials median residues (STMRs) are used to calculate international estimated daily intake (IEDI) and national estimated daily intake (NEDI) values. The STMR values are the median residues from residue field trials. Similarly, in the United States, mean residues from field trials are often used in revised dietary intake estimates.

RESIDUE MONITORING PROGRAMS

Many countries monitor their food supplies for pesticide residues (SCPH, 1998). Although the objective of many of the monitoring programs is to insure that pesticide residues do not exceed legal limits, monitoring data can also be used to calculate dietary intake estimates. Although monitoring programs provide data that can be used to calculate more refined dietary intake estimates, few programs report the data in ways that can be readily used for such calculations. Therefore, monitoring data are infrequently used for dietary intake estimates, even though several documents have recommended that such data can be used (FAO/WHO, 1995, 1997; PSD, 1999a).

In the United States, the Pesticide Data Program (PDP) is conducted by the US Department of Agriculture (USDA) and is designed to provide data for dietary intake calculations. The PDP is the only program that provides pesticide residue data in electronic format which also supplies specific information about the detection limits of the analytical method used to determine the residue concentration (USDA, 1996b, 1997c, 1998, 1999).

ANCILLARY DATA

That residue concentrations may decrease is commonly known. Residues may decrease as a result of simple aging, as a result of processing, or as a result of consumer preparation. Furthermore, many dietary intake calculations are made under the assumption that 100 % of the crop is treated, although it is rare that a pesticide is used on 100 % of the crop.

FOOD PROCESSING DATA

Processing studies should be conducted under conditions that replicate, on an experimental scale, commercial processing practices. The objective of a processing study is not to simply calculate the residue levels in processed foods, but to determine the ratio of residues in processed food relative to residues in the raw agricultural commodity. Such processing factors may then be used to adjust residue levels in tolerances or MRLs, mean residue or STMR values from field trials, or residue levels from monitoring data.

Food consumption data for processed foods are required if processing factors are to be used to any real advantage. Even if information is not directly available in the food consumption database, such as food balance sheets or the GEMS/Food global diets, proportions may be used. For example, we might know that apple consumption is 35 g/day. Agricultural or commerce statistics might indicate that 35 % of apples produced are used for the production of apple juice. We could then partition the food consumption data so that 65 % of the consumption is multiplied by the residue in the raw agricultural commodity and 35 % of the consumption is multiplied by the residue in the raw agricultural commodity and by an apple-juice processing factor.

PESTICIDE USE DATA

Information about the amount of a crop that is treated may be compiled by government agencies, by private businesses or by pesticide producers. Pesticide use data should be expressed in terms of 'base acres (or hectares) treated' instead of 'total treated acres (or hectares)'. In other words, data about the total area treated at least once are of use in a dietary intake assessment. However, total treated area, which includes crop area having multiple treatments, is of less utility.

Percent crop treated information can be used as a simple adjustment to residue values for chronic dietary intake calculations. Some countries incorporate the proportion of the crop that is treated in estimating chronic intake. The rationale is that chronic intake is measured over a long period of time, and that sometimes people consume treated foods and sometimes they consume untreated foods. In contrast, some countries restrict the use of percent crop treated data to highly blended food commodities, such as grains and vegetable oils. As will be discussed elsewhere in this chapter, percent crop treated information is used differently in probabilistic Monte Carlo assessments of acute dietary intakes.

Regardless of the specific data utilization rules with respect to percent crop treated information, it is generally prudent to consider the most recent data, and to verify that trends are properly taken into account, because use practices may change.

PESTICIDE USE PATTERNS

As discussed elsewhere in this chapter, field trials are often conducted under maximum 'label' conditions. However, in practice, farmers may apply the compound at a lower application rate, using less than the maximum number of applications, or with a PHI longer than the minimum. Frequently, pesticide manufacturers also analyze samples from plots treated at less than the maximum rate or longer PHIs. If information is available about the proportion of the crop that is treated according to less than maximum conditions, residue values can be calculated which take actual pesticide use patterns into account.

REGULATORY CONSIDERATIONS

Most regulatory bodies consider dietary intake to be of little concern if initial TMDI- or TMRC- type point estimates indicate that dietary intake is below toxicologically significant levels (EPA, 2000b). However, if the deterministic estimate exceeds levels deemed to be safe, refinements to the analysis are required if registration of the compound in question is to continue. It is important to remember that refinements to the risk assessment do not have any impact upon actual risk. Any risk assessment, regardless of its simplicity or sophistication, merely provides an estimate of potential risks in the real world. Changing the assumptions under which an assessment is conducted does not affect the parameters that effect real risks. A more refined risk assessment provides a more accurate estimate of actual risk.

PRELIMINARY TIERS

Assessments conducted under preliminary tiers frequently follow worst-case assumptions. In the United States, the worst-case acute dietary intake estimate is known as a Tier 1 assessment. Tier 1 assessments are conducted using tolerance level residues in the entire supply of registered crops. In short, residues at tolerance levels are assumed in 100 % of the crop. Likewise, at the international level, the TMDI calculation for chronic dietary intake is conducted assuming maximum residue limit residues in 100 % of commodities with existing or proposed MRLs.

REFINED ASSESSMENTS

Instead of worst-case assumptions, dietary intake assessments may be conducted by using more realistic assumptions. For example, instead of using tolerance level residues, a Tier 2 acute dietary intake assessment in the United States distinguishes between blended foods, such as juices, oils and grains, and 'single-serving' foods, such as apples or oranges, in which the entire treated commodity might be consumed. In the Tier 2 assessment, mean field residues may be used for blended foods, and either tolerance levels or the maximum value observed in residue field trials may be used for single-serving foods.

Internationally, chronic dietary intake may be calculated with the STMR value instead of the MRL values. STMR stands for 'supervised trials median residue', a value which represents central tendency, as does the mean. The STMR formerly was used in the second term of Case 2 international estimates of acute dietary intake. However, the report of the 2000 Joint Meeting on Pesticide Residues (JMPR, 2001) changed this term in the IESTI equation to the highest residue in a composite sample (HR).

Another parameter that is over-stated in preliminary assessments is the percent of the crop that is treated. In the UK, only blended foods may be adjusted by percent crop treated. The UK authorities believe that it is appropriate to apply percent crop treated adjustments to account for the potential pooling of crops from treated and untreated fields. However, the UK authorities believe that some people may consistently purchase single-serving crops from the same market and therefore may consistently consume treated produce. However, it seems unlikely that all of the various foods that a person would consume would consistently and regularly contain residues. In the United States, refined chronic intake estimates adjust the residue value by the percent of the crop that is treated. For example, if the residue value on a crop is 1 mg/kg, but only 30 % of the crop is treated, the effective residue used in the assessment would be 0.3 mg/kg. The rationale for this approach is that chronic intake represents intake over an extended time period, perhaps even during the course of a lifetime. Over the course of a long exposure period, sometimes people would consume treated food and sometimes they would consume untreated food. Not all authorities follow this viewpoint.

Refined chronic dietary intake assessments may be called international estimated daily intake (IEDI) or national estimated daily intake (NEDI) assessments. Both statistics represent a refinement away from the worst-case assumptions of the TMDI calculation. The difference in the two statistics is based on the different food consumption data used in the calculation. The general form of the calculation, for both the IEDI and the NEDI, is as follows:

$$NEDI = \Sigma(F_i \times RL_i \times K) \tag{6.8}$$

where F_i is the food consumption data for a given food commodity, RL_i the appropriate residue level for that commodity, and K the correction value for reduction or increase in the residue due to processing or storage.

The source of data for RL_i may include supervised trials median residue (STMR) values, values at or below the limit of determination (LOD), residues in edible portions only, residues adjusted for the effects of processing, storage, or cooking; monitoring or surveillance data, proportion of crop treated, and proportions of commodity imported and grown domestically. A complete discussion of the NEDI calculation is provided in a recent PSD guidance document (PSD, 1999a).

Percent crop treated may be used in a different manner in an assessment of acute dietary intake. The percent crop treated data may be used to represent the probability of occurrence of residues. Thus, information about the percent of the crop that is treated may be used to indicate the likelihood of consumed food containing residues in a Monte Carlo probabilistic assessment. A Monte Carlo assessment is based upon repetitive sampling from the available data. Generally, there must be enough repetitions (also called iterations) to ensure that the entire data distribution is adequately sampled. Therefore, if 40 % of the crop is treated, the Monte Carlo assessment should select a discrete residue value from the residue distribution approximately 40 % of the time and a zero the remaining 60 % of the time. It is important to note that the Monte Carlo approach does *not* modify the magnitude of the residue, only the likelihood of selecting a positive residue sample at all.

Monte Carlo approaches have been followed to refine acute distributional analyses. Probabilistic assessments conducted in the UK are conducted for consumers only, i.e. people who have consumed the food in question on the day or during the dietary period. They also may be conducted for single commodities. In other words, individuals who do not consume any of the foods on which the compound is registered are excluded from the assessment. Consequently, the UK approach generally takes a more conservative view on the utilization of data and the interpretation of the results. In the United States, Monte Carlo dietary intake assessments are conducted to calculate *total* daily intake from all foods on which the pesticide is registered. The US EPA has published a series of science policy papers that describe how to conduct probabilistic dietary intake assessments, how to use data in probabilistic dietary intake assessments, and advanced topics related to such assessments. The various policy papers have provided relatively sophisticated concepts for using data in probabilistic assessments. In the UK, probabilistic assessments are typically conducted for one food at a time. In contrast, intake assessments conducted in the United States typically evaluate intake for the total population, including individuals who do not consume any of the foods on which the compound is registered.

Regardless of the specific characteristics of the probabilistic assessments conducted in the United States and the United Kingdom, the percent of crop treated is used in both cases to provide the probability of selecting a positive residue for a particular iteration. The net result is that residues are placed in the context of likelihood of occurrence instead of assuming maximum residues for all consumers, a situation that clearly does not occur in practice.

DIETARY DATA CONSIDERATIONS

Different types of data support different types of assessments. For example, the ability to access more than 30 000 individual daily food consumption records in the USDA CSFII database moves Monte Carlo assessments from the realm of the theoretical to that of the possible. However, such comprehensive databases are the exception rather than the rule. Therefore, Monte Carlo assessments of acute dietary intake will not be possible for most of the food consumption databases that are available. However, as shown in preceding sections, it is still possible to conduct acute dietary intake estimates and to refine estimates of chronic dietary intake. This section will discuss some of the methods that might be used to refine dietary intake assessments according to the data and computational methodology that might be available.

WORLD HEALTH ORGANIZATION

As we have seen, acute distributional intake assessments, especially those that incorporate Monte Carlo techniques, require individual food consumption records.

The food consumption data provided via GEMS/Food are organized in a way that supports calculation of point estimates of chronic intake. WHO computations will be conducted by using the GEMS/Foods regional diets. Although the WHO is considering reconfiguring the regions (Barraj and Petersen, 1997), the consumption data will retain their roots in national food balance sheets, without estimates of maximum consumption values. Therefore, unless the regional diets undergo a fundamental change, it is not expected that assessments of acute dietary intake will be possible using WHO methodology. However, the IEDI and NEDI types of adjustments based upon refined residue values may still be used. One must exercise caution, however, when adjusting residue values for the effects of processing, and ensure that information about the consumption of processed commodity is available. Food utilization statistics may provide sufficient data to estimate consumption of processed commodities. For example, assume that we know that *per capita* apple consumption is 40 g/d (0.04 kg/d), the MRL in apples is 2 mg/kg, 40 % of the apple consumption is consumed as juice, and residues in apple juice are 90 % lower than in apples. It would then be possible to calculate intake of residues in apples as shown in Table 6.1.

Clearly, having some information about apple juice consumption and residue levels in apple juice results in almost a 46 % reduction in the intake estimate. Again, this refinement was made with the type of consumption data available via the GEMS/Foods regional diets, but does require additional information about the amount of apple juice consumed and about the reduction of residues in juice.

GENEVA AND YORK CONSULTATIONS

Recognizing the importance of dietary intake estimates in evaluating the public safety of pesticide residues, the 26th and 27th sessions of the Codex Committee on Pesticide Residues (CCPR) requested a Consultation (FAO/WHO, 1995). The 1995 York Consultation convened in response to that request. The primary goal of this Consultation was to review the existing guidelines and to consider new methods that would improve dietary intake methods. Because dietary intake estimates are such an important part of evaluating exposure to pesticides, it was decided that accurate dietary intake estimates promote better decisions at the international level, as well as for individual Member Countries.

Table 6.1 Dietary intake of residues in apples

Dietary intake component	Consumption component	Residue component	Dietary intake[a]
TMDI calculation	0.04 kg/d	2 mg/kg	0.001 143 mg/kg bw/d
Intake from apples	0.6 × 0.04 kg/d	2 mg/kg	0.000 686 mg/kg bw/d
Intake from apple juice	0.4 × 0.04 kg/d	0.1 × 2 mg/kg	0.000 046 mg/kg bw/d
Apples plus juice	–	–	0.000 732 mg/kg bw/d

[a] Assuming a body weight (bw) of 70.1 kg.

The Consultation confirmed the TMDI methodology previously described. In doing so, however, the Consultation acknowledged that an intake exceeding the ADI did not necessarily mean that people actually were exposed to excessive residue levels via the diet.

The Consultation discussed several factors that might be used to refine a dietary intake assessment, including the following:

- using median residue values instead of MRLs;
- excluding non-toxic moieties from the residue definition;
- incorporating samples with residues at or below the limit of determination (LOD);
- considering residues only in the edible portion instead of the whole commodity;
- accounting for reduction (or increase) of residue concentration due to storage, processing or cooking;
- considering proportion of crop or commodity actually treated;
- accounting for relative proportion of domestic and imported commodity;
- using more accurate consumption data;
- using residue data from monitoring programs instead of field trial studies;
- using data from total diet studies.

Using factors such as those listed above allows the best use of available data in the dietary intake assessment. Since the TMDI overestimates intake, the use of such factors provides a more accurate estimate of dietary intake. If a refined dietary intake estimate still exceeds the ADI, then a risk management issue must be resolved.

The product of the Consultation was a set of twelve recommendations, as follows:

- Revise the existing *Guidelines for Predicting Dietary Intake of Pesticide Residues* to take into account the Consultation report. Dietary risk assessment should then be conducted according to the revised guidelines.
- In collaboration with the Joint FAO/WHO Meeting on Pesticide Residues (JMPR), GEMS/Food should assess pesticide residues using the 'revised' *Guidelines* methods for calculating TMDI and IEDI estimates.
- National authorities should use the NEDI methodology outlined in the 'revised' *Guidelines*.
- Additional data should be supplied by industry and other sponsors of the pesticide if the intake estimate exceeds the ADI after all available factors are appropriately applied. If the intake estimate still exceeds the ADI, then the dietary intake situation becomes a risk management concern.
- JMPR deliberations regarding MRLs should, as appropriate, develop separate residue definitions and should identify median residue values.
- The fate of pesticides during processing needs to be better understood. Therefore, the FAO and WHO should conduct a comprehensive review of *international* level food processing information.

- The CCPR must develop additional monitoring and surveillance data on pesticide residues for dietary intake assessments at the national level.
- A consideration of acute toxicity must be part of the JMPR's routine assessment of a pesticide's toxic potential. As appropriate, JMPR should consider establishing an acute reference dose when it sets the acceptable daily intake.
- The national food balance sheets must be grouped appropriately into a set of 'cultural' diets by GEMS/Food. The cultural diets should be updated approximately every 10 years.
- Food balance sheets are restricted in their ability to supported refined dietary intake assessments. Therefore, all countries, especially developing countries, should conduct food consumption surveys to better address questions of dietary intake. Population sub-groups, as well as the general population, should be included in such surveys. Data on large portion weights should be obtained, as feasible.
- GEMS/Food should develop an international database of large portion weights for fruits, vegetables and other commodities for use in assessing acute dietary intake of pesticide residues.
- After gaining experience in the application of the 'revised' *Guidelines*, FAO and WHO should review them.

In 1997, another Consultation was convened in Geneva (FAO/WHO, 1997). This consultation had five main objectives. The five objectives and the Consultations general conclusions were as follows:

1. *Review the five regional GEMS/Food diets and revise them if necessary.* The Consultation agreed to increase the number of diets from five. Initially, 13 unique diets were to be developed, based upon a cluster analysis of all available FAO food balance sheet data. Subsequently, it was considered that nine regional diets would provide a proper balance between increased specificity and costs of obtaining the necessary data and devising the diet clusters.
2. *Further develop data required for acute dietary exposure assessment, particularly a database on large portion weights.* A procedure considered appropriate for all chemical contaminants was agreed upon by the Consultation. This procedure requires that larger portion weights be determined for each food commodity having international MRLs. Large portion weights should be developed for children as well as adults.
3. *Harmonize the international and national approaches to dietary exposure assessment across different chemical contaminants.* The Consultation recognized that the principles of dietary intake assessment are essentially the same at both the national and international level and from one country to another. The differences in methods are based in differences in the data that are available for conducting such assessments. Given the similarity in underlying principles and desiring to promote a high level or scientific rationale and consistency, the Consultation *recommended* that all dietary

exposure assessments considered by all relevant Codex use the terminology from the Consultation report.

4. *Encourage governments to practise consistency and transparency in the way they conduct dietary exposure assessments.* The Consultation correctly described dietary exposure assessment as an iterative process. It is also a highly technical process and effective communication between the risk assessor and the risk manager is essential. To promote effective transfer of information and correct decision making, the Consultation stated that such communication should be detailed and address the quality and quantity of data underlying the decisions.

5. *Consider the special needs of developing countries with respect to dietary exposure assessment.* In its recommendations, the Consultation addressed the needs of developing countries and integrating developing countries into the Codex process.

The Consultation made a total of 36 recommendations, roughly classified as relating to general considerations, chronic dietary exposure assessment, acute dietary exposure assessment, developing countries and harmonization. The most significant recommendations were as follows:

- Data should be used in the best way in dietary exposure assessments, exercising caution to avoid laborious assessments that add little to the assessment process.
- Dietary exposure assessments should not be relied upon indefinitely but should be reviewed and updated as necessary on a regular basis.
- Exposure from multiple routes should be considered when appropriate.
- The quality and quantity of the data should be considered when interpreting the results of a dietary exposure assessment.
- If data are sufficiently robust, FAO/WHO should investigate the use of probabilistic techniques for dietary exposure assessment.
- Population groups with different food consumption patterns or different sensitivities to the chemical should be specifically considered.
- Food consumption data should be expressed in units of gram per kilogram body weight.
- Five-year average consumption should be recalculated from the GEMS/Food regional diets at least every 10 years.
- JMPR and the Joint FAO/WHO Expert Committee on Food Additives (JECFA) should use more realistic residue values, such as the supervised trials median residue, in estimating dietary exposure.
- JMPR and JECFA should establish, when appropriate, acute reference doses.
- Member Countries and manufacturers should conduct research on the question of unit-to-unit variability of residue concentration; the body of the report contains recommendations for variability factors to use in acute exposure assessments.
- WHO should develop a database of large portion weights for the general population and for individuals aged six years and younger.

- International level exposure assessments should use data that reflect the differences in dietary patterns among countries and within countries (if appropriate).
- Quality criteria for data used in dietary exposure assessments should be developed and made known.

A workshop addressing pesticide variability and acute exposure was held in York in the UK in 1998 (PSD, 1999b). Debate about the level of variability factor to be used in dietary intake estimates was inconclusive although it was agreed that it should be in the range 5 to 10. This workshop also discussed the issues in establishing acute reference doses and probabilistic techniques as applied to dietary exposure assessments.

UNITED STATES METHODOLOGIES

The EPA's dietary intake model is DEEM™ – the Dietary Exposure Evaluation Model. This model is based upon either the 1991–1992 Continuing Survey of Food Intake by Individuals (CSFII) or the 1994–1996 CSFII, augmented by data for children from the 1998 Supplemental Children's Survey (USDA, 2000). This model is also capable of probabilistic assessments using Monte Carlo techniques. In Monte Carlo sampling, a residue distribution is repeatedly sampled numerous times, usually many hundreds or thousands of times. One of the most significant features of the probabilistic acute dietary exposure model is that the residue distribution includes zero residues in proportion to the percent of the crop that is *not* treated. Thus, if 25 % of a particular crop is treated and the Monte Carlo assessment specified 1000 iterations, approximately 750 of the iterations would draw a zero and approximately 250 would draw from the residue values.

Probabilistic dietary intake assessments were a new area for the EPA, and in 1996 the EPA published a policy for conducting acute dietary exposure assessments (EPA, 1996). This initial policy paper established the EPA's tiered system for acute dietary exposure assessments. Tiers with higher numbers do not have to be conducted unless assessments conducted at lower tiers yield unacceptable results. All of the tiers, however, use the entire distribution of consumption. The differences lie in the type of residue data and other mitigating factors that are applied. The four tiers in EPA's scheme of acute dietary exposure assessments are as follows:

- Tier 1 – 100 % of the crop is assumed to contain tolerance level residues. Adjustments may be made for the effects of processing.
- Tier 2 – 100 % of the crop is assumed to be treated. Single-serving commodities[4] are assumed to contain tolerance level residues. Exposure from residues

---

[4]Single-serving commodities are those in which the entire commodity is consumed at one meal, such as an apple or an orange, or for which several units are typically consumed at one meal, such as strawberries.

in blended and partially blended commodities[5] are estimated by using mean residue values from field trials or the 95th percentile value from monitoring data. The Tier 2 assessment incorporates the effects of processing.

- Tier 3 – Residues in single-serving commodities are sampled probabilistically from field trial residue distributions. Residue data from monitoring programs may be used if the data from composite samples have been 'decomposited' to single-serving residues.[6] The residue distributions incorporate the percent of the crop that is not treated so that sometimes a zero residue is used in the individual exposure calculation. Exposure from residues in blended and partially blended commodities is estimated by using mean residue values from field trials or the distribution of residues from monitoring data.[7] The Tier 3 assessment incorporates the effects of processing.

- Tier 4 – This Tier analysis is similar to the Tier 3 analysis except that residue data distributions from individual single serving commodities are used. These data would typically be from specific market basket surveys conducted for the purpose of acute dietary exposure assessment. The Tier 4 assessment also incorporates the effects of processing.

The passage of the Food Protection Act (US Congress, 1996) imposed stringent requirements on the EPA, particularly in the area of risk assessment. The law required a great number of changes in a relatively short period. Therefore, the EPA identified nine science policies that needed to be addressed for them to properly comply with the requirements of the US Food Quality Protection Act (FQPA). The science policies directly related to dietary exposure assessment, with the guidance papers written about the policy being shown in Table 6.2.

Generally, incorporating the percent of crop treated has a tremendous impact upon the outcome of the acute dietary risk assessment. When an assessment includes this feature of the probabilistic assessment with processing factors, refined residues in blended commodities, and other realistic adjustments, the risk assessment may result in acceptable levels of risk, or even virtually no risk at all.

A probabilistic assessment, however, is not a veil that hides risk. If a relatively high proportion of the crop is treated, the risk estimate may not be mitigated to acceptable levels, even if several other crops in the assessment use highly refined data. Furthermore, if the crop is one that is a significant part of the diet, or if residue levels are high, the statistical nature of the probabilistic assessment will *not* mask high exposure values. The values will be there, and it will be an issue

---

[5]Blended commodities are those, such as grains, which typically are consumed after the individual units have been extensively mixed. Partially blended commodities are those which typically are mixed, such as juices, but the mixing may occur on a regional or local scale.

[6]'Decompositing' is described later in the Food Consumption Data section under Issues.

[7]The procedures for residue data utilization in the Tier 3 assessment were stated differently in the original policy document (EPA, 1996). This description describes EPA guidance as published in the various science policy documents, standard operating procedures and other guidance documents.

Table 6.2 EPA science policies directly related to dietary exposure assessment

Science policy	Papers describing the policy
Dietary exposure assessment – whether and how to use Monte Carlo analyses and the 99.9th percentile issue	• *Guidance for the submission of probabilistic human health exposure assessments to the Office of Pesticide Programs* • *Choosing a percentile of acute dietary exposure as a threshold of regulatory concern* • *Use of the Pesticide Data Program (PDP) in acute dietary assessment*
Exposure assessment – interpreting 'No residues detected'	• *A statistical method for incorporating non-detected pesticide residues into human health dietary exposure assessments* • *Assigning values to non-detected/non-quantified pesticide residues in human health dietary exposure assessments* • *Threshold of regulation policy – deciding whether a pesticide with a food use pattern needs a tolerance (revised)*
Dietary exposure estimates	• *A user's guide to available OPP TS[a] information on assessing dietary (food) exposure to pesticides*
Drinking water exposures	• *Estimating the drinking water component of a dietary exposure assessment (revised)*
Additional papers relating to dietary exposure	• *The role of use-related information in pesticide risk assessment and risk management* • *Data for refining anticipated residue estimates used in dietary risk assessments for organophosphate pesticides* • *Guidelines for the conduct of bridging studies for use in probabilistic risk assessment* • *Guidelines for the conduct of residue decline studies for use in probabilistic risk assessment* • *Quantitative assessment of uses of concern for drinking water[b]* • *Factoring drinking water treatment into drinking water assessments for pesticides[b]*

[a]OPPTS, Office of Prevention, Pesticides and Toxic Substances (USA).
[b]Not published at the time this chapter was written.

for risk managers to decide if the risk really poses a threat to public health. This is the primary role of the probabilistic risk assessment – to put relatively high exposure values into context, not to mask them.

ISSUES IN ESTIMATING DIETARY EXPOSURE AND RISK

As the preceding sections demonstrate, more stringent statutory requirements have prompted the development of new techniques for estimating dietary exposure and risk. These new techniques, particularly the probabilistic methods for evaluating acute dietary risk, rely heavily on data. The improper use of data, improper models, mis-interpretations, or invalid assumptions all can make even the most sophisticated assessment of little practical worth. This section describes some of the important issues regarding dietary risk assessment, particularly in the new era of assessments that rely on so much high-quality data and sophisticated methods of using the data.

VARIABILITY AND UNCERTAINTY

Variability is an inherent property of life. A simple stroll down a city street reveals a vast array of genetic variability. Similarly, the data used in dietary risk assessments are variable. Residue levels on crop plants vary. Toxicological responses vary. The changes of residue concentrations when raw commodities are processed vary. One of the significant strengths of the probabilistic approach is putting the inherent variability of the data into proper context, thus providing the risk assessor with an understanding of what constitutes high exposure levels and the relative importance of the residue level.

Uncertainty, however, arises from errors in the models or the data. Uncertainty arises when a sample is not analyzed properly, when a sample is collected improperly, or when a mathematical model does not include the appropriate variables. Essentially, variability is a normal part of the risk assessment process, whereas the risk assessor should try assiduously to reduce uncertainty.

Variability is both the source of much of the error in worst-case assessments, and a quality that can be used to calculate better estimates of risk. The Monte Carlo assessments described in previous sections use the variation in residue concentration to simulate different intake levels by the population of interest. This contrasts drastically with the worst-case assessment that assumes a uniformly high residue level in all foods consumed. As appropriate with the available data, computer models should be able to account for variability in the amount of food that is consumed, the residue levels on that food, the effects of processing, and the likelihood that the food is treated at all. One of the next areas of research may be the impact of variability in the toxicity information upon the interpretation of the risk assessment.

OUTLIERS

Outliers are observations which are outside the usual limits of a data distribution. Statistical tests are available that can test questionable values to determine whether they might be outliers. Outliers can legitimately be excluded from a data set.

In the United States, the rapid expansion of probabilistic assessments for acute dietary risk led to tremendous scrutiny of the data used for dietary risk assessments, particularly the food consumption data. For the most part, the data have withstood this scrutiny, but the question of outliers in the data is still an important one. One question that risk managers must weigh is the extent to which a small number of atypical observations in a data distribution should determine the outcome of a risk assessment, and possibly the decision of the risk manager. Most regulatory authorities have chosen to regulate at the extreme tail of the acute exposure distribution. The magnitude of the exposure estimate at these points can be affected tremendously by extreme values in the data.

The issue of outliers typically relates only to questions of acute dietary exposure, because chronic exposure is evaluated over a relatively long time period, perhaps as long as a lifetime. For such long time periods, the significance of sporadic spikes in either consumption or residue data is dampened by the times when exposures are much lower.

FOOD CONSUMPTION DATA

Clearly, the types of calculations for probabilistic dietary exposure estimates require a great deal of data. In addition, many regulatory agencies are becoming more interested in various sub-groups of the population, particularly infants and children. Consequently, the types of analyses that regulatory authorities will require in the future will be increasingly complex. However, in some cases, the food consumption data used to calculate the dietary exposure estimates might not have very many observations. Even the CSFII data of the United States had fewer than 1000 food consumption estimates for infants, and a marginal number for older children. This situation prompted the execution of the 1998 Supplemental Children's Survey (USDA, 2000) which provided approximately 12 000 additional observations for children between infancy and nine years of age. The amount of data was enough of a concern that the USDA conducted a supplemental survey of consumption by young children to augment the basic CSFII data set.

Data quality has been an extremely important issue, particularly as the data are used for applications beyond those for which they were originally intended. Consequently, the organizations that collect the data typically take great care to guard the integrity of the data. For example, in the 1994–1996 CSFII, USDA scientists confirmed data reports of unusual food consumption amounts (USDA, 1997a).

Additionally, questions have been raised about the extent to which the data provided by surveys represent actual consumption patterns. Often, the proportion of samples for a particular population group in the survey corresponds to the proportion of that group in the actual population. Nevertheless, the actual numbers of individuals sampled can be very small. In such situations, the impact of a few extreme consumption values can have a tremendous impact upon the outcome of the risk assessment.

The exposure analyst should also consider the possible impact of periodicity upon the dietary exposure analysis. Conducting national surveys of food consumption is an ambitious undertaking that consumes the time of many highly trained scientists. As a result, such surveys generally are conducted every several years. It is not uncommon for more than 10 years to pass between surveys. In addition to the time that transpires between collecting the survey data, it often takes two or three years to process the data. Thus, even though the food consumption data remain static between survey periods, actual food consumption patterns are dynamic and changeable. Perhaps fruit juices tend to be consumed more than whole fruit. Perhaps intake of fats decreases as a result of government nutritional education. For whatever reason, it is a challenge for any food consumption survey to provide estimates of food consumption that are in synchrony with actual consumption patterns.

Another area that can have a significant impact upon the outcome of a dietary exposure assessment is the degree of specificity in the food consumption data. Most food consumption data, as was previously discussed, are collected from interviews, diaries or recording forms. Pesticide residues are regulated on the basis of raw agricultural commodities, such as tomatoes, wheat, mushrooms, peppers and pork. The food consumption data, in contrast, records how much pizza an individual consumed. Therefore, a major task in converting food consumption data for use in dietary risk assessment is to transform the raw data into a form consistent with the dietary exposure models that will be used. As described in previous sections, many models organize the food consumption data in a relatively aggregated state. Such organization of the data is good for efficiency, but usually does not allow for the full utilization of the available residue data, in particular, data on the impact of processing on residue concentration.

Finally, the food consumption data should be collected and processed in such a way that comparisons among population groups can be made. If one government objective is to ensure that pregnant women are not at risk from ingesting pesticide residues via the diet, then the food consumption database should report consumption estimates for adult women.

RESIDUE DATA

A major issue regarding residue data is that of variability. The residue concentration in food is the result of many factors, including the amount of pesticide that is applied and the minimum length between the last application and harvest. Residue concentration may also be influenced by the amount of UV radiation, moisture, the crop canopy and various edaphic factors. Even in a controlled field trial, pesticide levels do vary.

One of the major concerns regarding residue data has been the apparent lack of correspondence between presumed residue levels on individual commodity samples and the composite data that typically are analyzed in a residue field trial or monitoring program (PSD, 1997a, 1997b). The basic concern is that a composite sample may dilute the effect of an individual component of the composite that has a high residue. Follow-on work in the UK on pesticide variability has not revealed any consistent relationships to variability and any of the parameters examined (Harris, 2000). The most extreme example would be a situation in which the entire residue observed in a composite sample was derived from one unit in such a sample.

In the United States, monitoring data were typically excluded from consideration of acute dietary exposure because of the concern over composite samples. Recently, however, the US EPA has suggested that a procedure termed 'decompositing' may allow monitoring data to be used in assessments of acute dietary exposure. Basically, decompositing techniques make use of the observed variability of the composite samples to estimate the variability among the individual units within the composites (EPA, 2000a).

The technique of decompositing, in all likelihood, will be used as a surrogate until results of specific market basket surveys on individual unit samples increase our understanding of variability among individual units within a composite sample.

PESTICIDE USE DATA

One of the most powerful adjustments that can be made to a dietary exposure estimate is the adjustment for the percent of the crop that is actually treated. For assessments of chronic exposure, the adjustment is typically a simple multiplication of the crop residue by the percent crop treated estimate. Thus, for chronic dietary assessment, the percent crop treated estimate functions as a coefficient. For example, if the residue value is 0.5 mg/kg, and 25 % of the crop is treated, then, in effect, the residue value used in a refined chronic dietary risk assessment would be 0.13 mg/kg (0.5 mg/kg × 25 %).

There are some differences in practice depending on the actual situation in the country that is carrying out the assessment. For example, in the UK, adjustments for percent crop treated are restricted to estimates of exposure to residues in blended commodities, such as juices, oils or grains. In the United States, in contrast, adjustments for percent crop treated may be applied to any food, even those that are not blended.

For probabilistic acute dietary risk assessments, percent crop treated is *not* used as a simple coefficient. Instead, it is used to indicate the probability that a particular crop would be treated. If the model determines that the crop is treated on a particular pass through the model, then a residue value is sampled from the residue distribution. If the model determines that the crop is not treated on that pass, then the exposure contribution of the crop on that pass is zero.

Pesticide use data encompass more than the amount of the crop that is treated, however. Pesticides are applied according to the conditions permitted on the product labels. However, the limits on the label are maximum values. Individual growers may find, for example, that the maximum application rate permitted by the product label is not suitable for his or her situation; different pests, for example, often require different application rates on the same crop. In such cases, the grower may choose to apply the pesticide at a lower rate (never at a rate higher than that permitted by the product label). If the risk analyst knows the proportion of the crop that is treated at the lower rate, and also has data demonstrating residue levels on the crop that is treated at the lower rate, it is possible to adjust the residue value in the exposure assessments to account for the (presumably) lower residue concentration expected if the pesticide is applied at a lower rate.

AGGREGATE AND CUMULATIVE EXPOSURE

In the United States, the Food Quality Protection Act (FQPA) requires the EPA to consider the potential health effects of combinations of routes of exposure and compounds. The term 'aggregate' is applied to exposure from multiple routes or uses. For example, consider a pesticide used in crop production, to control residential termites, and to control insects on turf. An aggregate exposure assessment for such a compound would include a dietary component from the agricultural use, an inhalation component from the termiticide use, and possibly three components from the turf use, i.e. inhalation, dermal, and incidental ingestion.

Some pesticides may be categorized with other compounds with respect to mechanism of toxicity. Such compounds, according to the FQPA, must be evaluated in combination to assess the total impact of exposure to all of the compounds.

At the time of this writing,[8] the policies and methodologies for conducting such assessments were still under development and discussion in the United States. Although the requirement to conduct such assessments exists only in the United States at this time, the concepts of aggregate and cumulative exposure and risk are attracting attention in the international regulatory arena. In time, other regulatory authorities in addition to the US EPA may require aggregate and cumulative risk assessments.

Given the uncertainty at this time regarding methodologies and interpretation of such assessments, this chapter will not go into any further detail regarding this topic.

WORKED EXAMPLES OF ESTIMATES FOR RESIDUE DIETARY INTAKE

As an example[9] of dietary intake point estimates, consider Chemx – this is a hypothetical compound used on the crops shown in Table 6.3. There are Codex MRLs, US tolerances and UK MRLs. Note that the crop list in the United States differs from that in the United Kingdom. The summary statistics were calculated

---

[8]2002.

[9]In this chapter, Chemx will be used as an exemplary compound for illustrating the various types of dietary intake calculations. Chemx is a hypothetical compound created for purpose of illustration only. The residue information, processing factors, use information and toxicology end-points are all hypothetical and are not intended to represent any real pesticide that is either currently registered, formerly registered and now cancelled, or in development.

Table 6.3   Maximum residue levels and adjustment factors for Chemx[a]

Crop	Codex MRL (mg/kg)	US tolerance (mg/kg)	UK MRL (mg/kg)
Apple	1	1	0.5
Pear	1	1	0.5
Peach	5	2	5
Tomato	5	5	_[b]
Strawberry	3	_[b]	3

Crop	'Global' STMR (mg/kg)	US anticipated residue (mg/kg)	UK STMR (mg/kg)
Apple	0.1	0.04	0.04
Pear	0.15	0.09	0.06
Peach	0.35	0.6	0.45
Tomato	0.5	0.7	–
Strawberry	1.2	–	0.9

Crop	Crop treated, 'global' %	Crop treated, US %	Crop treated, UK %
Apple	35	25	10
Pear	15	5	15
Peach	15	2	10
Tomato	30	25	–
Strawberry	–	–	50

Processed food	Proportion of food consumed as processed	
	UK	Global
Apple juice	0.35	0.2

	Processing factor
Apple juice	0.05
Tomato juice	0.15
Tomato paste	1.7
Tomato puree	0.85

[a]Note that these data are for the purpose of illustration only and do not represent actual information for any real pesticide chemical.
[b]No data available.

from the raw field trial data shown in Table 6.4. All of the data are hypothetical for illustrative purposes.

In this example, we will compare chronic dietary intake estimates calculated using international level food consumption data (WHO, 1998), as well as country-specific consumption data from the UK (Anon (no date); Gregory et al., 1990, 1995; Mills and Tyler, 1992) and the United States (USDA, 1992, 1993, 1994, 1995, 1996a, 1997b). We will also consider acute dietary estimates calculated using the NESTI methodology, as well as probabilistic Monte Carlo techniques,

Table 6.4 Hypothetical field trial residue data (mg/kg) for 16 trials for each crop in each country representing the residue levels found when Chemx was used at the maximum allowed by the label[a]

United States data				United Kingdom data			
Apple	Pear	Peach	Tomato	Apple	Pear	Peach	Strawberry
0.005	0.015	0.05	0.05	0.007	0.022	0.14	0.45
0.005	0.016	0.09	0.13	0.008	0.031	0.24	0.52
0.005	0.018	0.12	0.29	0.01	0.032	0.32	0.72
0.008	0.02	0.14	0.5	0.015	0.042	0.37	0.72
0.013	0.022	0.24	0.52	0.024	0.044	0.4	0.77
0.02	0.023	0.37	0.7	0.024	0.053	0.42	0.8
0.02	0.027	0.4	0.72	0.03	0.055	0.44	0.8
0.03	0.042	0.44	0.72	0.032	0.058	0.45	0.9
0.039	0.06	0.5	0.77	0.044	0.063	0.45	0.9
0.042	0.075	0.6	0.77	0.05	0.069	0.5	0.94
0.052	0.078	0.62	0.8	0.061	0.072	0.6	1
0.072	0.09	0.68	0.8	0.062	0.088	0.6	1.2
0.072	0.09	0.93	0.9	0.066	0.09	0.6	1.2
0.077	0.12	0.98	0.92	0.088	0.14	0.68	1.3
0.09	0.15	1.4	1.2	0.097	0.17	0.8	1.4
0.092	0.6	1.8	1.4	0.11	0.66	0.9	1.6

[a]Note that these data are for the purpose of illustration only and do not represent actual information for any real pesticide chemical.

as described above in the section 'United States Methodologies' (EPA, 2000b). All exposure estimates calculated with the US food consumption data used the Dietary Exposure Evaluation Model (DEEM™) (Novigen, 1996). The probabilistic estimates using the UK consumption data were calculated by using a developmental version of DEEM™ based upon the UK food consumption data.

In these examples of dietary risk assessments, Chemx has a chronic ADI of 0.01 mg/kg bw day and an acute reference dose (ARfD) of 0.025 mg/kg bw. These numbers are hypothetical and are used for sake of illustration only.

THEORETICAL MAXIMUM DAILY INTAKE

Table 6.5 shows the standard TMDI intake estimate for adults using the European regional diet (WHO, 1998) and the MRL value. Consumption data for population sub-groups, such as children, are not available with the GEMS/Foods data at this time.[10] One modification has been made to the food consumption data. The GEMS/Food data only report apple consumption, without differentiating consumption for apple juice. However, Table 6.5 provides some information about the proportion of apples that are consumed as apple juice (Note: that this figure is purely hypothetical). The apple consumption has been adjusted to partition

---

[10]2002.

Table 6.5 TMDI estimates for Chemx using GEMS/Foods consumption data for the European region

Commodity	Consumption (kg food/d)	Consumption (kg food/kg bw/d)	Residue (mg/kg food)	Dietary intake (mg/kg bw/d)	ADI[a] %
Apple	0.0320	0.000 533	1	0.000 533	5.3
Apple juice	0.0080	0.000 133	1	0.000 133	1.3
Pears	0.0113	0.000 188	1	0.000 188	1.9
Strawberry	0.0053	0.000 088	5	0.000 442	4.4
Peaches	0.0125	0.000 208	3	0.000 625	6.3
Peaches, dried	0.0001	0.000 002	3	0.000 005	0.1
Tomato, fresh	0.0382	0.000 637	5	0.003 183	31.8
Tomato, juice	0.0020	0.000 033	5	0.000 167	1.7
Tomato, paste	0.0040	0.000 067	5	0.000 333	3.3
Tomato, puree	0.0020	0.000 033	5	0.000 167	1.7
Tomato, peeled	0.0040	0.000 067	5	0.000 333	3.3
			Totals:	0.006 110	61.1

[a] ADI is 0.01 mg/kg bw/d.

total apple consumption into consumption of apple juice and consumption of apple, *per se*. The consumption data are in the form of mean *per capita* consumption on a daily basis. The basic consumption data are divided by a standard body weight of 60 kg to yield consumption in kg of food/kg bw/d. Although the use of Chemx varies slightly between the US and the UK, the 'global' intake assessment considers all uses of the compound.

UK THEORETICAL MAXIMUM DAILY INTAKE

The UK version of the TMDI calculation for toddlers (children aged $1\frac{1}{2}$ to $4\frac{1}{2}$ years), described in the Point Estimates section under 'Models', is shown in Table 6.6. The UK TMDI estimate was calculated by using the UK Consumer Model available from the UK PSD web site (PSD, 1996). The UK TMDI estimate, and the associated dietary risk, is almost nine times higher than that calculated with the European regional diet. The difference is probably attributable to the UK methodology, which considers the two highest 97.5th percentile intake values.

THEORETICAL MAXIMUM RESIDUE CONTRIBUTION

The Point Estimates section under 'Models' also describes the US version of the TMDI, termed the TMRC. The TMRC estimates for Chemx are displayed in Table 6.7. As can be seen from this summary table for the US TMRC, the US food consumption data provide rather more detailed information than some of the other consumption databases. Although this might not have much of an impact on a TMRC calculation, the ability to sub-divide the dietary exposure estimate with greater detail can have a tremendous impact upon the refined assessment.

Table 6.6 TMDI estimates for Chemx using the UK consumer exposure model for toddlers

Commodity	UK TMDI (mg/kg bw/d	ADI[a] %
Apple	0.007 530	75
Peach	0.037 034	370
Pear	0.003 359	33
Strawberry	0.006 621	66
Totals:	0.054 544	545

[a]ADI is 0.01 mg/kg bw/d.

Table 6.7 Theoretical mean residue contributions (TMRCs) calculated by using the US consumption data and DEEM™ model

Commodity	TMRC (mg/kg bw/d)	ADI[a] %
Apple	0.001 564	15.6
Apple, dried	0.000 078	0.8
Apple, juice	0.005 110	51.1
Apple, juice concentrate	0.000 364	3.6
Peach	0.000 595	6.0
Peach, dried	0.000 018	0.2
Pear	0.000 235	2.3
Pear, dried	<0.000 001	<0.1
Pear, juice	0.000 076	0.8
Tomato	0.004 197	42.0
Tomato, juice	0.000 043	0.4
Tomato, puree	0.001 767	17.7
Tomato, paste	0.002 352	23.5
Tomato, catsup	0.000 766	7.7
Tomato, dried	—[b]	—[b]
Totals:	0.017 170	172

[a]ADI is 0.01 mg/kg bw/d.
[b]No data available.

It is important to recognize the differences in techniques that have an impact even on screening level assessments such as the TMDI or TMRC. As shown in Table 6.3, MRLs have been established for Chemx on five crops. However, only four of the crops are registered in the US and the UK, with three of the crops – apple, pear, and peach – in common (see Table 6.4). The TMDI calculation using the GEMS/Foods methodology, therefore, calculates the TMDI incorporating all five foods, because this TMDI is truly a *global* dietary exposure estimate. TMDI values calculated with the GEMS/Food methodology use mean food consumption data, which are normalized to a standard body mass of 60 kg. The TMDI calculated using the UK PSD Consumer Model is calculated with the two highest 97.5th percentile intake estimates, using food consumption data that have been normalized to a standard body mass of 14.5 kg. The US TMRC is calculated for children aged one to six using mean *per capita* consumption data. The US database does not use a standard body mass, because the food consumption database includes information about the individual body mass of the survey participants. The TMDI value calculated using the 97.5th percentile values is approximately 10-fold higher than the TMDI calculated by either the GEMS/Food or US methodologies. Although all of these dietary intake estimates are 'worst-case' preliminary estimates, the actual values differ because the consumption data, the way of using the consumption data, and some of the methods used to perform the calculation differ.

INTERNATIONAL ESTIMATED DAILY INTAKE

Table 6.8 shows a refined dietary exposure estimate based upon the GEMS/Food consumption data and the international methodology, namely an example of the international estimated daily intake (IEDI) calculation. In this evaluation, four refinements have been made. First, the GEMS/Food apple consumption has been modified to estimate exposure from apple juice, as described above for the GEMS/Food TMDI calculation. In addition, the global STMR value from Table 6.3 is used instead of the MRL for the residue portion of the dietary exposure equation. Finally, adjustments have been made for the percent of the crop that is treated. Percent-crop-treated adjustments have only been applied to crops that are blended, such as apple juice and processed tomato products, as is the

Table 6.8 IEDI calculation using the GEMS/Food consumption data for the European region

Commodity	Consumption (kg food/d)	Consumption (kg food/kg bw/d)	Residue[a] (mg/kg food)	Dietary intake (mg/kg bw/d)	ADI[b] %
Apple	0.0320	0.000 533	0.1	0.000 053	0.5
Apple juice	0.0080	0.000 133	0.0035[c]	0.000 000 3	0.0
Pears	0.0113	0.000 188	0.15	0.000 028	0.3
Strawberry	0.0053	0.000 088	1.2	0.000 106	1.1
Peaches	0.0125	0.000 208	0.35	0.000 073	0.7
Peaches, dried	0.0001	0.000 002	0.35	0.000 001	0.0
Tomato, fresh	0.0382	0.000 637	0.5	0.000 318	3.2
Tomato, juice	0.0020	0.000 033	0.022[c]	0.000 001	0.0
Tomato, paste	0.0040	0.000 067	0.26[c]	0.000 017	0.2
Tomato, puree	0.0020	0.000 033	0.13[c]	0.000 004	0.0
Tomato, peeled	0.0040	0.000 067	0.5	0.000 033	0.3
			Totals:	0.000 636	6.4

[a] STMR, supervised trials median residue.
[b] ADI is 0.01 mg/kg bw/d.
[c] Residue values for these processed commodities were: (STMR × percent crop treated × processing factor).

practice in the UK. Finally, the residue values for processed commodities have
been adjusted for the effect of processing. For example, the STMR for apples is
0.1 mg/kg. For estimating dietary exposure from apple juice, the residue value
in effect is the (STMR × percent crop treated × apple juice processing factor),
which results in a residue of 0.0005 mg/kg (0.1 mg/kg × 10 % × 0.05). Thus,
the IEDI for apple juice is calculated with a residue value of 0.0005, a value
more than 500-fold lower than the value used to calculate the TMDI value for
apple juice.

NATIONAL ESTIMATED DAILY INTAKE

The PSD Consumer Model was used to calculate a national estimated daily intake
(NEDI) value for toddlers. The UK STMR values were used for the residue, and
residues in apple juice were adjusted for the effects of processing and the percent of the crop that is treated. A comparison of Tables 6.6 and 6.9 will show
that the UK NEDI calculation included apple juice, whereas the TMDI calculation did not. The PSD Consumer Model includes only consumption data for
apples. However, the estimated proportion of apples consumed as apple juice
(a hypothetical value) was used to partition consumption of apple juice from
overall consumption of apples. The NEDI is approximately 9-fold lower than the
TMDI value.

ANTICIPATED RESIDUE CONTRIBUTION

The US food consumption data and methodology were used to calculate an Anticipated Residue Contribution (ARC), as shown in Table 6.10. In the US, the term
'anticipated residue' is used to describe the value that is intended to represent
residue levels on foods as they are eaten. The tolerance value, as is the MRL,
is a legal enforcement value, which is not expected to occur regularly in foods.
The anticipated residue, in practice, pertains to any residue other than a tolerance
value which is used to estimate dietary exposure. Thus, the mean value from

Table 6.9 NEDI calculation using the UK consumer exposure model
for toddlers

Commodity	UK TMDI (mg/kg bw/d)	ADI[a] %
Apple	0.000392	3.9
Apple juice	0.000001	<0.1
Peach	0.003333	33.3
Pear	0.000403	4.0
Strawberry	0.001986	19.9
Totals:	0.006115	61.1

[a] ADI is 0.01 mg/kg bw/d.

Table 6.10 Anticipated residue contributions (ARCs) calculated by using the US consumption data and the DEEM™ model

Commodity	ARC (mg/kg bw/d)	ADI[a] %
Apple	0.000 094	0.9
Apple, dried	0.000 005	0.8
Apple, juice	0.000 047	0.5
Apple, juice concentrate	0.000 003	<0.1
Peach	0.000 179	0.2
Peach, dried	0.000 005	0.1
Pear	0.000 021	0.2
Pear, dried	<0.000 001	<0.1
Pear, juice	0.000 076	0.8
Tomato	0.000 588	5.9
Tomato, juice	0.000 006	0.1
Tomato, puree	0.000 247	2.5
Tomato, paste	0.000 329	3.3
Tomato, catsup	0.001 07	1.1
Tomato, dried	_[b]	–
Totals:	0.0001639	16.4

[a] ADI is 0.01 mg/kg bw/d.
[b] No data available.

field trials, the maximum value from a monitoring program, or a residue adjusted for the percent of crop treated or the effects of processing would all be types of anticipated residues. Conceptually, these are the same types of adjustments that were applied in the IEDI calculations. IEDI values were calculated by using the GEMS/Foods data, while the NEDI calculations used the PSD Consumer Model. The details of how these adjustments were used differ, as do the consumption values. Nonetheless, the concept of using more refined data to estimate dietary exposure is generally applicable. As for the TMRC, mean *per capita* consumption values were used to calculate dietary exposure. However, instead of using the tolerance values, the ARC is calculated by using the mean value from field trials (Tables 6.3 and 6.4) and adjustments for processing and percent crop treated. In the US, percent-crop-treated adjustments may be applied to all foods, blended and unblended.

Estimating acute dietary exposure is a process quite different from estimating a TMDI or NEDI. The acute dietary risk assessment seeks to evaluate the potential impact of extreme intake levels on human physiology. The general methodologies are described under the section 'United States Methodologies'. Table 6.11 illustrates the NESTI and distributional approaches to acute dietary risk assessment, while Table 6.12 provides an example of probabilistic acute dietary risk assessments. In all cases, the toxicity end-point against which estimated exposure is evaluated is an acute reference dose (ARfD) of 0.025 mg/kg bw. The various exposure calculation methodologies used the data handling conventions

Table 6.11 Preliminary acute dietary exposure assessments[a]

Type of estimate	Estimated exposure for type or commodity	97.5th Percentile value	
		Exposure (mg/kg bw/d)	ARfD[b] %
US Tier 2	Total daily exposure	0.041 437	166
UK NESTI	Apples	0.027 274	109
	Apple juice	0.000 540	2.2
	Peaches	0.266 572	1066
	Pears	0.036 741	147
	Strawberries	0.006 890	27.6

[a]Note that it is inappropriate to add the NESTI values together because each value is based upon a high-end consumption estimate, namely the 97.5th percentile.
[b]ARfD is 0.025 mg/kg bw/d.

Table 6.12 Probabilistic acute dietary exposure assessments[a]

Type of estimate	Estimated exposure for type or commodity	97.5th Percentile value	
		Exposure mg/kg bw/d	ARfD[b] %
US Tier 3	Total daily exposure	0.003 304	13.2
UK Monte Carlo	Apples plus apple juice	0.001 123	4.5
	Peaches	0.002 882	11.5
	Pears	0.002 497	10.0
	Strawberries	0.002 635	10.5

[a]Note that it is inappropriate to add the UK Monte Carlo values together because each value is based upon a high-end exposure estimate, namely the 97.5th percentile.
[b]ARfD is 0.025 mg/kg bw/d.

commonly permitted by the pertinent national regulatory authority, such as the US EPA or the UK PSD.

The US Tier 2 assessment is a non-probabilistic assessment based upon the USDA CSFII data. For the UK NESTI calculation, data from the toddler survey are used. The US database groups children between the ages of 1 and 6 years, whereas the UK database groups children between the ages of $1\frac{1}{2}$ and $4\frac{1}{2}$ years.

The US Tier 2 assessment is a distributional analysis, although it does not use a probabilistic approach. The distribution originates in the food consumption data. Unlike the TMRC or ARC calculations, the acute assessment uses the individual daily consumption values for all of the data meeting the selection criteria, in this case children between the ages of 1 and 6 years. The tolerance value is used for *all* unblended commodities. Although consumption of apples varies, a residue value of 1 mg/kg would be used in the assessment for each data record reporting consumption of apples. Processed commodities are assumed to be blended, and so the residue value for apple juice was the US anticipated residue, adjusted for processing. Thus, the residue value used for apple juice in the US Tier 2 assessment would be 0.003 mg/kg (0.06 mg/kg × 0.05) for all data records of apple juice consumption. No adjustment is made for percent of crop treated in the US Tier 2 assessment. If monitoring data were available, the maximum residue value observed in the monitoring database could be used as the residue for apple juice instead of the field trial mean anticipated residue. As shown in Table 6.11, exposure estimated using the US Tier 2 method is unacceptable. Exposure is evaluated relative to the RfD, while in the case of acute exposure, an acute RfD (ARfD). If the estimated exposure exceeds the ARfD, exposure is considered to be too high and the risk is unacceptable. If the estimated exposure is less than the ARfD, then exposure is not too high and the risk is considered to be acceptable.

It should be remembered that the US methodology results in a *total* daily exposure estimate. One person's estimated exposure could arise from consuming moderate amounts of all of the pertinent foods, whereas another person in the survey database could reach the same level of exposure by eating a relatively large amount of only one of the pertinent foods. Unlike the chronic exposure estimate represented by the TMRC or ARC, the acute exposure estimate *can not* be partitioned into components that neatly combine to produce the total exposure. The total exposure is determined by the overall consumption *pattern* of the individual.

Acute exposure was estimated for toddlers with the UK consumption data by using the NESTI approach, and these values are shown in Table 6.11. Apple juice and strawberries were treated as Case 1 commodities for which the residue value in the composite sample represents the residue. The residue was adjusted for the effect of processing, but not for the percent of crop that was treated. All other commodities were treated as Case 2 commodities, without any adjustment for processing or for the percent of crop that was treated. As can be seen in Table 6.11, several of the commodities exceeded the ARfD, but, in particular, peaches and pears.

Since the initial acute exposure estimates were unacceptable, acute exposure was then estimated by using probabilistic techniques. It is important to remember, however, that the judgment of unacceptability is based upon a simple comparison to the ARfD. In this example, the initial exposure estimates exceed the ARfD, and so risk is considered to be unacceptable. Actual exposure levels are usually much lower than an initial estimate would indicate.

The US Tier 3 assessment incorporates Monte Carlo sampling. Just as for the Tier 2 assessment, in the Tier 3 assessment total daily exposure was estimated using the general procedure described above under the section 'United States Methodologies'. Percent crop treated was used in the assessment, but not as a residue coefficient as in the chronic ARC estimate. In the acute Tier 3 assessment, percent crop treated is used to determine the probability of a given pass of the exposure algorithm (called an iteration) to sample a treated or untreated sample. If the simulation selects an untreated sample for a commodity, then that commodity does not contribute to total daily exposure on that iteration. If the simulation selects a treated sample for a commodity, then a residue value is selected at random from the residue distribution. The major difference between the Tier 2 and Tier 3 acute assessments is that the residue value in the Tier 3 assessment varies as determined by the rules of probability. In the Tier 2 assessment, all commodities are assumed to contain the same high residue concentration. Table 6.12 shows that the probabilistic assessment reduces the estimated risk by almost 13-fold, thus resulting in acceptable levels of risk, again based upon a comparison to the ARfD.

Similar results are observed with the probabilistic approach when using the UK data. At this time,[11] probabilistic assessments have not undergone much scrutiny by regulators outside of the United States. Although the UK version of the DEEM™ is able to calculate a total daily estimate of exposure calculations, the probabilistic approach was followed on a commodity-by-commodity basis. This facilitates comparison to the NESTI calculation. Furthermore, given the uncertainty about how percent crop treated might be incorporated into a probabilistic assessment outside of the United States, the probabilistic sampling used only the residue values shown in Table 6.4. All pertinent commodities were assumed to be treated, but the magnitude of the residue did vary. Exposure from apple juice was combined with exposure from apples. Table 6.12 shows that acute dietary exposure did not exceed acceptable levels when calculated in this way.

CONCLUSIONS

A reading of this chapter demonstrates that the calculation of dietary risk can follow many different procedures. Furthermore, the methods that are used to estimate dietary risk are often a function of the data that are available to the scientist. It is important to remember that it is virtually impossible to actually *know* what the level of dietary exposure is, either on an individual or population basis. The preliminary risk estimates provided by the TMDI or TMRC calculations clearly grossly overestimate dietary risk. It is virtually impossible that anyone would consistently consume foods that always contain the highest possible concentration of pesticides allowed. Yet, this is exactly what is assumed in the TMDI and TMRC calculations. Knowing that the TMDI or TMRC calculations grossly overestimate dietary risk, we can be very well assured that the populace will not be adversely affected if such calculations do not exceed safe levels. However, TMDI or TMRC estimates that do exceed safe levels (the chronic ADI) do not necessarily mean that people are actually ingesting such high levels of pesticides in the foods that they eat.

When the preliminary risk calculations exceed acceptable levels, more refined estimates using better data and less outrageous assumptions may be calculated. As shown in the worked examples, such calculations often result in tremendous

---

[11] 2002.

reductions in the estimated dietary exposure and risk. Even so, the refined calculations do not necessarily reflect actual risk. The presumption, and hope, of the risk assessor is that actual risk is much lower than even the most refined risk estimate.

REFERENCES

Anon, (no date). *The Diets of British Schoolchildren*, Report on health and social subjects, No. 36, Her Majesty's Stationery Office, London.

Barraj, L. and Petersen, B. (1997). *A Method for Revising and Redefining Regional Diets for Use in Estimating the Intake of Pesticides*, Working Paper CONSUMP CON 2, Joint FAO/WHO Consultation on Food Consumption and Exposure Assessment of Chemicals, Geneva, 10–14 February, 1997 (available from the World Health Organization, Geneva, on request).

CAST (1992). *Pesticides – Minor Uses/Major Issues*, Council for Agricultural Science and Technology, Ames, IA, USA.

EC (1991). *Council Directive 91/414/EEC*, of 15 July 1991 concerning the placing of plant protection products on the market, OJ L 230, European Commission, (19 August, 1991).

EPA (1996). Final office policy for performing acute dietary exposure assessment, memorandum from D. Edwards to EPA OPP Staff, US Environmental Protection Agency, (13 June, 1996).

EPA (2000a) *Office of Pesticide Programs Comparison of Allender, RDFgen, and MaxLIP Decompositing Procedures*, Background paper for the Scientific Advisory Panel Meeting, 28 February–3 March, 2000, US Environmental Protection Agency.

EPA (2000b). *Available Information on Assessing Exposure from Pesticides in Food: A User's Guide*, US Environmental Protection Agency, (June 21, 2000).

FAO/WHO (1995). *Recommendations for the Revision of the Guidelines for Predicting Dietary Intake of Pesticide Residues*, Report of an FAO/WHO consultation, WHO/FNU/FOS/95.11, Food and Agriculture Organization of the United Nations, Rome.

FAO/WHO (1997). *Food Consumption and Exposure Assessment of Chemicals*, Report of an FAO/WHO consultation, Geneva, Switzerland, 10–14 February, 1997, WHO/FSF/FOS/97.5, Food and Agriculture Organization of the United Nations, Rome.

Federal Register (1998). *Guidance for Submission of Probabilistic Human Health Exposure Assessments to the Office of Pesticide Programs*, **63**(214): 59780–59783, (5 November, 1998).

Food Marketing Institute (1994). *Trends in the United States – Consumer Attitudes and the Supermarket*, Food Marketing Institution, Washington, DC, USA.

Gregory, J., Foster, K., Tyler, H. and Wiseman, M. (1990). *The Dietary and Nutritional Survey of British Adults*, Her Majesty's Stationery Office, London.

Gregory, J., Collins, D., Davis, P., Hughes, J. and Clarke, P. (1995). *The National Dietary and Nutritional Survey: Children Aged $1\frac{1}{2}$ to $4\frac{1}{2}$*, Her Majesty's Stationery Office, London.

Harris, C. (2000). How the variability issue was uncovered: the history of UK residue variability findings, *Food Additives Contaminants*, **17**, 492–495.

Harris, C. A., Mascall, J. R., Warren, S. F. P. and Crossley, S. J. (2000). Summary report of the international conference on pesticide residues variability and acute dietary risk assessment, *Food Additives Contaminants*, **17**, 481–485.

JMPR (2001). *Pesticide residues in food-2000*. Report of the Joint Meeting of the FAO Panel of Experts on Pesticide residues in food and the environment and the WHO Core assessment group on pesticide residues Geneva, Switzerland 20–29 September 2000. FAO Plant Production and Protection Paper 163, Rome.

Korb, P. and Cochrane, N. (1989). World food expenditures, *Nat. Food Rev.*, **12**(4), p. 26.

Mills, A. and Tyler, H. (1992). *Food and Nutrient Intakes of British Infants Aged 6–12 Months*, Her Majesty's Stationery Office, London.

Novigen (1996). *Dietary Exposure Evaluation Module (DEEM™)*, Novigen Sciences, Inc.

NRC (1991). *Sustainable Agriculture Research and Education in the Field*, National Research Council, National Academy of Sciences, National Academy Press, Washington, DC, USA.

NRC (1993). *Pesticides in the Diets of Infants and Children*, National Research Council, National Academy of Sciences, National Academy Press, Washington, DC, USA.

PSD (1996). *Consumer Exposure Model* (Version 3.02), Pesticides Safety Directorate, UK.

PSD (1997a). *Unit to Unit Variation of Pesticide Residues in Fruit and Vegetables – the Probabilistic Approach to Risk Assessment*, Advisory Committee on Pesticides, 16 January, 1997, Pesticides Safety Directorate, UK.

PSD (1997b). *Organophosphorus Residues in Carrots: Monitoring of UK Crops in 1996/97 and Carrots Imported Between November and May 1996*, Pesticides Safety Directorate, UK.

PSD (1999a). *Guidance on the Estimation of Dietary Intakes of Pesticides Residues*, Pesticides Safety Directorate, UK (4 November, 1999).

PSD (1999b). *Report of the International Conference on Pesticide Residues Variability and Acute Dietary Risk Assessment, 1–3 December, 1998*, Pesticides Safety Directorate, UK (15 February, 1999).

SCPH (1998). *Monitoring for Pesticide Residues in the European Union and Norway, Report 1996*, Standing Committee on Plant Health.

Tomerlin, J. R. (2000). The US Food Quality Protection Act policy implications of variability and consumer risks, *Food Additives and Contaminants*, **17**, 641–648.

US Congress (1938). *Federal Food, Drug, and Cosmetic Act (FFDCA)*, 21 USC 301, *et seq.*, US Congress.

US Congress (1947). *Federal Insecticide, Fungicide and Rodenticide Act (FIFRA)*, 7 USC s/s 135, *et seq.* (1972), US Congress.

US Congress (1996). *Food Quality Protection Act (FQPA) Public Law*, 104–170, US Congress, (3 August, 1996).

USDA (1992). *Nationwide Food Consumption Survey: Continuing Survey of Food Intakes by Individuals 1989–1990*, Human Nutrition Information Service, United States Department of Agriculture, (Dataset).

USDA (1993). *Nationwide Food Consumption Survey: Continuing Survey of Food Intakes by Individuals 1990–1991*, Human Nutrition Information Service, United States Department of Agriculture, (Dataset).

USDA (1994). *Nationwide Food Consumption Survey: Continuing Survey of Food Intakes by Individuals 1991–1992*, Human Nutrition Information Service, United States Department of Agriculture, (Dataset).

USDA (1995). *Nationwide Food Consumption Survey: Continuing Survey of Food Intakes by Individuals 1993–1994*, Food Survey Research Group, Agricultural Research Service (ARS), United States Department of Agriculture, (Dataset).

USDA (1996a). *Nationwide Food Consumption Survey: Continuing Survey of Food Intakes by Individuals 1994–1995*, Food Survey Research Group, Agricultural Research Service (ARS), United States Department of Agriculture, (Dataset).

USDA (1996b). *Pesticide Data Program (USDA PDP), Annual Summary Calendar Year 1994*, Agricultural Marketing Service, Science and Technology Division, United States Department of Agriculture.

USDA (1997a). *Design and Operation: Continuing Survey of Food Intakes by Individuals 1994–1996*, Food Survey Research Group, Agricultural Research Service (ARS), United States Department of Agriculture.

USDA (1997b). *Nationwide Food Consumption Survey: Continuing Survey of Food Intakes by Individuals 1995–1996*, Food Survey Research Group, Agricultural Research Service (ARS), United States Department of Agriculture, (Dataset).

USDA (1997c). *Pesticide Data Program (USDA PDP), Annual Summary Calendar Year 1995*, Agricultural Marketing Service, Science and Technology Division, United States Department of Agriculture.

USDA (1998). *Pesticide Data Program (USDA PDP), Annual Summary Calendar Year 1996*, Agricultural Marketing Service, Science and Technology Division, United States Department of Agriculture.

USDA (1999). *Pesticide Data Program (USDA PDP), Annual Summary Calendar Year 1997*, Agricultural Marketing Service, Science and Technology Division, United States Department of Agriculture.

USDA (2000). *CSFII 1994–1996, 1998 Dataset*, National Technical Information Service Accession Number PB2000–500027, Food Survey Research Group, Agricultural Research Service (ARS), United States Department of Agriculture, (Dataset on CD-Rom).

WHO (1997). *Food Consumption and Exposure Assessment of Chemicals*, Report of an FAO/WHO consultation, Geneva, Switzerland, 10–14 February, 1997, WHO/FSF/FOS/97.5, World Health Organization, Geneva.

WHO (1998). *GEMS/Food Regional Diets: Regional* per capita *Consumption of Raw and Semi-processed Agricultural Commodities*, WHO/FSF/FOS/98.3, World Health Organization, Geneva.

7 Chronic Intake

LES DAVIES, MICHAEL O'CONNOR and SHEILA LOGAN[1]
Australian Department of Health and Ageing, Canberra, Australia

INTRODUCTION

This chapter outlines some of the methods used by regulatory authorities to look at the possible health risks to the general population from long-term exposure

[1]Currently (2003) United Nations Environment Programme (UNEP), Geneva, Switzerland.

Pesticide Residues in Food and Drinking Water: Human Exposure and Risks. Edited by Denis Hamilton and Stephen Crossley
© 2004 John Wiley & Sons, Ltd   ISBN: 0-471-48991-3

to pesticides which are used in agriculture and in or around residential areas. The first part of the chapter outlines some of the different types of pesticides, which can be classified by chemical structure, target pest or mode of activity. An understanding of the chemistry, mode of action and use-pattern of a pesticide is an important precursor to conducting a health risk assessment since it may provide an indication of the intrinsic toxicity of the compound and some idea about how widespread the public exposure to the chemical is likely to be.

A BRIEF OVERVIEW OF SOME DIFFERENT TYPES OF PESTICIDES

The use of synthetic chemicals to control pests and diseases has become widespread in the 20th century. An increase in food production and quality over this time has been attributed to the proper use of agricultural and veterinary chemicals. Some chemicals were initially developed for different purposes, while others were specifically designed to mimic a natural pesticide or to interrupt a particular metabolic pathway in a target pest. Synthetic pesticides cover a very wide range of chemical structural types; for a useful guide to the majority of synthetic pesticides ever developed, both superseded and in current use, the reader is referred to *The Pesticide Manual* (Tomlin, 1994).

Pesticides may be classified or sub-classified by commonality of chemical structure. They can also be grouped according to the target on which they act (e.g. herbicides or insecticides) or by their known or assumed biochemical mode of action (e.g. insect growth regulators, acetylcholinesterase inhibitors, etc.). A useful guide to pesticides is *Pesticide Profiles: Toxicity, Environmental Impact, and Fate* (Kamrin, 1997).

Herbicides

Herbicides are pesticides used for weed control. They are frequently applied early in the growing season, when weed growth can inhibit the germination or early growth of a crop; application may be either pre-emergent (i.e. prior to emergence of the seedling), particularly in dry areas, or early post-emergent. They may also be used at the end of the growing season in crops such as cotton, to promote desiccation of the vegetative growth and increase the ease of harvesting. They can be targeted to a particular type of plant (i.e. broad leaf versus grasses).

Some crops have now been genetically engineered to be resistant to the effects of herbicides, thus allowing the use of cover spraying of these herbicides to remove competing weeds without affecting the crop. Most modern synthetic herbicides have low mammalian toxicity because they are designed to mainly affect specific metabolic pathways within plants. Important chemical classes of herbicides are the 1,3,5-triazines (e.g. atrazine and simazine), the ureas (e.g. diuron and isoproturon) and the sulfonylureas (e.g. chlosulfuron and tribenuron).

Insecticides

There is a large range of synthetic insecticides, including important chemical classes commonly referred to as organochlorines (now largely obsolete), organophosphorus compounds (or 'organophosphates'), carbamates, pyrethroids (synthetic analogues of the natural pyrethrums), insect growth regulators, and the relatively recent nicotinyl/chloronicotinyl compounds (related to naturally occurring nicotine).

The organochlorines (a commonly used term referring to the persistent organochlorine pesticides which include the cyclodienes, DDT and related compounds, lindane and the hexachlorocyclohexanes, and toxaphene) were the first widely used group of synthetic insecticides, coming into use after World War II. These chemicals were generally long-acting, controlling pests for an extended period of time. Unfortunately, their high fat solubility and chemical stability means that they can bioaccumulate over time, with concentrations increasing in animals higher in the food chain. Their ability to volatilize in warm regions means they also can spread over quite long distances, with measurable concentrations being found near the Arctic Circle and alpine areas where they have not been used. These organochlorines have been phased out of agricultural use in most countries because of the concerns about environmental persistence, bioaccumulation and trans-boundary movement. Nevertheless, they need to be considered in the context of chronic intake of pesticide residues since they can still be found at low levels (generally decreasing with time) in a limited number of products, particularly of animal origin.

Organophosphorus compounds (OPs) are commonly used as insecticides in a variety of crops, and as ectoparasiticides in animal husbandry. As well as being highly toxic to insects, they generally have quite high acute toxicity in mammals. They act by inhibiting acetylcholinesterase, an enzyme which breaks down acetylcholine, a neurotransmitter chemical in both the central and peripheral nervous system. When this enzyme is inhibited, acetylcholine can remain in the gap or synaptic junction, between two nerves, or between nerve and muscle, causing persistent nerve or muscle stimulation. This can produce gross signs including tremors and convulsions. Some chemicals in this class have also been implicated as having chronic effects, including delayed neurotoxic effects in the nervous system and possible ocular toxicity (effects on the eye).

Carbamates (e.g. aldicarb and methiocarb) form an important group of insecticides. Like the organophosphorus compounds, they inhibit acetylcholinesterase but their effects are quicker in onset and more rapidly reversible. Chemicals which are structurally related to these carbamates have also been developed as fungicides, herbicides and molluscicides.

The synthetic pyrethroids, which mimic the structure and action of naturally occurring pyrethrins, are very widely used as insecticides. Like many other insecticides, they act on the nervous system of insects; in particular, they affect sodium channels in the membranes of nerve cells, thus disrupting such cells and the transmission of electrical signals along nerve cell axons. They are not as toxic to mammals as they are to target insects because they are quickly broken down into readily excretable metabolites.

For the past several decades, approximately 80% of the insecticide market has taken up by organophosphorus compounds, carbamates and pyrethroids (Yamamoto and Casida, 1999). However, other types of compounds are taking an increasing portion of the market. Synthetic nicotinoids and neo-nicotinoids (related to the natural compound nicotine as the pyrethroids are related to pyrethrum) are gaining increasing importance. These compounds interact with nicotinic acetylcholinergic receptors in the central nervous system of insects. Another class of insecticides of growing importance is the so-called insect growth regulators (IGRs) which kill insects by interfering with the normal process of juvenile development, either by disrupting hormonal processes or exoskeleton development. IGRs, from several different chemical classes, are relatively selective to specific pests, provide a reasonably long period of protection, and have not shown resistance problems. However, they generally kill insects more slowly than, say, the organophosphorus compounds, they are developmental stage-specific (which means they may work only when applied at the proper growth stage), and must be applied before pest populations reach economic thresholds. They are generally more costly to produce than conventional pesticides and there are concerns about their effects on aquatic organisms.

Fungicides

Next to herbicides and insecticides, fungicides are an economically very important group of pesticides. One of the major chemical classes of fungicides include the azole compounds (e.g. propiconazole and fenbuconazole), but as for herbicides and insecticides, there are a diverse range of chemical types. Fungicides are of two types, i.e. protectants and systemics. Protectant fungicides, e.g. the dithiocarbamates, protect the plant or fruit against infection at the site of application and do not penetrate the tissue. Systemic fungicides penetrate the plant and prevent disease from developing on parts of the plant away from the site of application. They often control disease by eradication of the fungus.

ASSESSING THE HEALTH RISKS FROM LONG-TERM EXPOSURE TO PESTICIDES AND THEIR RESIDUES

SOCIETAL CONCERNS ABOUT THE POSSIBLE DANGERS POSED BY PESTICIDES

Of all the chemicals to which humans are potentially exposed, pesticides are unique by reason of their deliberate use to kill or otherwise control macro-organisms considered detrimental to human welfare. In view of the economic problems that pests have caused since the advent of agriculture, some quite toxic pesticides, both naturally occurring (e.g. strychnine and arsenic) and synthetic compounds, have been used in crop production and animal husbandry. It is because of concerns about the possible adverse effects of pesticidal compounds that most countries have intro-duced quite stringent regulatory systems to ensure that chemicals introduced into agriculture do not pose an unacceptable health and environmental risk. These days, before new agricultural or veterinary chemicals can be approved for sale and use, regulatory agencies are required to assess, among a range of other investigations usually conducted by sponsor companies, the following studies:

- detailed and extensive toxicological testing in laboratory test animals (as surrogates, or models, for possible toxic effects in humans);
- toxicological investigations on organisms and animals in the environment, to assess possible unintended effects on beneficial insects, earthworms, fish, birds, etc.;
- field trials to determine likely dermal and inhalation exposure of agricultural workers using the chemical, as well as estimates of possible bystander exposure;
- field trials to determine likely residues in crops, in order to estimate the possible dietary exposure of the general population to the chemical.

The task of assessing the toxicity of a chemical and the potential for human expo-sure to it (either occupationally, from ingestion of residues in food, or through domestic use in the home and the garden) is called *chemical risk assessment*. This chapter concentrates particularly on the issue of estimating long-term expo-sure to pesticides, especially dietary intakes of pesticide residues in food, and describes how regulators compare this estimate with an acceptable (or 'safe') health standard determined from the package of toxicology studies conducted on animals *in vivo* and on cell and tissue systems *in vitro*.

POSSIBLE ROUTES OF PUBLIC EXPOSURE TO PESTICIDES

The most common route of exposure to pesticides for the general public is by ingestion of treated food commodities containing residues. Additionally, peo-ple living near agricultural areas may be accidentally exposed by inhalation or dermal absorption, particularly if pesticides are applied without due regard to weather conditions, spray droplet size or adequate buffer zones. For example, fine ultra-low volume sprays arising from aerial crop-spraying can be carried for considerable distance under windy conditions.

People may also be exposed through the use of pesticides in and around the home, e.g. termiticide treatments, household insect sprays, lawn grub treatments, etc. Exposure from termiticides and home–garden products is mainly by der-mal absorption (entry through the skin) and inhalational uptake (entry through the lungs).

Acute poisoning by the oral route can occur accidentally, most commonly in children, and also deliberately in cases of suicide or attempted suicide.

Where people such as farmers, farm workers and pest control operators are employed to apply pesticides, they are mainly exposed by absorption through the skin or by breathing in vapours, spray mists or dusts.

HAZARD ASSESSMENT AND THE ESTABLISHMENT OF NOELs AND ADIs

The capacity of a chemical to cause harm depends on its intrinsic properties, i.e. its toxicity (capacity to interfere with normal biological processes) and its ability to burn, explode, etc. The **hazard** posed by a chemical directly relates to its intrinsic properties. The concept of **risk** (that is, the likelihood of harm occurring) is introduced when the extent of **exposure** is considered in conjunction with the hazard data. The basic approach to risk assessment can be expressed by the following simple formula:

$$Risk = Hazard \times Exposure \qquad (7.1)$$

Thus, if either the hazard or the extent of exposure can be reduced or minimised, the risk or likelihood of harm can be reduced or minimised. The WHO Environmental Health Criteria monograph on *Principles for the Assessment of Risks to Human Health from Exposure to Chemicals* discusses the issues relating to hazard and risk assessment and risk management for chemicals in some detail (WHO, 1999).

The possible effect on human health is the main consideration when assessing the risks posed by pesticide residues in foods. When assessing the possible health effects arising from the ingestion of residues of a particular pesticide, it is important to determine the level of intake of that pesticide that is considered to be 'safe' over a lifetime exposure, that is, without any apparent adverse effects on health when ingested daily over a lifetime or most of a lifetime. Consideration of this is usually based on information derived from toxicology studies carried out in experimental animals. The studies generally considered most suitable for the derivation of an Acceptable Daily Intake (ADI)[1] are long-term studies in experimental animals, with the test compound admixed with the diet; these studies, involving regular daily intake of the chemical under investigation, most closely mimic the chronic exposure resulting from ingestion of food containing low levels of pesticide residues.

No-Observed-Effect Levels

A comprehensive package of toxicology studies allows the determination of the daily dose of a pesticide or test chemical which can be given over a certain period of time by a particular dose route, at which no effects are observed. This is known as

---

[1] The ADI or 'Acceptable Daily Intake' may be compared with the TDI ('Tolerable Daily Intake') and RDI ('Recommended Daily Intake'). Because pesticides are deliberately applied to crops to improve food production, low residues which may result can be considered 'acceptable' if they are not of toxicological concern. In contrast, the TDI is a health limit for food contaminants which should **not** be present but which may be unavoidable, e.g. aflatoxins. The RDI refers to the recommended daily intake of required nutrients, e.g. vitamins.

the No-Observed-Effect Level (NOEL). This level[2] has been simply defined (WHO, 1990) as the highest dose of a substance which causes no changes distinguishable from those observed in normal (control) animals. The No-Observed-Adverse-Effect Level (NOAEL) is the highest dose of a substance at which no toxic (i.e. adverse) effects are observed (WHO, 1990). Whether a NOEL or NOAEL is used will depend on technical policy considerations in different regulatory agencies. Sometimes, there will also be differences of scientific opinion about whether a particular finding in a toxicology study is necessarily 'adverse'. In reviewing the toxicity of a particular chemical, it is customary to set a NOEL or NOAEL for each repeat-dose toxicology study conducted on the chemical. Then, an overall NOEL/NOAEL for the chemical is selected, either from the most appropriate study or, most commonly, the lowest value is selected from the package of studies.

An example of the establishment of a NOAEL is shown in Table 7.1; this is taken from the evaluation of the organophosphorus pesticide, ethoprophos, by the Joint FAO/WHO Meeting on Pesticide Residues (JMPR, 1999b).

The JMPR took the value of 0.04 mg/kg bw as an overall NOAEL for ethoprophos, to be used in establishing an acceptable daily intake (see discussion below); this NOAEL was based on findings in a two-year rat dietary study and a rat two-generation reproduction toxicity study (gavage). In this case, the lowest NOAEL of 0.025 mg/kg bw seen in a one-year dog study was not used because the marginal effects seen in the liver were not seen in other studies in dogs or in other test species, and furthermore, there was poor dose selection in this particular study, with a 40-fold difference between the dose at which the NOAEL was established and the next highest test dose.

Acceptable Daily Intakes

Once established, the lowest (or most appropriate or relevant) NOEL (or NOAEL) derived from animal toxicology testing is used to set an ADI (or in the USA, the Reference Dose (RfD)) for humans. This is done by dividing the NOEL or NOAEL (expressed in terms of mg test chemical/kg body weight/day) by a safety factor, which is conservatively chosen to allow what is considered to be a more than adequate margin of safety, as follows:

$$ADI = \frac{NOEL \ (or \ NOAEL)}{SF} \qquad (7.2)$$

where SF is a safety factor.

---

[2]The 'No-Observed-Effect Level' (or 'No-Observable-Effect Level') can be defined in more detail as the highest dose of a substance administered to a group of experimental animals at which there is an absence of observable effects on morphology, functional capacity, growth, development or life span, which are observed or measured at higher dose levels used in the study. Thus, dosing animals at the NOEL should not produce any biologically significant differences between the group of chemically exposed animals and an unexposed control group of animals maintained under identical conditions. The NOEL is expressed in milligrams of chemical per kilogram of body weight per day (mg/kg bw/d).

Table 7.1 No-observed-adverse-effect level (NOAEL) values recorded in the toxicology evaluation of the organophosphorus pesticide ethoprophos (JMPR, 1999b)

Study	NOAEL (mg/kg bw/d)[a]	Effect at next highest dose tested
Mouse, 2-year (dietary)	0.25	Brain acetylcholinesterase inhibition
Rat, 2-year (dietary)	0.04	Brain acetylcholinesterase inhibition
Rat, 2-generation reproduction study (gavage)	0.04	Reduced bodyweight gain in parental animals and brain acetylcholinesterase inhibition
Rat, single-dose neurotoxicity (gavage)	5.0	Behavioural changes and inhibition of erythrocyte acetylcholinesterase
Rat, developmental study (gavage)	2.0[b]	Soft stools, faecal staining, lower body weight gain
	18[c]	No fetotoxic effects seen at highest dose tested of 18 mg/kg bw
Rabbit, developmental study (gavage)	2.5	No maternotoxicity or fetotoxicity seen at the highest dose tested of 2.5 mg/kg bw
Dog, 1-year (capsule)	0.025	Liver effects – vacuolation and pigment deposition

[a] mg/kg body weight/day.
[b] Maternotoxicity NOAEL.
[c] Fetotoxicity NOAEL.

Safety factors are not rigidly applied and can vary from 100 to 2000, depending on the supporting toxicological database. When NOELs are based on studies in animals, the usual safety factor used to derive an ADI is 100, made up of a factor of 10 for inter-species extrapolation, and an extra factor of 10 to allow for variations between individuals in human populations. A safety factor of only 10 may apply if it is possible to derive a NOEL from an appropriate test conducted in humans; while tests in human volunteers were once relatively common, these days ethical concerns have been raised by some regulatory agencies about testing pesticides in this way. Further safety factors may be incorporated (1) to provide additional protection for special risk groups (e.g. infants), or (2) where the toxicological database is not complete or there are some concerns about its quality, or (3) the nature of the potential hazard(s) indicate(s) the need for additional caution. The further safety factors may lead to an overall safety factor up to 5000 in some cases.

It should always be borne in mind that, in any toxicology study, the NOEL (or NOAEL) will be determined by the doses selected for testing in that particular study. Sometimes, no clear NOEL is established, i.e. effects may be seen even at the lowest dose selected for testing in a study. The toxicologist may consider that the low dose is a lowest-observed-effect level (LOEL) (that is, the lowest dose of a substance which causes changes distinguishable from those observed in normal (control) animals) (WHO, 1990) if there is a reasonable justification for concluding that if a lower dose had been used, the effect would not have occurred. In this case, an ADI may be derived, but using an extra safety factor (usually 10) on top of the factor of 100 usually employed when extrapolating from animal studies to humans.

An example in which an extra safety factor on a LOEL was used to derive an ADI is provided by a year 2000 registration consideration by Australian authorities. Dogs were the most sensitive test animals but a clear NOEL could not be established because somewhat reduced body weight gain was noted even at the lowest dose used in a one-year toxicity study; while this finding was not linked with any other observations at this dose, some concern that it could possibly be related to thyrotoxic effects seen at higher doses led to the use of an extra safety factor in establishing an ADI for this compound. The size of the safety factor chosen in any particular case will depend on the perceived severity of the toxicology end-point observed; however, if the end-point is of significant concern, regulatory agencies are more likely to request the sponsor company to conduct a further study (or studies) using lower doses in order to establish a clear NOEL or NOAEL, in preference to using a LOEL and extra safety factor.

Using the safety factor approach to derive ADIs is based on the assumption that exposure at less than the ADI is without appreciable risk, but no attempt is made to quantify the level of risk, i.e. it is not a quantitative risk assessment approach but based on a presumption of some threshold dose, below which no effects of concern are likely to occur.

Although it is commonly taken to be synonymous with the 'ADI', the US RfD is distinctly defined. Developed by the US Environmental Protection Agency (EPA) for assessment of risks associated with systemic toxicity, the RfD is an estimate of a daily exposure to the human population (including sensitive subgroups) that is likely to be without appreciable risk of deleterious effects during a lifetime. However, it does not assume that all doses below the RfD are 'acceptable' (or risk-free), nor that all doses which exceed the RfD are necessarily 'unacceptable' (i.e. result in adverse effects). The equation for its derivation is as follows:

$$\text{RfD (mg/kg bw/d)} = \frac{\text{NOAEL}}{\text{UF} \times \text{MF}} \tag{7.3}$$

where UF is an Uncertainty Factor, and MF is a Modifying Factor.

In practice, the standard uncertainty factors (UFs) used are as for safety factors (see above). In addition, a modifying factor (MF), of greater than zero but less than or equal to 10, is sometimes also applied, based on a professional judgement of the entire database of the chemical. For all intents and purposes, the ADI and RfD are quite similar, but for a detailed explanation of the RfD concept and how it varies from the ADI, and its role in risk management, the reader is referred to the relevant US EPA Background Document (US EPA, 1993).

MAXIMUM RESIDUE LIMITS – ESTABLISHMENT AND USE

Governments have a responsibility to regulate the food supply to ensure that foods offered to consumers are safe and wholesome. The occurrence of pesticide residues in foods is an unavoidable consequence of their intended use. In order to protect the health of the consumer, governments establish Maximum Residue Limits (MRLs) or tolerances, to ensure that dietary exposure to pesticide residues is kept to a minimum and within acceptable standards. Accordingly, MRLs are established for all raw agricultural commodities where pesticides are to be used directly on food crops or on crops intended to be fed to food-producing species. They are also established for animal products consumed by humans, e.g. meat, milk and eggs. In some cases, they are set for processed foods if the pesticide is likely to concentrate during processing (e.g. milling, cooking and dehydrating) or for primary processed foods such as flour and vegetable oils.

The MRL is defined as the **maximum** concentration of a pesticide residue (expressed as mg/kg of the food commodity) resulting from the officially autho-rised safe use of a pesticide that is legally permissible or acceptable in or on food commodities and animal feeds (FAO, 1986). MRLs are set as low as pos-sible consistent with Good Agricultural Practice (GAP).[3] The levels at which MRLs are set are either finite (i.e. detectable by an acceptable analytical method) or at, or about, the limit of analytical determination (i.e. when residues are not detectable by an acceptable analytical method). GAP requires that the minimum amount of pesticide be applied in such a manner as to achieve effective pest or disease control while ensuring the smallest practicable residue level. MRLs are not based on any health criteria, but are established only where the known toxicological risks do not constitute an undue human health hazard and where the amount of pesticide residue that may be consumed by any individual is not likely to exceed the ADI (or RfD) for the pesticide over a prolonged period of time.

Accordingly, the process of setting MRLs is separate from the evaluation of the pesticide's toxicity. MRLs are set on the results of field trials, usually conducted by the manufacturer, in several geographical regions that typify areas in which the crop is produced, so that different climatic conditions, cultural practices and soil types are represented. The field trials are designed to determine the maximum amount of pesticide residue which will remain in or on a food commodity under the most extreme conditions likely to be encountered in commercial practice. Trials utilising application rates in excess of those required to control the target pest or disease may also be conducted in order to more conveniently determine the rate of depletion of the pesticide from the target plant or animal tissues.

---

[3] Good Agricultural Practice (GAP) is formally defined as nationally authorised safe uses of pesticides under actual conditions necessary for effective and reliable pest control. It encompasses a range of levels of pesticide applications up to the highest authorised use, applied in a manner that leaves a residue which is the smallest amount practicable. Authorised safe uses include nationally registered or recommended uses, that take into account public and occupational health and environmental safety considerations. Actual conditions include any stage in the production, storage, transport, distribution and processing of food commodities and animal feed (IPCS, 1989).

While MRLs are based on maximum application rates and maximum number of applications, in practice this level of usage would not always occur. Additionally, given the expense involved in purchasing and applying pesticides, it is unlikely that they will be used when or where they are not required.

In establishing MRLs for a pesticide, consideration needs to be given to a number of factors related to the potential use of the chemical. It needs to be established how frequently the chemical will be used, and at what dose rate to ensure adequate control of the target pest. This will be determined by a number of characteristics of the chemical, including its environmental persistence, as well as the sensitivity of target organism(s) to the pesticide, the characteristics of the host crop or farm animal and the normal cultural practices. Where pesticides are designed to control pests or diseases which are active close to harvest time, for example, fungi and fruit fly, the timing of application to give optimum pest control but to leave a residue at harvest no higher than necessary must be given careful consideration. The interval between the final application and harvest is legally prescribed as the pre-harvest interval.

In summary, an MRL is more appropriately considered as a legal limit to control pesticide residues in or on foods sold in commerce by providing a mechanism to measure compliance with label directions with respect to application rates, withholding periods, and whether the pesticide is approved for use on that particular crop or commodity. While it is not a health standard *per se*, food safety is assured when the residue is at or below the MRL. It is the ADI (or RfD) which is the health guideline level. The acceptability of an MRL from the public health point of view is determined by comparing the ADI (or RfD) with the estimated dietary intake of pesticide residue following use in accordance with GAP.

There are a number of factors which can complicate the calculation of dietary intake of pesticide residues. For example, much of the available agricultural produce will not have been treated with the chemical under consideration, or the produce may have been treated early in the growing season and may not contain detectable residues at the time of harvest or consumption. In cases in which residues in food commodities are detectable, they will most commonly be well below the MRL because of withholding periods or use of pesticide products at lower than maximum allowable application rates.

DETERMINATION OF INTAKES OF PESTICIDE RESIDUES FROM FOOD, AND COMPARISON WITH THE ADI

As discussed above, the most likely route of long-term exposure of the public to pesticides is via intake of pesticide residues in food. Dietary exposure to a pesticide depends on both the actual residue in or on foods and food consumption patterns. In determining the health risks associated with chronic intake, it is necessary to determine the quantity of the pesticide residue likely to be consumed over a prolonged period of time and compare this estimate with the ADI (or RfD).

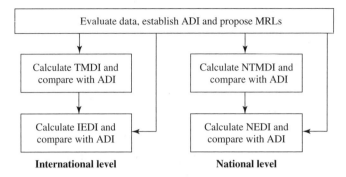

Figure 7.1 Scheme for the assessment of long-term dietary intake of pesticide residues. Reproduced from *Guidelines for Predicting Dietary Intake of Pesticide Residues* (revised), document WHO/FSF/FOS/97.7, Global Environmental Monitoring – Food Contamination Monitoring and Assessment Programme in collaboration with the Codex Committee on Pesticide Residues, Figure 7.1, World Health Organization, Geneva, Switzerland (1997), with permission of the World Health Organization. Abbreviations defined in text.

Dietary intake calculations are carried out to determine whether consumption of food commodities containing pesticide residues would result in consumption of residues which would exceed the ADI for that pesticide. The World Health Organization (WHO) *Guidelines for Predicting Dietary Intake of Pesticide Residues* (revised), published by the Programme of Food Safety and Food Aid of the World Health Organization (WHO, 1997) details internationally accepted procedures to estimate dietary exposure and determine acceptability of proposed MRLs from a public health viewpoint both at the national and international level.

The approach to dietary intake assessments at the national and international level are summarized schematically in Figure 7.1.

NATIONAL THEORETICAL MAXIMUM DAILY INTAKE

National Theoretical Maximum Daily Intake (NTMDI) calculations can be used as an initial screening tool for estimating dietary intakes, and are based on the assumption that food commodities contain residues at the **maximum** permitted level. NTMDI calculations can also be used to provide an estimate as to whether proposed national MRLs (for a new pesticide or for additional uses of an existing pesticide) are likely to provide food which is suitable for consumption. That is, once MRLs resulting from the use of a pesticide according to label recommendations (with respect to application rate, frequency and method) have been established, it is possible to consider the likely intake of residues by the population or particular sub-populations, assuming that all of the commodities on which the pesticide has been approved for use contain residue levels at the MRL. (Needless to say, median residue levels data from supervised trials provide a much better basis for estimating likely population intakes – see below.)

Similarly, in the absence of actual residue level data from relevant field trials, NTMDI calculations can be used to estimate whether international Codex MRLs (set by the Codex Alimentarius Commission's Codex Committee on Pesticide Residues) may be acceptable to national authorities.

The usual way of determining an NTMDI is to consider the national daily intake of each food commodity for which an MRL has been set, and assume that the pesticide is present at the MRL in all of the food consumed. It is relatively simple to calculate the total amount of the pesticide consumed by multiplying the amount of the food commodity consumed (in kg/d) by the MRL (in mg/kg of the commodity), thus giving the total amount of pesticide consumed (in mg/d). Once the pesticide intake from each of the food commodities consumed has been determined, and the sum of all intakes calculated, the quantity of pesticide consumed per unit body weight (mg/kg bw/d) can be calculated.

Thus, the NTMDI is determined according to the following equation:

$$NTMDI = \Sigma MRL_i \times F_i \qquad (7.4)$$

where MRL_i is the national maximum residue limit for a given food commodity, and F_i is the national mean consumption of that food commodity per person (kg/d).

The NTMDI calculation serves a useful function as a screening tool but scientifically it is a poor estimate of actual intake. It allows regulatory agencies to estimate roughly the upper limits of consumption of pesticide residues, although it is recognised as producing a gross overestimate of pesticide intake. Agencies ask the question 'Is the NTMDI greater or less than the ADI?'. Should the total pesticide consumption calculated by this method be less than the ADI, regulatory agencies can be quite confident that there will not be a risk posed to consumers from the chronic dietary intake of residues of the chemical. However, the NTMDI approach has been criticised as contributing to the poor credibility of regulators in the eyes of the public – the further refinement of the calculation to make the intake estimate more realistic appears to the consumer as if regulators are fiddling with the calculations until they reach the desired result!

A worked example using Australian data on chlorfenvinphos is presented in Table 7.2 to illustrate the use of the WHO *Guidelines for Predicting Dietary Intake of Pesticide Residues* (revised) (WHO, 1997). Chlorfenvinphos is a broad-spectrum organophosphate pesticide which has been registered for use in Australia and other countries for over 30 years. In 1999, the National Registration Authority for Agricultural and Veterinary Chemicals (NRA)* undertook a comprehensive review of chlorfenvinphos under the Existing Chemicals Review Program, a systematic re-registration program which reassesses older registered chemicals to determine whether they continue to meet contemporary registration standards (NRA, 1999).

*Now the Australian Pesticides and Veterinary Medicines Authority (APVMA).

Table 7.2 Chlorfenvinphos: national theoretical maximum daily intake (NTMDI) calculations (modified from NRA, 1999)

Commodity	Food consumption (kg/person/d)	MRL (mg/kg)	NTMDI (mg/adult)
Cattle milk[a]	0.603	0.008	0.0048
Cattle (edible offal of)	0.000 11	0.1[b]	0.000011
Cattle meat (in the fat)[c]	0.0145	0.2	0.0029
Sheep (edible offal of)	0.000 01	0.1[b]	0.000 001
Sheep meat (in the fat)[c]	0.002 79	0.2	0.000 56
Potato	0.0658	0.05[b]	0.0033
		Total mg/person	0.0116
		mg/kg body weight	0.000 173
		% of ADI	35[d]

[a] Cattle milk (in the fat) MRL is 0.2 mg/kg but has been adjusted to 0.008 mg/kg for expression on a 'whole-milk' basis (assuming 4 % fat content).
[b] At or about the limit of determination.
[c] For meat (in the fat) in cattle and sheep, the intake calculations have taken account of a 20 % fat content in the meat (the intakes for meat in cattle and sheep are 0.0728 and 0.0140 kg, respectively).
[d] Rounded value.

Recommendations arising from the chlorfenvinphos review included a downward revision of the Australian ADI to 0.0005 mg/kg body weight/d and revocation of a number of Australian MRLs for agricultural commodities that were no longer linked to registered uses patterns. To determine the chronic dietary intake risk, the NRA used consumption data for a 67 kg body weight adult[4] from a dietary modelling system developed by the Australia New Zealand Food Authority (ANZFA).

The predicted intake of chlorfenvinphos residues (Table 7.2) is approximately 0.000 173 mg/kg body weight which is equivalent to 35 % of the ADI (0.0005 mg/kg body weight/d). However, it should be noted that NTMDI calculations represent an over-estimate of pesticide dietary exposure as it is assumed that the pesticide is present at the MRL (i.e. the **maximum** permitted residue limit, and not the more realistic but still conservative **median** value from the set of residue trials) in the food commodity. This NTMDI intake does not take into account the fact that not all produce will be treated with the pesticide, and that, of the produce treated, the majority is likely to have residues well below the MRL. Furthermore, this method does not consider the fact that MRLs are set on residues in the whole commodity, including the inedible portions. In many fruits, particularly, the residues in the inedible portion (citrus and banana peel, for example) may constitute the majority of the residues present. The NTMDI calculation also does not include the effect of any processing on residue levels. Common processing occurring in the home includes washing fruit and vegetables, peeling and cooking. Most processing is likely to reduce pesticide levels, although

---

[4] It may be noted that at the international level, the Codex Committee on Pesticide Residues assumes the average body weight to be 60 kg.

a few may act to concentrate the pesticides in the edible portion of the commodity (e.g. bran and wheat germ).

The WHO publication *Guidelines for Predicting Dietary Intake of Pesticide Residues* (revised) (WHO, 1997) provides further discussion on the calculation and use of NTMDIs.

NATIONAL ESTIMATED DAILY INTAKE

As the NTMDI calculation for chlorfenvinphos is equivalent to 35 % of the ADI, there would normally be no need to undertake National Estimated Dietary Intake (NEDI) calculations which provide a more refined or 'best estimate' of dietary intake. However, for the purpose of illustration, NEDI calculations will be performed by using chlorfenvinphos as an example.

Better estimates of actual consumption of a pesticide can be determined by introducing a number of the factors ignored in calculating the NTMDI. If national data are available on the proportion of the crop or the commodity which is treated with the pesticide under consideration, this can be used to calculate the percentage of the crop which would be expected to have residues. However, caution is required since in some areas, all of the produce consumed may come from local growers who may be required to use the pesticide on a regular basis, due to prevailing conditions; thus, some individuals may always be exposed on this basis.

The importation of a large percentage of the crop or commodity consumed may affect the consumption of residues. Depending on the residue levels in imported produce, it may either increase or decrease pesticide residue intake. Information on the use of the pesticide in question in the exporting country will be useful; if the pesticide is not used in the producing country, then residues can be expected to be zero. However, accurate figures on application rates and the percentage of crops treated in other countries may be difficult to obtain. In Australia, it is possible to estimate the percentage of imports contributing to national food commodity intake from apparent consumption data collected by the Australian Bureau of Statistics (e.g. ABS, 1999).

Increasingly, field trial data are being collected to provide so-called Supervised Trials Median Residues (STMRs) data to better reflect actual residue levels in commodity crops. However, it must be recognised when using these data that they have been obtained from a relatively small sample of the produce and may not reflect levels which are present overall. By necessity, most monitoring and surveillance schemes use limited numbers of samples, which may not accurately reflect overall residue patterns across the country.

Whereas MRLs are established for residues in the whole commodity, including the inedible portions, estimated dietary intake calculations should, if possible, utilise residue data on edible portions of fruit. For fruits with inedible skin, such as bananas, citrus and melons, supervised trials commonly measure residues in the whole commodity, on skin, and in the edible portion. Where residue data on edible portions are available, they are used directly as the starting point for intake estimation.

Likewise, consideration should be given in any refinement of dietary intake calculations to the effect of transport, storage, commercial processing and cooking on residue levels. Residues are commonly dissipated during these processes, although sometimes they may concentrate in processed fractions, thus resulting in higher levels than in the raw commodity. Washing and cleaning will often reduce residue levels, particularly for those pesticides which are not significantly absorbed by the commodity. Milling of cereals to flour and polishing of rice result in significant lowering of residues. However, extraction of oils from oil seeds and conversion of fruits to pomace may result in concentration of residues. Unfortunately, for some older pesticides there is a paucity of data on the effects of storage, processing and cooking on residues, but for new pesticides or those that have been through a thorough re-registration process the effect of food processing on residues has usually been well studied, which allows more realistic estimates of the dietary intake of residues.

While there are a number of factors which can be used to refine estimates for predicting long-term dietary intake of pesticide residues, only STMRs have been utilised in the NEDI calculations for chlorfenvinphos.

The NEDI calculations for chlorfenvinphos (Table 7.3) result in a dietary intake estimate of 0.000 077 07 mg/kg body weight, which is equivalent to 15 % of the ADI and approximately a 55 % reduction of the NTMDI estimate. It can be concluded that the risk posed by chronic dietary exposure to chlorfenvinphos in Australia is acceptable. This conclusion is consistent with the 1996 monitoring data and the 1994 and 1996 Australian Market Basket Surveys (ANZFA, 1994, 1998) which indicated no detectable chlorfenvinphos residues were found in food commodities.

Table 7.3 Chlorfenvinphos: national estimated daily intake (NEDI) calculations (modified from NRA, 1999)

Commodity	Food consumption (kg/person/d)	MRL (mg/kg)	STMR (mg/kg)	NEDI (mg/adult)
Cattle milk[a]	0.603	0.008	0.0028	0.001 69
Cattle (edible offal of)	0.000 11	0.1[b]	0.1	0.000 011
Cattle meat (in the fat)[c]	0.0145	0.2	0.01	0.000 145
Sheep (edible offal of)	0.000 01	0.1[b]	0.1	0.000 001
Sheep meat (in the fat)[c]	0.002 79	0.2	0.01	0.000 0279
Potato	0.0658	0.05[b]	0.05	0.003 29
			Total mg/person	0.005 16
			mg/kg body weight	0.000 077
			% of ADI	15[d]

[a] Cattle milk (in the fat) STMR is 0.07 mg/kg but equivalent to 0.0028 mg/kg when expressed on a 'whole-milk' basis (assuming 4 % fat content).
[b] At or about the limit of determination.
[c] For meat (in the fat) in cattle and sheep, the intake calculations have taken account of a 20 % fat content in the meat (the intakes for meat in cattle and sheep are 0.0728 and 0.0140 kg, respectively).
[d] Rounded value.

ESTIMATING DIETARY INTAKE AT THE INTERNATIONAL LEVEL

At the international level, estimates of dietary intake can be conducted centrally with data provided for MRL evaluation. MRLs set by the Codex Committee on Pesticide Residues may be used to make a first estimate of pesticide residue intake, the Theoretical Maximum Daily Intake (TMDI). As already discussed above, this estimate can be used to separate those pesticides for which there are no concerns for long-term intake from those that require further consideration. However, recent changes at the international level mean that TMDIs are now infrequently calculated, except for those compound–commodity situations which have not yet been re-evaluated by Codex and related scientific supporting bodies. There are a number of reasons for this, with an important one being that the TMDI is not an intermediate step in the process, i.e. there is no factor to apply to the MRL (a legal limit) to produce an actual supervised trial median residue (STMR) level. Furthermore, the residue definition established for MRL enforcement purposes may not necessarily be the ideal definition for dietary intake assessment. For dietary intake purposes, it is desirable (if not always possible) to factor in any metabolites which have toxic effects similar to, overlapping with, or possibly greater than, that of the parent compound. However, for testing of food consignments for compliance with MRLs, it is not desirable to include metabolites if they are present as only a minor part of the residue or are in a relatively constant proportion to levels of the parent compound, since this only adds to the cost and complexity of what should be a routine and regular monitoring program. Similarly, the issue of metabolites common to different pesticides would lead to difficulties and anomalies in MRL enforcement; in contrast, dietary intake assessments should take into account toxicologically relevant metabolites regardless of their source (Hamilton *et al.*, 1997).

National dietary surveys for food consumption should be used where available to predict intakes, particularly where there are dietary preferences which may alter consumption patterns. The WHO is currently in the process of revising and expanding regional diets, and has tentatively assigned countries to thirteen regional or cultural dietary groups (CCPR, 1999a, 1999b). The regional diets are a close approximation of dietary patterns but are based on food balance sheet data (apparent consumption *per capita*) and may overestimate mean intakes (see Chapter 6 – *Diets And Dietary Modelling for Dietary Exposure Assessment*).

As for the NEDI (see above), the International Estimated Daily Intake (IEDI) incorporates additional factors to provide a 'best estimate' of dietary intake; calculations include median residues in edible portions from supervised trials and effects of processing and cooking, as follows:

$$IEDI = \Sigma STMR_i \times F_i \times P_i \qquad (7.5)$$

where $STMR_i$ is the supervised trial mean residue level for a given food commodity, P_i the processing factor for that food commodity, and F_i the food regional consumption of that commodity.

Examples of TMDI and IEDI calculations for fenamiphos, taken from the work of the 1999 report of the Joint Meeting of the FAO Panel of Experts on Pesticide Residues in Food and the Environment, and the WHO Core Assessment Group on Pesticide Residues (JMPR, 1999a) have been included in the appendices to this chapter. It may be noted that the refinement using the IEDI calculation significantly lowers the estimated intake to more realistic levels than the TMDI.

ESTIMATING FOOD CONSUMPTION

While beyond the scope of this present chapter (see Chapter 6 – *Diets and Dietary Modelling for Dietary Exposure Assessment*) it is useful to consider some of the difficulties in estimating consumption of different foods by a population. It is almost impossible to get a selected survey group to keep accurate records of what was eaten over the course of a day for more than a few days. For this reason and for reasons of the effort and expense of conducting prolonged surveys, data from short-term dietary surveys are often used in the estimation of chronic intake of pesticide residues. Such short-term surveys tend to overestimate high consumption levels of foods.

In using the results of food intake estimates, there is the option of basing calculations on the mean intake of a particular food for **all respondents** in the survey, or a different mean intake can be used, based only on the **consumers** of that particular food. A range of issues need to be considered in conducting food intake surveys, and interpreting and using the data from such surveys; a useful discussion of some of the issues may be found *In Food Consumption and Exposure Assessment of Chemicals* (FAO/WHO, 1997).

EXPOSURE TO PESTICIDES FROM SOURCES OTHER THAN RESIDUES IN FOOD

PESTICIDE CONTAMINATION OF DRINKING WATER

Pesticides can contaminate water supplies, either from run-off from catchment areas into streams and watercourses, from aerial spraying, or from percolation through soil into ground water. In considering total pesticide intake, chemical risk assessment needs to take into account possible dietary intake from drinking water as well as from residues occurring on or in food.

In addition to legal upper limits on pesticide residues in a large range of foods, most countries have also established limits for pesticide levels which may occur in drinking water.[5] Control authorities use many approaches for regulatory

---

[5]In a number of countries (e.g. Australia, Canada, New Zealand and the USA), a determination about which pesticides to regulate in drinking water is made on the basis of likely health risks (i.e. the toxicological hazard posed by the chemical) and the likelihood that the contaminant could occur in drinking water at a level of concern. In the European Union, rather than setting limits on a case-by-case basis dependent on the toxicological hazard, limits for pesticides in drinking water as per Directive 98/83/EC are set at a general cut-off of 0.1 µg/l for each pesticide, with a maximum of 0.5 µg/l for total pesticides.

limits in water; one approach, as used in Australia, is to establish two different guideline values for those pesticides which have the potential to contaminate water (either because of their use pattern around water catchment areas or because their physicochemical properties mean that they are reasonably stable and can also readily leach through soil and percolate into ground water). One value is a health-related guideline value, based on the calculated acceptable daily intake, or ADI (see above). The other is referred to as a 'Guideline Value' (or 'Action Level') and is commonly the analytical limit of detection of the pesticide in water using the most appropriate modern assay method. This means that if a pesticide is detected in drinking water, action should be taken by the appropriate water supply authorities to identify the source of contamination and take action to prevent further contamination, even though the level may be well below that causing any health concerns.

How are Health Guideline Values for Drinking Water Established?

The health guideline value is calculated by assuming that intake from water will comprise a proportion, commonly 10 %, of the total daily intake of the pesticide from the diet (i.e. food and water). Thus, the health guideline value is given by the following:

Health guideline value (mg/l)

$$= \frac{\text{ADI (mg/kg bw/d)} \times \text{Average weight of a person (kg)} \times 10 \ \%}{\text{Average water consumption per person (l/d)}} \quad (7.6)$$

Values for the average body weight of an adult will differ slightly between regulatory agencies in different countries; it commonly is between 60 to 70 kg, while 2 l/d is the estimated (maximum) amount of water consumed by an adult. The value of 10 % for the contribution of pesticide residues in water to total daily dietary burden of residue intake is a commonly used value (WHO, 1993), although this may be varied if it is considered justified to do so.

An actual example follows: this is for the herbicide atrazine which can contaminate water supplies because of its use pattern and its reasonable mobility in many soil types.

The Australian health-based guideline value of 0.04 mg/l for atrazine was determined as follows:

$$0.04 \text{ mg/l} = \frac{0.005 \text{ mg/kg body weight per day} \times 70 \text{ kg} \times 0.5}{(2 \text{ l/d}) \times 2} \quad (7.7)$$

In the above equation:

- 0.005 mg/kg body weight per day is the ADI determined by the Therapeutic Goods Administration (NRA, 1997);

- 70 kg is taken as the average weight of an adult (a figure used by the Australian National Health and Medical Research Council in setting drinking water guidelines);
- 0.5 is a proportionality factor based on the conservative assumption that at least 50 % of the ADI will arise from the consumption of drinking water (atrazine has never been detected in the Australian food supply);
- 2 is an extra safety factor which takes into consideration the likely presence of metabolites of atrazine which have a similar toxicity profile to the parent atrazine and which may constitute about 50 % of the total atrazine-derived compounds;
- 2 l/d is the estimated amount (maximum) of water consumed by an adult;
- the ADI value includes a safety factor of 100 on the no-observed-effect level (NOEL) obtained from toxicology studies in test animals (10 for interspecies variation, and 10 for human variability).

With respect to estimating the intake of pesticides from drinking water, it should be pointed out that health values calculated from an ADI cannot be used in a TMDI-like calculation since they bear no relation to possible residue levels in water. Only limited monitoring data for pesticides in water are currently available and then only for a few compounds, targeted because of their known mobility in soil or their particular agricultural use, e.g. in aerial spraying near waterways or catchment areas. Regulatory agencies are considering ways of best estimating total pesticide burden from food, drinking water and other residential and bystander exposures (see section below on 'Aggregate and Cumulative Risk').

LONG-TERM EXPOSURE TO PESTICIDES FROM INHALATION AND DERMAL CONTACT

Oral exposures may result from dietary consumption of food and water, and from incidental exposure from residential uses (e.g. children sucking their hands after playing in a treated yard). Dermal and inhalation exposure of home-owner applicants and residents could possibly result from residential pesticide applications either in the yard or inside the home. For some pesticides, other non-occupational exposures are possible such as those resulting from applications to school buildings, parks, recreational areas or adjacent agricultural crops.

In a number of these situations, possible exposures are not likely to occur on a long-term basis, since the pest treatments are not likely to occur frequently. Nevertheless, in conducting a hazard assessment on a pesticide, toxicologists generally aim to determine both the likely extent of its dermal absorption, its physicochemical volatility and its proposed use pattern.

If there is evidence of measurable penetration of a pesticide through the skin, product labels will be required to carry safety directions indicating the need to use gloves and, possibly, other protective clothing.

The volatility of a pesticide will determine whether it readily forms vapours which may be inhaled. One example of a pesticide which is somewhat volatile is the organophosphorus compound, dichlorvos (see Table 7.4 for a list of vapour pressures of dichlorvos, several other OPs, a gas and an organic solvent). It is because of this property that dichlorvos is used in pest strips and in fogging machines for fumigating buildings.

To prevent exposure, strict precautions should be taken to keep humans and pets away from buildings undergoing fumigation until such time as the pesticide has had a chance to act and the building is then adequately ventilated with fresh air before re-entry.

For pest strips which are commonly hung in living areas of homes or work places, particularly on farms or in tropical areas where flies, mosquitoes or other biting insects are a problem, it is possible to estimate the likely human exposure to pesticide vapour, provided that an estimate has been made of the rate of release of the active ingredient from the inert matrix of the pest strip. The following calculation can be made:

$$\text{Equivalent oral dose (mg/kg bw)} = \frac{C \times EL \times MV \times AF \times 10^{-6}}{BW} \qquad (7.8)$$

where C is the concentration of substance in the air (mg/m^3), EL the exposure period (min), MV the minute volume (ml/min) (this value is species-specific – for humans, a mean of 7,400 ml/min is commonly used (US EPA, 1988; Derelanko, 2000)), AF the absorption factor, i.e. fraction of inhaled substance which is absorbed (default $= 100\%$), 10^{-6} the m^3-to-ml conversion, and BW the body weight (kg).

In this way, inhalational exposure over one day can be compared with the ADI.

Table 7.4 Vapour pressures (at 25 °C) of several organophosphorus pesticides (and comparator compounds)

Chemical	Vapour pressure (mPa)
Methyl bromide[a]	227 000 000
Acetone[b]	30 800 000
Ethanol	7 870 000
Dichlorvos	2100
Chlorpyrifos	2.7
Fenthion	0.74
Parathion-methyl	0.41
Azinphos-methyl	0.18

[a] Gas at room temperature.
[b] Volatile organic solvent.

PERSISTENT ORGANIC PESTICIDES AS ENVIRONMENTAL CONTAMINANTS

The so-called organochlorine pesticides (OCPs), which were the first widely used group of synthetic insecticides (see the introductory section, 'A Brief Overview of Some Different Types of Pesticides'), raise particular concerns with respect to estimating chronic human intake. Even though very few of these are still used, their chemical stability and good fat solubility means that some persist in the environment and are measurable in some commodities.[6] Nevertheless, available monitoring data suggest that residues of OCPs are declining with time and are now only detected in a relatively small proportion of foods sampled, e.g. elevated intakes of OCPs such as dieldrin and hexachlorobenzene during the 1970s declined at an exponential rate over the next 20 years (Miller *et al.*, 1999).

However, it appears that some infants may be exposed to OCP residues during the early breastfeeding period. This arises from the fact that stable compounds stored in adipose tissue of mothers are released into breast milk at the onset of lactation. The most recent testing of limited breast milk samples for OCPs in Australia indicated that the DDT metabolite DDE predominates, with much lower and less frequent occurrence of heptachlor epoxide and BHC (ANZFA, 1998). Data collected during the 1990s suggested a low but relatively stable level of total DDT (DDE plus DDT) over the period, albeit at levels significantly lower than those reported in the 1970s. While there are concerns that generation and publication of such monitoring data might discourage mothers from breastfeeding their infants, the WHO is keen to point out that the advantages of breast milk to the infant far outweigh any risks from the potential hazards of OCP residues (JECFA, 1990).

RISK ASSESSMENT AND RISK MANAGEMENT

Risk assessment, as outlined in FAO/WHO consultations (e.g. FAO/WHO, 1995), consists of the following:

(1) hazard identification;
(2) hazard characterisation (dose–response assessment);
(3) exposure assessment;
(4) risk characterisation, on the basis of hazard characterisation and exposure assessment.

The two main data collection and analysis steps in chemical risk assessment are hazard characterisation (i.e. determining whether a chemical has toxic effects and what doses are likely to cause these effects) and exposure assessment. Assessing

---

[6]A similar situation exists with some chlorinated organic compounds which are byproducts of industrial processes or waste incineration (e.g. dioxins) or were manufactured for industrial uses (e.g. polychlorinated biphenyls (PCBs)).

the likely level of exposure of an individual or the population as a whole to a particular chemical is an important step since if there is no exposure, a toxic chemical poses no risk. This may be illustrated by the following: imagine a perfect containment system which absolutely prevents any exposure of humans and the environment to a toxic substance. Since exposure is zero, the risk it poses to humans and the environment is also zero, although the toxicity of the substance remains unchanged.

What are the options for food regulators when the risk assessment indicates that, for a significant number of people, the dietary intake of a particular pesticide residue is likely to exceed the ADI? If it is apparent that, in the calculation of dietary intake the best available residue level and food consumption data have been used, then it is at this stage that risk management steps need to be taken. There are a number of options to reduce the dietary load, e.g. the use pattern of the pesticide could be changed, i.e. fewer applications to the crop per season or application at a lower rate, regulatory approval for use of the pesticide could be withdrawn from one or more crops, or the pesticide could be withdrawn from the market.

In a few cases, if new or additional toxicology data are available, it may be appropriate to review the ADI and the basis on which it was set. However, ADIs are health standards and their establishment should in no way be influenced by subsequent steps in the risk assessment process, be it exposure assessment, risk characterisation or risk management.

AGGREGATE AND CUMULATIVE RISK

Several factors must be considered when conducting a chemical risk assessment with respect to the general population. As has been discussed above, all possible routes of exposure (i.e. oral, dermal and inhalation) should be taken into account.

The US EPA, in particular, is refining methods for estimating what is termed aggregate risk (US EPA, 1999). Aggregate risk assessments consider exposures from three major sources, namely food, drinking water and non-dietary and non-occupational (typically residential) exposures. In an aggregate assessment, exposures from these sources would be added together and compared to a quantitative estimate of hazard (e.g. a NOEL) or, where quantitative or semi-quantitative estimates are made for each exposure type, the risks themselves can be aggregated.

Currently, most regulatory agencies consider the potential use patterns of pesticides when conducting pesticide risk assessments and consider exposures by each possible pathway. Thus, for example, for a compound like the herbicide atrazine for which it is apparent from regular residue surveys on foodstuffs that residues don't occur but that it can leach into water supplies, the risk assessment will focus on intake via drinking water. On the other hand, for a volatile insecticide like dichlorvos which is used in indoor pest strips and in fogging machines for fumigation of pest-infested buildings, the regulatory assessment will pay particular attention to the risk of inhalation exposure.

Another issue related to chronic intake of pesticide residues is that of how to add together the health risks posed by exposure to different pesticides with common mechanisms of toxicity; this is often referred to as 'cumulative risk assessment'. Regulatory agencies are beginning to conduct such assessments but there are problems in determining on what basis pesticides should be grouped. For example, one could consider organophosphorus compounds and carbamate compounds together as they both act to inhibit acetylcholinesterase in the nervous system (albeit with significantly different time-courses of action), thus causing the accumulation of acetylcholine which then acts on nicotinic and muscarinic acetylcholine receptors to produce characteristic adverse effects. However, if the cascade of events following acetylcholine receptor stimulation is considered, then any compound that stimulates the cholinergic synapse or induces 'nicotinic' actions (such as nicotine and the synthetic nicotinoids and neo-nicotinoids) could also be considered to have a common mechanism! In addition, what about compounds which might act at the same receptor sites in an insect nervous system, but some 'activate' while others 'block' the receptor, i.e. have agonist or antagonist interactions? Would these be considered to have a common mechanism of action because they targeted the same receptor? It appears that issues related to cumulative risk will be under discussion by toxicologists for some time to come.

THE FUTURE FOR PESTICIDE INTAKE ASSESSMENT?

As discussed, much attention in the next few years will be placed upon practical ways of estimating aggregate exposure (that is, exposure to chemicals by the dermal and inhalation routes as well as by ingestion) and cumulative exposure (that is, exposure to different chemicals which have a common mechanism of action and whose combined effects are additive).

With respect to dietary intake assessment, the availability of inexpensive but powerful computers will allow more and more countries to expand and improve their national dietary intake tables, not only to cover the average population intake of particular foodstuffs but to consider the intakes of special subgroups, etc. One such computer model which has been developed by Food Standards Australia New Zealand is Dietary Modelling of Nutritional Data (DIAMOND). This program is based on individual dietary records from a very large number of respondents. It contains a range of databases, including the following:

- Food consumption data from single 24-h recall surveys (three national dietary surveys conducted in Australia between 1983 and 1995 (2–70 year olds surveyed), plus a 1997 New Zealand dietary survey (15 + years old)).
- Food chemical concentration data (MRLs for pesticide residues, Maximum Permitted Concentrations for contaminants, Maximum Permitted Levels for food additives, pesticide residue levels in food commodities from supervised field trials, market basket surveys for levels of pesticide residues and

contaminants in prepared and processed food, and manufacturers' data for levels of food additives).

- A recipe database (used to split prepared food into its constituent raw commodities).

Utilising its databases, this program can rapidly calculate, for different population sub-groups, the likelihood of dietary intake of any pesticide residue exceeding the established health standard. Similar programs have been developed by other agencies, e.g. the UK (MAFF), the US EPA and Health Canada.

There are a number of areas where more information will improve our ability to estimate dietary intakes of pesticide residues. With respect to hazard identification and assessment, there is a need for toxicologists to continue to investigate whether there are population sub-groups which may be more vulnerable than normal to particular pesticides. On the exposure side, there is a need for more extensive residue data which is representative of a country's food supply, both domestically produced and imported. There is a need for long-term dietary surveys (e.g. food frequency surveys) to take into account day-of-the-week and seasonal changes in eating habits, in order to construct model diets that represent habitual food consumption levels.

It has been argued by some that, since the likelihood of human poisoning is very much greater from microbiological contamination of inadequately prepared or stored food rather than from the presence of low levels of pesticide residues, the need for detailed hazard and dietary exposure assessment for synthetic pesticides will decrease. However, this argument does not take into account the perception among a significant portion of the community that the health risks posed by synthetic chemicals, particularly those finding their way into food, are of greater concern than contamination from naturally occurring chemicals. Because of this, regulatory toxicologists and food scientists are likely to come under increasing public pressure to adopt more conservative assumptions and approaches in the risk assessment of chemical pesticides. On the other hand, by virtue of advances in biological knowledge about target pest species, in chemical synthetic methods and in high-throughput screening tests, chemical pesticides are likely to become more specific for particular pests and less toxic to non-target organisms. In addition, pressures on the agricultural industry to reduce their reliance on chemical pesticides will encourage broader moves to Integrated Pest Management (IPM) programs, thus expediting the development and use of biological pesticides or other pest control measures which do not leave residues in agricultural produce.

ACKNOWLEDGEMENTS

The authors would like to thank Drew Wagner (National Occupational Health and Safety Commission) and Janis Baines (Food Standards Australia New Zealand) for their valuable comments on the draft manuscript.

REFERENCES

ABS (1999). *National Nutrition Survey: Foods Eaten, Australia, 1995*, Australian Bureau of Statistics, Canberra Table 1 – Mean daily intake: Fine age group, by sex, pp. 14–22.

ANZFA (1996). *The 1994 Australian Market Basket Survey*, Australia New Zealand Food Authority, Commonwealth of Australia, Canberra, Australia.

ANZFA (1998). *The Australian Market Basket Survey 1996*, Australia New Zealand Food Authority, Commonwealth of Australia, Canberra, Australia.

CCPR (1999a). *Consideration of Intake of Pesticide Residues*, Report of the Joint FAO/WHO Expert Consultation on Food Consumption and Exposure Assessment, Document CX/PR 99/3, Codex Committee on Pesticide Residues, 31st Session, The Hague, The Netherlands 12–17 April, 1999 (prepared February, 1999).

CCPR (1999b). *Consideration of Intake of Pesticide Residues: Report of Intake of Pesticide Residue Intake Studies at International and National Level Based on Revised Guidelines for Predicting Dietary Intake of Pesticide Residues*, Progress report by WHO on prediction of dietary intake of pesticide residues, Document CX/PR 99/4, Codex Committee on Pesticide Residues, 31st Session, The Hague, The Netherlands, 12–17 April, 1999 (prepared January, 1999).

Derelanko, M. J. (2000). *Toxicologist's Pocket Handbook*, CRC Press, Boca Raton, FL, USA, Table 108, p. 132.

FAO (1986). *International Code of Conduct on the Distribution and Use of Pesticides*, Food and Agriculture Organization of the United Nations, Rome.

FAO/WHO (1995). *Application of Risk Analysis to Food Standards Issues*, Report of the Joint FAO/WHO Consultation, Document WHO/FNU/FOS/95.3, Geneva, Switzerland, 13–17 March, 1995, Food and Agricultural organization of the United Nations, Rome.

FAO/WHO (1997). *Food Consumption and Exposure Assessment of Chemicals*, Report of an FAO/WHO Consultation, Document WHO/FSF/FOS/97.5, Geneva, Switzerland, 10–14 February, 1997, Food and Agriculture Organization of the United Nations, Rome.

Hamilton, D. J., Holland, P. T., Ohlin, B., Murray, W. J., Ambrus, A., DeBaptista, G. C. and Kovacicová, J. (1997). Optimum use of available residue data in the estimation of dietary intake of pesticides, *Pure Appl. Chem.*, **69**, 1373–1410.

IPCS (1989). *Glossary of Terms for Use in IPCS Publications*, International Programme on Chemical Safety, World Health Organization, Geneva.

JECFA (1990). *Evaluation of Certain Food Additives and Contaminants*, 35th Report of the Joint FAO/WHO Expert Committee on Food Additives, WHO Technical Report Series No. 789, World Health Organization, Geneva.

JMPR (1999a). *Pesticide Residues in Food 1999 – Report*, FAO Plant Production and Protection Paper No. 153, Food and Agriculture Organization of the United Nations, Rome.

JMPR (1999b). *Pesticide Residues in Food – 1999 evaluations. Part II, Toxicological*, World Health Organization, WHO/PCS/00.4 (2000), World Health Organization, Geneva.

Kamrin, M. A. (Ed.) (1997). *Pesticide Profiles: Toxicity, Environmental Impact and Fate*, CRC Press/Lewis Publishers, Boca Raton, FL, USA.

Miller, G., Anderson, S. and Connell, D. (1999). *Report on Levels of Persistent Organochlorine Pesticides (OCP) in Australia (with Particular Reference to ANZECC's Schedule X)*, Report 99118/Envirotest, prepared for Environment Australia by Envirotest, Brisbane, Queensland (November, 1999).

NRA (1997). *The NRA Review of Atrazine*, Existing Chemicals Review Program, The National Registration Authority for Agricultural and Veterinary Chemicals, Canberra, Australia (November 1997) [http://www.affa.gov.au/nra/prsat.html].

NRA (1999). *The NRA Review of Chlorfenvinphos* (draft report), Existing Chemicals Review Program, National Registration Authority for Agricultural and Veterinary Chemicals, Canberra, Australia (September, 1999) [http://www.affa.gov.au/nra/cfvp.html].

Tomlin, C. (Ed.) (1994). *The Pesticide Manual*, 10th Edn, Crop Protection Publications, British Crop Protection Council and The Royal Society of Chemistry, Cambridge, UK and Farnham, UK.

US EPA (1988). *Reference Physiological Parameters in Pharmacokinetic Modeling*, Office of Risk Analysis, EPA Document 600/6-88/004, US Environmental Protection Agency, Springfield, VA, USA.

US EPA (1993). *Reference Dose (RfD): Description and Use in Health Risk Assessments*, Background Document 1A, US Environmental Protection Agency, Springfield, VA, USA (March, 1993) [http://www.epa.gov/ngispgm3/iris/rfd.htm].

US EPA (1999). *Guidance for Performing Aggregate Exposure and Risk Assessments*, Office of Pesticide Programs, US Environmental Protection Agency, Springfield, VA, USA (October 29, 1999) [http://www.epa.gov/fedrgstr/EPA-PEST/1999/November/Day-10/6043.pdf].

WHO (1990). *IPCS Environmental Health Criteria 104: Principles for the Toxicological Assessment of Pesticide Residues in Food*, World Health Organization, Geneva.

WHO (1993). *Guidelines for Drinking Water Quality – Volume 1: Recommendations*, World Health Organization, Geneva.

WHO (1997). *Guidelines for Predicting Dietary Intake of Pesticide Residues* (revised), Global Environment Monitoring – Food Contamination Monitoring and Assessment Programme in collaboration with the Codex Committee on Pesticide Residues, Document WHO/FSF/FOS/97.7, World Health Organization, Geneva.

WHO (1999). *IPCS Environmental Health Criteria 210: Principles for the Assessment of Risks to Human Health from Exposure to Chemicals*, World Health Organization, Geneva.

Yamamoto, I. and Casida, J. E. (Eds) (1999). *Nicotinoid Insecticides and the Nicotinic Acetylcholine Receptor*, Springer-Verlag, Tokyo.

APPENDICES

APPENDIX 1

Table 7.A1 Fenamiphos: – theoretical maximum daily intake (TMDI) calculations at the international level (ADI = 0.0008 mg/kg body weight or 0.048 mg/person (60 kg body weight))

Code	Commodity	MRL (mg/kg)	STMR or STMR–P (mg/kg)	Middle Eastern Diet (g/d)	Middle Eastern TMDI (mg/d)	Far Eastern Diet (g/d)	Far Eastern TMDI (mg/d)	African Diet (g/d)	African TMDI (mg/d)	Latin American Diet (g/d)	Latin American TMDI (mg/d)	European Diet (g/d)	European TMDI (mg/d)
FP 0226	Apple	0.05	–	7.5	0.0004	4.7	0.0002	0.3	0	5.5	0.0003	40	0.0020
FI 0327	Banana	0.05[a]	–	8.3	0.0004	26.2	0.0013	21	0.0011	102.3	0.0051	22.8	0.0011
VB 0402	Brussels sprouts	0.05	–	0.5	0	1	0.0001	0	0	1.1	0.0001	2.7	0.0001
VB 0041	Cabbages (head)	0.05	–	4.5	0.0002	8.7	0.0004	0	0	9.5	0.0005	24.1	0.0012
VR 0577	Carrot	0.2	–	2.8	0.0006	2.5	0.0005	0	0	6.3	0.0013	22	0.0044
OC 0691	Cotton seed oil (crude)	0.05	–	3.8	0.0002	0.5	0	0.5	0	0.5	0	0	0
MO 0105	Edible offal (mammalian)	0.01[a]	–	4.2	0	1.4	0	2.4	0	6.1	0.0001	12.4	0.0001
PE 0112	Eggs	0.01[a]	–	14.6	0.0001	13.1	0.0001	3.7	0	11.9	0.0001	37.6	0.0004
FB 0269	Grapes	0.1	–	15.8	0.0016	1	0.0001	0	0	1.3	0.0001	13.8	0.0014
MM 0095	Meat (mammalian)	0.01[a]	–	37	0.0004	32.8	0.0003	23.8	0.0002	47	0.0005	155.5	0.0016
VC 0046	Melons (except watermelon)	0.05[a]	–	16	0.0008	2	0.0001	0	0	2.8	0.0001	18.3	0.0009
ML 0106	Milks	0.005[a]	–	116.8	0.0006	32	0.0002	41.8	0.0002	160	0.0008	294	0.0015
SO 0697	Peanut	0.05[a]	–	0.3	0	0.2	0	2.3	0.0001	0.3	0	3	0.0002
OC 0697	Peanut oil (crude)	0.05[a]	–	0	0	1.8	0.0001	3.5	0.0002	0.5	0	1.8	0.0001
VO 0051	Peppers	0.5	–	3.4	0.0017	2.1	0.0011	5.4	0.0027	2.4	0.0012	10.4	0.0052
FI 0353	Pineapple	0.05[a]	–	0	0	0.8	0	10.2	0.0005	3.1	0.0002	15.8	0.0008
PO 0111	Poultry (edible offal of)	0.01[a]	–	0.1	0	0.1	0	0.1	0	0.4	0	0.4	0
PM 0110	Poultry meat	0.01[a]	–	31	0.0003	13.2	0.0001	5.5	0.0001	25.3	0.0003	53	0.0005
VO 0448	Tomato	0.5	–	81.2	0.041	7	0.0035	16.5	0.0083	25.5	0.0128	63.9	0.0320
VC 0432	Watermelon	0.05[a]	–	49.3	0.0025	9.5	0.0005	0	0	5.5	0.0003	7.8	0.0004
			Total:		0.050		0.0087		0.0134		0.0236		0.0538
			% of ADI:		105		18		28		49		112

[a]MRL at or about the limit of determination.

Table 7.A2 Fenamiphos – international estimated daily intake (IEDI) levels (ADI = 0.0008 mg/kg body weight or 0.048 mg/person (60 kg body weight))

Code	Commodity	MRL (mg/kg)	STMR or STMR–P (mg/kg)	Middle Eastern		Far Eastern		African		Latin American		European	
				Diet (g/d)	IEDI (mg/d)	Diet (g/d)	IEDI (mg/d)	Diet (g/d)	IEDI (mg/d)	Diet (g/d)	IEDI (mg/d)	Diet (g/d)	IEDI (mg/d)
FP 0226	Apple	–	0.01	7.5	0.0001	4.7	0	0.3	0	5.5	0.0001	40	0.0004
JF 0226	Apple juice	–	0.008	0.1	0	0.1	0	0.1	0	0.1	0	0.1	0
FI 0327	Banana	–	0.02	8.3	0.0002	26.2	0.0005	21	0.0004	102.3	0.0020	22.8	0.0005
VB 0402	Brussels sprouts	–	0.01	0.5	0	1	0	0	0	1.1	0	2.7	0
VB 0041	Cabbages (head)	–	0.01	4.5	0	8.7	0.0001	0	0	9.5	0.0001	24.1	0.0002
VR 0577	Carrot	–	0.02	2.8	0.0001	2.5	0.0001	0	0	6.3	0.0001	22	0.0004
OC 0691	Cotton seed oil (crude)	–	0.01	3.8	0	0.5	0	0.5	0	0.5	0	0	0
MO 0105	Edible offal (mammalian)	–	0	4.2	0	1.4	0	2.4	0	6.1	0	12.4	0
PE 0112	Eggs	–	0	14.6	0	13.1	0	3.7	0	11.9	0	37.6	0
FB 0269	Grapes	–	0.02	15.8	0.0003	1	0	0	0	1.3	0	13.8	0.0003
MM 0095	Meat (mammalian)	–	0	37	0	32.8	0	23.8	0	47	0	155.5	0
VC 0046	Melons (except watermelon)	–	0.02	16	0.0003	2	0	0	0	2.8	0.0001	18.3	0.0004
ML 0106	Milks	–	0	116.8	0	32	0	41.8	0	160	0	294	0
SO 0697	Peanut	–	0	0.3	0	0.2	0	2.3	0	0.3	0	3	0
OC 0697	Peanut oil (crude)	–	0	0	0	1.8	0	3.5	0	0.5	0	1.8	0
VO 0051	Peppers	–	0.055	3.4	0.0002	2.1	0.0001	5.4	0.0003	2.4	0.0001	10.4	0.0006
FI 0353	Pineapple	–	0.01	0	0	0.8	0	10.2	0.0001	3.1	0	15.8	0.0002
PO 0111	Poultry (edible offal of)	–	0	0.1	0	0.1	0	0.1	0	0.4	0	0.4	0
PM 0110	Poultry meat	–	0	31	0	13.2	0	5.5	0	25.3	0	53	0
VO 0448	Tomato	–	0.05	81.2	0.0041	7	0.0004	16.5	0.0008	25.5	0.0013	63.9	0.0032
VJ 0448	Tomato juice	–	0.03	0.3	0	0	0	0	0	0	0	2	0.0001
VC 0432	Watermelon	–	0.02	49.3	0.0010	9.5	0.0002	0	0	5.5	0.0001	7.8	0.0002
			Total:		0.0063		0.0014		0.0017		0.0040		0.0065
			% of ADI:		13		3		3		8		14

8 Acute Intake

**KIM Z. TRAVIS[1] DENIS HAMILTON[2], LES DAVIES,
MATTHEW O'MULLANE and UTZ MUELLER[3]**
[1] *Syngenta AG, Bracknell, UK*
[2] *Department of Primary Industries, Brisbane, Australia*
[3] *Australian Department of Health and Ageing Care, Canberra, Australia*

Pesticide Residues in Food and Drinking Water: Human Exposure and Risks.   Edited by Denis Hamilton and
Stephen Crossley
© 2004 John Wiley & Sons, Ltd   ISBN: 0-471-48991-3

INTRODUCTION

When a pesticide is used on food or feed crops, residues of the pesticide may appear in the food delivered to the consumer. Use on a food crop such as tomatoes or lettuce can lead directly to residues in those foods but it is not immediately obvious that use on pastures or cereals might produce residues in meat, milk or eggs. Before such uses of pesticides are approved by government authorities, a consumer risk assessment should be conducted.

It has long been the practice, for each pesticide, to determine the daily intake of residues which is considered to be 'safe' over lifetime exposure,[1] that is, an intake without any apparent adverse effects on health when ingested daily over a lifetime or most of a lifetime.

However, in the early 1990s it became apparent that account needed to be taken of the fact that in some food consumption situations, pesticide residues could pose acute hazards. Research on residues of acutely toxic pesticides (organophosphates and carbamates) in individual fruits and vegetables (which may be consumed as individual units during a single sitting or over one day) revealed random occurrences of comparatively high residue levels; while some variability was expected, the magnitude of the variability was not (PSD, 1998; Crossley, 2000). It was further realized that a proportion of the population which consume significant amounts of such foods are at a finite risk of ingesting such 'hot' commodity units.

In these cases, the reference health standard for acute dietary intake should not be the acceptable daily intake (ADI) but a more-appropriately established acute reference dose (ARfD).

The low level of residues occurring in water and the methods for chronic risk assessment (permitted health levels calculated from ADI) mean that it is difficult to imagine a situation where the legitimate use of a pesticide could lead to a short-term risk from consumption of drinking water.

This present chapter discusses the background to short-term (acute) exposure and risk and the differences from the more traditional chronic risk assessment. It then explains the framework and data requirements for acute risk assessment. National governments have tackled the issue in various ways. The issues at international level, as exemplified by the Joint FAO/WHO Meeting on Pesticide Residues (JMPR) approach, are slightly different from those faced at national level.

WHY CONSIDER ACUTE RISK?

Systems and processes for chronic dietary risk assessment have been in place for several decades, and have evolved slowly over that period. In contrast, the assessment of acute exposure and risk is more recent and several forces have driven its development. The importance of each has varied over time and among

---

[1]In the present context, 'dietary exposure' or 'exposure' or 'intake' all have the same meaning.

different countries, but all have been factors in the evolution of the science. The main driving forces have been the following:

- temporary departures over the ADI
- acute toxic effects
- variability of residues in single items
- variability of food consumption or diets

TEMPORARY DEPARTURES OVER THE ADI

Chronic toxicity studies are generally conducted in conditions where exposure is held constant over time. It has normally been assumed that brief periods when exposure exceeds the ADI are not likely to be a problem so long as the average exposure is still below the ADI. This is equivalent to assuming that it is the long-term average exposure or long-term total exposure that matters in terms of biological response. However, this conceptual model of biological response is not valid for all circumstances, i.e. for all pesticides and all biological end-points.

If the paradigm (set of underlying concepts) is not universally true, then some difficult questions can arise. How far above the ADI can exposure go, for how long, and how frequently, before we would expect to see effects? Are we observing a minor erosion of the standard 100-fold safety factor or a more significant problem?

The JMPR (FAO, 1999) noted that occasional exceedance of the ADI is generally considered of no toxicological concern provided that the ADI is not exceeded over the long term. However, for pesticides that present an acute hazard, such temporary excesses above the ADI may be of concern and an acute reference dose (ARfD) should be established to set an upper limit on such excursions. The JMPR also drew attention to certain acutely toxic compounds where the ARfD may have the same numerical value as the ADI; in such cases, fluctuations above the ADI should not occur.

The intake of residue on a single occasion or in a day can be somewhat higher than the long-term or typical daily average for two reasons. First, the specific food portion may contain higher residues than average and secondly, the consumer may eat more of that food on a particular day than average. The extreme case occurs where the two situations coincide, i.e. the consumer eats a large portion of food with the higher residue. In estimating short-term risk from pesticide residues, we need data on large portion consumption and the occurrence of high residue.

ACUTE TOXICITY EFFECTS

There are some reported incidents of adverse effects on human health following single dietary exposures to pesticide residues in food, e.g. following the illegal use of aldicarb on watermelons (Goldman et al., 1990). The great majority of such incidents involve misuse of the products concerned but they do remind us that acute hazards exist, and that they need to be managed to prevent them becoming acute risks.

The JMPR has established a number of acute reference doses. Among the most common acute toxic effects are acetylcholinesterase inhibition and neurotoxicity.

Some toxic effects are of more concern than others, particularly irreversible effects. Decisions on the safety factor to be applied based on the severity of the toxic effect were considered more an issue for risk management than for scientific assessment (PSD, 1998).

A conference report (PSD, 1998) envisaged that there may be situations where more than one ARfD could be established for different population sub-groups where there are different qualitative toxicological end-points. Dinocap is such an example where an ARfD of 0.008 mg/kg bw[2] applies to women of child-bearing age and an ARfD of 0.03 mg/kg bw applies to the general population with the exception of women of child-bearing age (FAO, 2001b). The ARfD of 0.008 mg/kg bw was based on developmental effects resulting from prenatal exposure; the ARfD for the general population was a conservative value from a long-term toxicity study in mice because no study specifically designed for setting an ARfD (apart from the developmental toxicity study) was available.

VARIABILITY OF RESIDUES IN SINGLE ITEMS

Extensive research in the UK and other countries has shown the considerable variation of residue levels between individual carrots or apples or a number of other fruits and vegetables within the one lot, i.e. from the same field with the same pesticide treatment. Even post-harvest treatments, e.g. dipping apples in the packing shed produces substantial variation in residue levels from one unit to the next (Roberts et al., 2002).

Field residue studies almost always involve the collection of a composite sample, i.e. a sample consisting of several or many units of the food commodity concerned. This is a deliberate practice which seeks to ensure that the residue in the sample is representative of the average for the particular crop as a whole. The averaging that results from this is entirely appropriate for chronic risk assessment.

The variability of residue levels in individual units within a lot may be described by a variability factor (v) defined as an upper percentile (97.5th percentile) of the residue levels found in individual units divided by sample mean (PSD, 1998).

Ambrus (2000) reported variability factors of 2.8, 3.1 and 4.3 for residues on individual apples from a single orchard. It is theoretically possible that in a composite sample of 10 apples all of the residue could be in a single apple, which would mean that the residue level in that single apple would be 10 times the residue level for the whole composite sample. In the context of chronic intake

---

[2]bw, body weight.

this is not important: a person may eat a high-residue apple one day and a low-residue one the next, and it will all average out over time. However, in the context of acute risk assessment this kind of variability is more of a concern. Acute risk assessment is often expressed as assessment of the risk of exposure over a period of one day, or for one meal, but ultimately the variability of residues on single items pushes the assessment towards a consideration of the risk from eating a single food item, e.g. one apple.

A preliminary way of looking at the short-term intake of residues is to calculate the residue level in a single potato or carrot or peach that would produce an intake equivalent to the acute reference dose when the single unit is consumed, and then ask 'are we likely to generate such a residue when the pesticide is used according to label instructions?':

$$\text{Residue level (mg/kg)} = \frac{\text{Acute RfD (mg/kg bw)} \times \text{body weight (kg)}}{\text{unit weight (kg)}} \quad (8.1)$$

Examples of such calculations are shown in Table 8.1. Of course, when data on residue levels and diets are available, better calculations are possible.

UK research has shown that residues on apples in a supermarket crate can be particularly variable. One reason for this is that it is very common for agricultural produce from different fields and from different growers to be mixed in the food supply chain, so that apples in a crate may have quite different pesticide application histories (Hill, 2000).

VARIABILITY OF FOOD CONSUMPTION OR DIETS

Besides the variability in residue levels, the consumption of a particular food may vary substantially from day to day. A range of factors contribute to this variability in food consumption including appetite, seasonal availability of food, cultural and regional eating patterns, and personal preference or habits. Some staple foods may show less variability on consumption levels from day to day, e.g. bread. However, others may be consumed in large quantities but with big

Table 8.1 Calculated residue levels (mg/kg) in edible portions of single units of fruits and vegetables required to give an intake by a 60 kg person equivalent to the ARfD

Acute RfD (mg/kg bw)	Residue level in edible portion equivalent to ARfD for 60 kg person consuming one unit[a]				
	Potato (160)	Carrot (89)	Tomato (123)	Peach (99)	Apple (127)
0.0008	0.3	0.5	0.4	0.5	0.4
0.003	1.1	2.0	1.5	1.8	1.4
0.03	11	20	15	18	14
0.1	38	67	49	61	47

[a] Figures in parentheses indicate unit weights (g) of edible portions.

gaps in between, e.g. cherries in season. The variability of food consumption supports the consideration of acute risk assessment to the same degree as the variability of residues among single items.

FRAMEWORK FOR ACUTE RISK ASSESSMENT

Acute risk assessment for pesticide residues has been implemented to varying extents and in different ways across the world. However, there are some common elements in national approaches that together form a framework for the acute risk assessment itself, and these will now be considered.

WHAT TRIGGERS AN ACUTE RISK ASSESSMENT?

Opinions are quite diverse regarding when an acute risk assessment should be undertaken. Some still consider that the classical chronic risk assessment has an excellent track record of adequately protecting public health, and acute risk assessment is an unnecessary complication. Toxicologists tend to consider that an acute risk assessment should be conducted if an acute hazard is evident in the toxicology database of the pesticide concerned (Marrs, 2000; Billington and Carmichael, 2000). Residue chemists would more likely support the conduct of an acute risk assessment if significant residues were detected on crops eaten fresh and whole.

Risk assessors might argue that it is not possible to determine whether a risk assessment is needed until after it has been conducted. If the acute risk assessment shows an unacceptable risk, then it was clearly needed, but if it did not, then the risk assessment did not add value. Thus, an acute risk assessment is always needed. This view is reflected in US regulatory practice.

ACUTE TOXICITY END-POINTS

An acute toxicity end-point (health standard) is needed as a reference dose for the dietary exposure estimate. Unlike chronic toxicity, the whole world seems to use a single term to describe this health standard – the acute reference dose (ARfD). This is formally defined (WHO, 1997b) as follows:

The estimate of the amount of a substance in food or drinking water, expressed on a milligram per kilogram body weight basis, that can be ingested over a short period of time, usually during one meal or one day, without appreciable health risk to the consumer on the basis of all the known facts at the time of the evaluation.

The 2002 JMPR changed the wording to ..... *that can be ingested in a period of 24 hours or less.....*

How can we estimate the acute reference dose? Many toxicity studies are performed (typically in laboratory animals) to evaluate the possible hazards posed by pesticides, but none is specifically designed for deriving an acute reference dose (Billington and Carmichael, 2000). It is a golden rule of risk assessment that the time-scale of the risk being evaluated must match the time-scale of the hazard and exposure components. It is not valid to use a study conducted over the lifetime of a laboratory animal (e.g. rat or dog) to set an acute toxicity standard for comparison with the dietary intake during one meal or one day. Yet studies involving a single dose, or dosing in a single day, followed by a period of observation and analysis of toxicological consequences (biochemical and histological assessment), are infrequently performed.

The current toxicology data package required by the authorities was not designed with acute dietary risk assessment in mind. Thus, the current procedure consists primarily of trawling through the entire toxicity data package looking for effects that might occur from one or a few doses (Dewhurst, 2000). This involves a great deal of expert judgement and interpretation and, in the absence of the ideal study, can result in quite conservative assumptions being made. For example, effects seen in a 28-day study (animals dosed every day) might be assumed to have occurred as the result of the dose administered on day one.

Clearly, this situation is less than ideal, and a sounder basis for setting an ARfD is needed. In fact, proposals have been made for the design of a study specifically for the purpose of setting an acute reference dose (FAO, 2001a). Ideally, such a study guideline would need thorough review before being used, in order to ensure that results are reliable, would be accepted by regulators, and would provide a proper basis for decision-making. The OECD is currently considering the proposal for an acute reference dose setting study within its guideline development and approval process.

RESIDUE DATA

Before a pesticide can be registered for use, information is required on all aspects of its toxicology, safety and behaviour in crops, animals and the environment. An important component of this information is the expected level of residue that might occur in food commodities offered for sale and in food prepared for consumption. Levels of residue are estimated from supervised residue trials where the pesticide is used as instructed on the label or proposed label. The trials are designed to cover the range of practical situations that occur during production.

The risk assessment for the pesticide residue takes into account the levels of residue occurring in the food items from the trials. The highest residues in the edible portion of food items from the trials are used in the short-term risk assessment.

CONSUMPTION DATA³

Food consumption data are collected for many purposes, primarily for public health reasons, including such diverse considerations as calorie, vitamin and cholesterol intake, and correlations between intake and health outcomes (e.g. sugar intake and dental caries). Pesticide residue intake is rarely a dominant factor in the design of such studies. Average food consumption data are adequate for many chronic risk assessment purposes. Such data may be averaged across different people or across different days. Usually, both forms of averaging are done, since it is exceptionally demanding to obtain detailed dietary consumption records for an individual that cover more than a few days. Clearly this averaging is not appropriate for the purposes of acute dietary risk assessment, for which records of the food consumed by individuals on single days, or at single meals, are needed.

Few involved in the process of pesticide risk assessment are experts in the field of dietary consumption databases. Discovering what databases exist, their contents, formats and suitability is a time-consuming and specialist task. For these reasons it is often said that a major barrier to the development of acute dietary risk assessment methodologies is the lack of adequate food consumption databases. Such databases often do exist, but the time and resources necessary to investigate and interpret them and make them available for dietary risk assessment purposes has not been committed.

TIERED APPROACH⁴

A practical basis for regulatory risk assessment is the use of a tiered approach (Figure 8.1). This generally begins with a screening-level assessment, Tier 1, which uses simplifying conservative assumptions, is quick and involves commonly available data. If a pesticide passes a risk assessment at Tier 1 we can be completely confident that it is safe. The higher tiers involve the progressive introduction of greater realism, but at the cost of increased complexity, cost, review time, and the need for more data to support the risk assessment.

If a pesticide fails at Tier 1, then this is not necessarily a 'black mark' – it simply means that a more detailed examination is needed before a judgement can be made. In theory, there may be any number of tiers, but ultimately the new refinements considered at the highest tier are quite specific to the properties of the pesticide in question and how it is to be used. Of course, if a pesticide fails at the highest tier then this implies it really is not safe, and the uses should be modified or the registration withdrawn. In reality, it is more common for a registration to be withdrawn, not because a pesticide is unsafe but because the cost of maintaining the registration, e.g. carrying out new studies to

³See also Chapter 6.
⁴See also Chapter 6 – section 'United States Methodologies'.

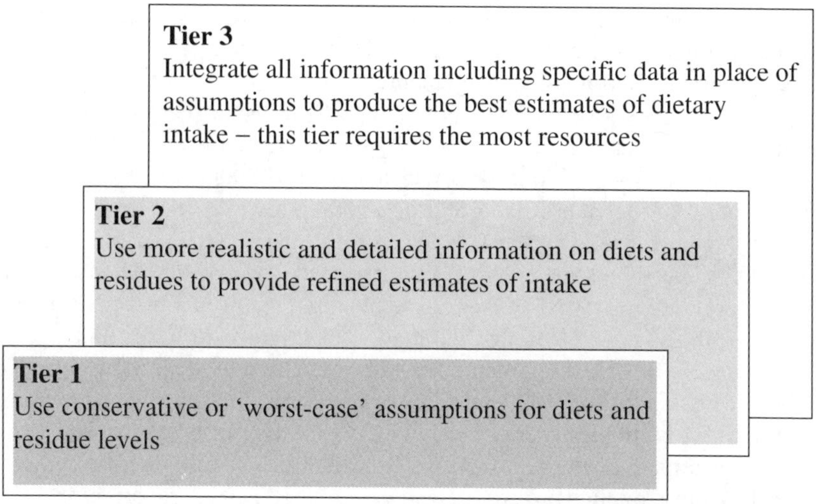

Figure 8.1 Illustration of the tiered approach for regulatory risk assessment

bring the data package up to the latest standards, is not justified by the sales or potential sales.

Numerous pesticide-specific factors will influence the true dietary risk of that pesticide to the entire population under real agricultural-use conditions. The world is a complex place, and it would be very onerous to find out precisely what is the true risk. Thus, the approach is to build simpler and simpler models of the real world, incorporating conservative default assumptions, e.g. all apples contain a high residue of the pesticide, none is cooked or peeled. The Tier 1 model is not realistic – it is conservative because it carries a series of regulatory worst-case assumptions on top of the normal 100 × safety factor. In subsequent tiers, we replace these worst-case assumptions with more realistic values that are directly relevant for the particular pesticide and use situation in question (e.g. residues are not always present at a high level, and apple juice may contain lower residues than whole apples). Ultimately, if we were to refine all of the assumptions in the model then we would discover the true risk, but until then we know we are always on the conservative side, that is, we will ensure that we are not underestimating the risk.

It is important to recognize that these regulatory models are only tools, and they have limitations. There is a danger that familiarity with these models could lead to some confusion between the models and reality. After many years positive experience of using a simple Tier 1 risk assessment model, it can become so comfortable that it becomes thought of as the 'gold standard'. If higher-tier approaches for refining risk estimates are not well developed or used, as has been the case in dietary risk assessment, then their proposed use can seem to threaten the established 'gold standard'. Regulators are then naturally concerned that the safety margins they have been accustomed to (i.e. with Tier 1) are being eroded. This is a misunderstanding of the tiered approach.

The purpose of tiering is to save resources (review time, money, test animals, etc.) by investing less effort into products with massive margins of safety. If the dietary intake of a pesticide residue reaches 99 % of the ARfD at Tier 1 (assuming the usual 100-fold safety factor is used), then the true margin of safety is not 100, but more like 1000 or 10 000. However, the conservatism of the Tier 1 approach allows the conclusion only that the margin of safety is *at least* 100. If the same pesticide at Tier 2 reached 20 % of the ARfD, the more refined, resource-intensive Tier 2 assessment would result in a more accurate risk estimate, with the conclusion that the margin of safety is *at least* 500.

The first tiers of risk assessment tend to be occupied by deterministic models, i.e. those with worst-case point estimates of food consumption and residue levels. By definition, the highest tiers of risk assessment approach reality, and since a major characteristic of the real world is its variability, it follows that higher-tier approaches must explicitly account for variability. This calls for probabilistic modelling, i.e. modelling that considers the full range of food consumption patterns and the full range of residue levels rather than single point estimates.

The tiered approach is not so appropriate at the international level, where the intention is to make best use of all available data. The JMPR evaluation takes place subsequent to national evaluation and registration. The relevant data already generated for national evaluations are provided to JMPR and it is more efficient in use of reviewer's time to evaluate all of the available data at one time (see discussion below under section 'International Assessment').

PROBABILISTIC METHODS[5]

In this approach, we combine the distribution of residue levels with the distribution of food consumption to produce a distribution of residue intake across the population. The modelling requires much more data than the deterministic method, e.g. single unit residue data and individual food consumption per day – not average or typical food consumption.

In the USA, the available monitoring and survey residue data are used in probabilistic modelling. The models and methods require increasing amounts of data as the estimates are refined. The percentage of crop treated is used as an incidence of residues occurring in the probabilistic modelling, i.e. it does not influence the residue levels, but it influences the likelihood of those levels occurring.

The perception is that a probabilistic approach 'leaves a proportion of the population unprotected'. In reality, some people are more exposed than others, and a very few people have far more exposure than the average, i.e. they are 'along the tail' of the exposure distribution. The deterministic approach decides initially

---

[5]See also Chapter 6.

how far along the distribution tail to regulate, i.e. the 97.5th percentile food consumption for eaters in combination with the 97.5th percentile residue level in a unit from a lot with a composite residue level equivalent to the highest found in the supervised trials. The probabilistic approach displays the whole distribution, and the risk manager then has a real job to do, namely to decide what percentile to regulate on. In reality, both methods decide on a cut-off working point, but the deterministic approach does not provide options for the risk manager.

NATIONAL ASSESSMENTS – EXAMPLE FROM AUSTRALIA

The most common route of exposure to pesticides for the general public is by ingestion of food commodities containing pesticide residues. In Australia, the methods for undertaking dietary exposure assessments are broadly outlined in a guideline document developed by the Australia New Zealand Food Authority (ANZFA)[6] and the National Registration Authority for Agricultural and Veterinary Chemicals (NRA), the ANZFA/NRA Protocol For Dietary Risk Assessments for Pesticide and Veterinary Drug Residues. The protocol, developed in consultation with staff from the Australian Department of Health and the Queensland Department of Primary Industries, covers procedures for estimating both acute and chronic dietary exposures to pesticides and veterinary drug residues in food as a result of the use of these chemicals on food crops and farm animals and the consumption of treated feed items by farm animals. The principles for dietary modelling outlined in this protocol are consistent with FAO/WHO guidelines for estimating intakes of pesticide residues (WHO, 1997b).

It should be noted that acute dietary risk assessments are conducted by only a few countries at this stage (2002), as the necessary data to undertake such assessments are limited.

ACUTE DIETARY EXPOSURE ESTIMATES

Acute dietary exposures to pesticide residues are normally only estimated for consumption of raw unprocessed commodities (fruit and vegetables) but may include consideration of meat, offal, cereal, milk or dairy product consumption, on a case-by-case basis. Food consumption data may be required for a single eating occasion, or for a single day, depending on the nature of the pesticide. A time scale of one day (24 h) was recommended for international calculations by the International Conference on Pesticide Residues Variability and Acute Dietary Risk Assessment (PSD, 1998).

Point or Deterministic Method of Acute Dietary Intake Estimation

Currently, Australia uses point estimate or deterministic methodologies to estimate acute intakes of pesticide residues. This non-probabilistic methodology

---

[6]Food Standards Australia New Zealand (FSANZ) since 2002.

assumes that there is a negligible level of risk if the single or point estimate of acute exposure for a given population does not exceed the acute reference dose. Point estimates are so called because single-point estimates are made for a range of factors in the dietary intake calculation, e.g. the amount of a food consumed, the residue level and the bodyweight of the consumer.

The procedure considers only one food commodity at a time, which is a reasonable approach since it is unlikely that an individual will consume more than one unit (i.e. a single fruit or vegetable) or more than one commodity containing a very high residue in the one meal or during one day.

Maximum Residue Limits (MRLs) and Supervised Trial Median Residues (STMRs) are based on analyses of composite samples of a number of commodity units (the number is chosen according to sampling protocols, but approximately 1–2 kg samples are commonly bulked for subsequent assay). It is clear that there will be some variation in residue levels between the individual commodity units which comprise the composite sample, i.e. the residue level in the composite sample will not reflect the actual range of residue levels in individual units, a finding that needs to be taken into account in assessing the risk of acute dietary exposure from consuming a single portion of the commodity.

The approach in Australia for estimating short-term intake of residues closely parallels the JMPR approach, but with nationally derived residue levels, body weights, diets and unit weights (see section below on 'International Assessment').

The first step is to determine if the commodity is homogenous or not in relation to consumption. For commodities that are basically 'homogeneous' when consumed because they are centrally processed like cereals or because there are a large number of individual units per portion (e.g. peas, beans and berries), individual unit variation is not considered to be of concern. However for commodities such as fruits and vegetables which are consumed whole or in large pieces (three or fewer commodity units per large portion), individual unit variability needs to be considered.

Three different cases are considered, depending on the type of commodity. The first of these covers those foods (such as peas, beans and berries, other than grapes for which a bunch is considered as the unit) in which the available composite residue data reflect the residue levels in the food portion consumed.

The second case covers foods (such as apples, bananas, melons and bunch of grapes) in which the available composite residue data may not reflect the residue levels in the food portion consumed, i.e. individual units which may provide a significant portion of the meal, could have a significantly higher residue level than that measured in a composite sample of the commodity. In this second case, two possibilities are identified: (1) the unit weight of the commodity is lower than the large portion weight (e.g. two or more apples or pears may be eaten at one sitting); (2) the unit weight of the commodity is higher than the large portion size (e.g. a third of a watermelon is eaten at one sitting).

Note that in the second case, a variability factor is included in the calculations. Variability is defined as the ratio of the 97.5th percentile single item residue to the mean residue level of the lot. Actual data from horticultural crops have shown that it is not uncommon to find 'hot' commodity units with high residue levels. For example, UK data showed that an individual carrot from a treated crop could contain up to 25 times the level in a composite sample from the same crop (PSD, 1998; Harris, 2000)

The third case relates to processed commodities where bulking and blending (e.g. breakfast cereal, flour, fruit juice, vegetable oil, etc.) mean that the median residue level derived from supervised trials (and adjusted for processing (STMR–P)) represents the highest likely residue.

Algorithms for these three different cases match those used for international estimated short-term intake (see section on 'International Assessment' later in this chapter).

Probabilistic Modelling[7]

With today's computing technology it is possible to utilize the full range of observations and data to simulate potential exposures to pesticide residues in food. By combining food consumption distributions with residue concentration distributions, these distributional models of exposure, most commonly referred to as Monte Carlo models (Callahan *et al.*, 1996) give a distribution of exposure, with estimates of the proportion of the population which may be at risk. A typical modelling approach would draw one food consumption figure at random from the distribution and then draw a concentration figure at random, combining them to generate the first point on an intake distribution. The process will be then repeated as many as 500 000 times to generate the intake distribution, before being repeated for each foodstuff containing the residue until a distribution of total intake from all sources is generated (Douglass and Tennant, 1997).

In contrast to the deterministic or point-estimate method which considers only one food commodity at a time, probabilistic methodology has an advantage in that it allows the consideration of more than one commodity at a time, thereby considering the consumer who may eat more than one commodity containing residues of a particular active constituent. The methodology may also allow consideration of residues of several different active constituents which have the same toxicological end-point.

For the probabilistic approach, single-unit residue data are required for a variety of chemical–crop combinations, as well as individual consumer diets. Monitoring and survey residue data may also be used, providing that they are derived from representative samples. For probabilistic modelling to be valid, the amount of data required is substantial; this includes extensive monitoring and survey data on residues as well as extensive food intake data across population sub-groups. In the USA, residues data generated at State and Federal level are used, with some indication of the percentage crop treated in each State. The USA also has

---

[7]See also Chapter 6.

a 'Continuing Survey of Food Intakes by Individuals' (CSFII) which provides ongoing collection of individual consumer records across four geographical areas by ethnic group, age and season.

At the time this chapter was written (2002), probabilistic modelling was being used by only the US EPA and, to some extent, by the UK Pesticides Safety Directorate and authorities in The Netherlands.

ACUTE DIETARY INTAKE ESTIMATION IN AUSTRALIA

In Australia, the National Registration Authority for Agricultural and Veterinary Chemicals routinely conducts acute dietary risk assessments for both new and review chemicals. Dietary consumption data are provided by FSANZ while toxicologists within the Therapeutic Goods Administration (TGA) establish the ARfD.

Although recognized as conservative (i.e. overestimating the likelihood that a proportion of the population will ingest residues above an estimated 'safe' level), the deterministic or point method is considered to be the only suitable approach for assessing acute dietary exposure in Australia until such time as more extensive residue and dietary consumption data are collected to allow probabilistic modelling: These data are currently not being generated in Australia at either State or Federal level, although expansion of the relevant databases is under consideration.

Australian data on residues levels, variability factors and processing factors are required for estimating the short-term dietary intake of pesticide residues, sometimes referred to as the national estimate of short-term intake (NESTI) (cf. the international estimate of short-term intake (IESTI) carried out by the JMPR).

Supervised field trials to collect residue data are conducted by the agricultural chemical industry or by specific grower associations, for the purposes of registering new 'agvet' chemicals or extending their uses to new crops. To date, the focus has been on measuring residue levels in composite samples for the purposes of establishing MRLs and STMRs, but as the industry becomes aware of the data requirements for acute dietary intake estimations, there will be increasing focus on measuring individual unit commodity residues in cases where refinement of an excessively conservative estimate is needed.

As noted above, relatively recent investigations have noted that in a particular treated crop, it is not all that uncommon to find commodity units with residue levels many-fold higher than the level in a bulked or composite sample of the crop. Acute dietary intake estimations based on composite-sample residue values include variability factors to account for this. In the absence of measured variability data, Australia utilizes the default factors established by the JMPR (see Table 8.3 below).

In Australia, processing factors for some commodities are available because processing trials have been undertaken. For example, there are sufficient data for selected agricultural chemicals on wheat to determine the residue concentration factor for wheat bran relative to wheat grain and the residue reduction factors for wheat flour relative to wheat grain.

The MRL is a compliance standard, set as low as possible consistent with good agricultural practice. Thus, the residue measured to monitor this compliance (most commonly, the parent compound only) has to be one which is relatively simple and inexpensive to extract and measure and may not include all of the residue components needed for the risk assessment. In some circumstances, the residue definition used for the ARfD may differ from that used for establishing ADIs, depending on whether the metabolite(s) that cause(s) the observed acute and chronic adverse effects are different. It is also conceivable that for a limited number of compounds, different population sub-groups could require different residue definitions for the acute risk assessment. (In practice, it is unlikely that the toxicology data would be sufficient to address these possibilities.)

In Australia, it is assumed that if residues are undetected, then for calculating NESTIs they are taken to be present at the Limit of Detection (LOD) unless evidence (e.g. from residue trials at multiples of the recommended rate) indicates that residues are essentially zero.

Dietary Consumption Data

For NESTI calculations, extensive data on dietary consumption habits for a cross-section of the population are required, as are large portion sizes and unit weight data for those commodities which may be consumed as whole units (e.g. fruit) or as parts of a unit (e.g. slice of watermelon).

The available Australian survey data (individual food consumption and body weight data from the 1995 National Nutritional Survey (NNS)[8] (ABS, 1999) is suitable for use in acute dietary exposure assessments. In some cases, however, the number of consumers of some commodities which are only occasionally consumed is too small to derive a statistically valid 97.5th percentile level of consumption for use in acute dietary exposure assessments. To derive a 97.5th percentile, 41 data points or more are required to ensure this figure is one reported by at least one 'real' person. This is more likely to occur for non-staple commodities and specific age groups (e.g. two to six years) where the number of total respondents in the age group is small (fewer than 3000).

It is possible that the limitations in the Australian food consumption database may be overcome in some dietary intake modelling situations, as follows:

(a) *Use of pooled data from a single food from a wider commodity group.* For example, if there are inadequate numbers of consumers of blueberries for any age group, the 97.5th consumption figure will be unreliable. The 97.5th level of consumption of berry fruits, including blackberries, strawberries

---

[8]The survey was conducted over a 13-month period during 1994–1995, with data collected in all States and Territories across different seasons and days of the week; food consumption data were reported for 13 858 respondents for the preceding 24 h (24-h recall method).

etc., could be used to represent the consumption of blueberries in an acute dietary exposure assessment of a pesticide used on blueberries. However, this approach may be invalid if the food commodity being considered is not consumed in the same way as other foods within the same commodity group, e.g. comparing limes with oranges would not be appropriate.

(b) *Use of consumption data from a similar commodity that has higher consumer numbers.* This could be useful in the situation where there are no consumption data at all for the commodity in question, or consumer numbers are too small. The limitation is that consumption of the related commodity may overestimate actual consumption of the commodity in question. However it allows a calculation to be undertaken, provided note is made of the assumptions used.

(c) *Use of international food consumption data.* Nutrition surveys have been undertaken in many countries, although not all have widely available summary results. In these surveys, there may be larger consumer numbers for given commodities, but if 97.5th percentile consumption levels are not published, access would be required to individual food consumption data.

The eating patterns of the country from which the data were sourced would also need to be considered; preferably the country would have similar eating patterns to those in Australia. Australia has been grouped with New Zealand, the USA, Canada, Chile and Argentina in the WHO regional diets; New Zealand and American national nutrition surveys are likely to yield the best data since Canada has not undertaken a national diet survey since 1991. There are limitations to this approach, including (1) that overseas data may not cover the whole population (e.g. there are no data for children in the 1997 New Zealand survey), and (2) that some food consumption data may already be the result of calculations and assumptions and further extrapolations from an extrapolated value are undesirable. Raw commodity data are available for New Zealand but are not directly available from US surveys.

(d) *Use of other percentile data.* In such cases, 90th, 95th or 99th percentile consumption figures could be used instead of 97.5th percentile data where such data are available.

There are some commodities which may be consumed either raw, cooked, or in prepared mixed food (e.g. carrots and fruits). One issue is whether the consumption figures used for an acute dietary exposure assessment should be estimated on the basis of the commodity consumed raw or on total commodity consumption (i.e. including the commodity consumed raw, cooked and as an ingredient in a prepared mixed food). For some commodities, this is not an issue, as it is known that the commodity is consumed either always raw or always cooked (e.g. lettuce is nearly always consumed raw, while potatoes are always consumed cooked). The food consumption values used in Case 2 calculations[9] should apply only to the form of the food being consumed, e.g. the

---

[9]See section on 'International Assessment' below.

consumption value for apples should be for fresh apples only and should not include apple juice or any form of processed apple or back-calculated equivalent of raw commodity. Intake calculations for apple juice and other processed apple commodities should be made separately from their own residue and food consumption values.

If a potentially sensitive target group was identified at the time that the toxicology data were assessed, food consumption data for this population sub-group should be used to provide a realistic estimate of dietary exposure to the compound being considered.

For situations in which the number of commodity units per large portion size is three or fewer, unit commodity weight data are required for NESTI calculations. Table 8.2 provides a list of fruit and vegetables in this category. Australian data are limited, although an Australian database is being developed (Bowles and Hamilton, 2000).

The current data set of individual food consumption and body weight data from the 1995 National Nutritional Survey (NNS) are suitable for use in probabilistic

Table 8.2 List of commodities with three or fewer commodity units that may comprise a large portion. The commodities listed are those for which a variability factor (v) should be used in calculating the NESTI if no residue data are available on an individual commodity basis (this list should not be considered exhaustive)

Citrus Fruits	Root and Tuber Vegetables	Cucurbits
Grapefruit	Beetroot	Cucumber[a]
Lemon[a]	Carrot	Courgette/Zucchini
Mandarin and other soft	Celeriac[a]	Melon[a]
citrus	Jerusalem artichoke	Watermelon[a]
Orange	Potato	Marrow[a]
Lime[a]	Parsnip	Pumpkin[a]
Pome Fruit	Swede[a]	Brassica
Apple	Sweet potato	Broccoli
Pear	Turnip[a]	Cauliflower[a]
Quince	Yam	Cabbage[a]
Stone Fruit	Bulb and Stem Vegetables	Chinese cabbage[a]
Apricot	Onion	Kohlrabi
Peach	Fennel bulb	Lettuce and Leaf Vegetables
Plum	Miscellaneous Fruit	Lettuce[a]
Nectarine	Avocado	Spinach
Berries	Banana	Chicory/Witloof
Table grapes (bunches)	Fig	Stem Vegetables
Fruiting Vegetables	Guava	Asparagus
Tomato	Kiwi fruit	Celery
Pepper (sweet)	Mango	Globe artichoke
Pepper (chilli)	Pawpaw/Papaya	Leek
Aubergine[a]	Pineapple[a]	Rhubarb

[a] A single portion of these commodities usually consists of less than one unit. For these commodities, acute intake would be calculated by using the equation given in Case 2b; otherwise, the equation given for Case 2a would apply (see section on 'International Assessment' below).

modelling for acute dietary exposure assessments, except for those commodities where consumer numbers in the adult or children sub-groups are less than 41.

ACUTE REFERENCE DOSE ESTIMATION

A critical component of any acute dietary risk assessment is the establishment of a suitable acute or short-term health standard, i.e. the acute reference dose (ARfD). The latter is a health standard set on the basis of toxicology findings from short-term studies in laboratory animals or possibly human trials. The ARfD is used as the reference dose against which the acute dietary intake estimate is compared. In theory, the dietary intake of pesticide residues below the ARfD should not lead to an adverse health effect in humans and therefore regulators would be prompted not to permit a product use where the ARfD is likely to be exceeded.

While the concept of health standards for acute exposures to chemicals is not new (e.g. short-term exposure limits or STELs for occupational settings), the consideration of acute effects arising from the dietary intake of pesticide residues is a relatively new concept.

Although the ARfD is ideally an upper limit or threshold value, there is currently a degree of conservatism built into the figure as a consequence of the type of end-points that are the basis for the no-observed-effect level (NOEL), as well as the use of conservative safety factors.

In principle, the process of setting an ARfD is relatively straightforward. The first step is to select a suitable study conducted in laboratory animals or humans in which a single dose or short-term dosing regime is employed. This could be an acute toxicity study, a developmental toxicity study, or a short-term repeat-dose study. The key factor in choosing the study is that observations or measurements (e.g. clinical signs, pathology, haematology, etc.) have been made following a single or short-term (i.e. one to two days) oral exposure. Once a suitable study has been selected, a toxicologically significant end-point indicative of an acute adverse effect is chosen. The dose at which this effect did not occur is the NOEL. The latter is then divided by an appropriate safety factor to yield the ARfD figure. Safety factors (or uncertainty factors) are used to extrapolate the findings observed in laboratory animals to humans. In Australia the Therapeutic Goods Administration currently uses default inter- and intra-species safety factors of 10. Therefore, a NOEL set on a human study would require a safety factor of 10 (to account for intra-species variability), while a NOEL set on an animal study (e.g. rats) would require a higher safety factor of 100 to account for both inter- and intra-species variability.

RISK CHARACTERIZATION

In chemical risk evaluation procedures, exposure estimates are compared with reference health standards to assess the likelihood of these standards being exceeded.

For an acute dietary risk assessment, estimates of dietary exposures from a single meal or over 24 h are compared to the ARfD. For a given pesticide, if the estimated chronic dietary exposure does not exceed the ADI and, where relevant, the estimated acute dietary exposure for a particular commodity does not exceed the ARfD, the proposed use may be registered (with MRLs being recommended for inclusion in the Australian Food Standards Code).

In cases in which the estimated acute dietary exposure for a particular commodity exceeds the ARfD, the following steps may be taken:

- The dietary exposure estimates can be checked and refined, ensuring that best use has been made of all available data. If the estimated dietary exposure still exceeds the ARfD, then:
  - if no relevant data are available that lead to a revision of residue levels, new data may need to be generated by the applicant for further assessment;
  - the toxicology data may be reviewed to confirm that the chosen endpoint is appropriate or a more suitable acute toxicity study may be requested from the applicant.

If the refined acute dietary exposure estimate still exceeds the ARfD (whether revised or not), risk management options are investigated.

RISK MANAGEMENT

If the acute dietary risk assessment for a particular pesticide indicates a potential unacceptable risk to public health, options for management of the risk include (1) not allowing the use of that pesticide on the particular crop or commodity, or (2) modifying the use pattern on the particular crop or commodity so that the pesticide is still effective but with reduced final residues.

Nevertheless, the point method as currently adopted provides a conservative over-estimate of acute exposure and thus risk management decisions need to be considered carefully.

ISSUES FOR CONSIDERATION IN ASSESSING ACUTE DIETARY RISKS

At a national level, some of the main issues under consideration with respect to acute dietary exposure assessments include the following.

Residue Data

- Cost and resources involved in generating residues data in Australia for use in point estimates of acute exposure (variability factors) or in probabilistic modelling (distributions of data).
- Investigation of the validity and use of overseas data, where available.

Food Consumption Data

- Use of alternative food consumption data to supplement situations for which there are small sample numbers in the existing database, particularly for the two to six year age group.
- Cost and resources involved in collecting additional food consumption data.

Acute Reference Dose Establishment

- Selection of appropriate toxicological end-points for setting acute reference doses.
- Establishment of criteria on whether or not it is necessary to set an ARfD for a particular pesticide.
- Determination of appropriate safety factors in setting ARfDs.
- The suitability and appropriateness of establishing an ARfD on the basis of studies conducted in humans.
- Whether there are situations in which more than one ARfD should be established for a pesticide, to allow for different effects in separate population sub-groups (e.g. women of child-bearing age, children, etc.).

Risk Management

- Development of a policy on regulatory action in situations where the acute reference dose is exceeded.
- Decisions regarding degrees of acceptable risk, and communication of that risk.

In conclusion, acute dietary risk assessment is an evolving regulatory science and the methods should be constantly reviewed so that intake modelling is as realistic as possible.

INTERNATIONAL ASSESSMENT – JMPR[10]

The Codex Alimentarius Commission (Codex) sets MRLs that apply to food commodities in international trade. Codex MRLs for pesticides are the responsibility of the Codex Committee on Pesticide Residues (CCPR). The JMPR evaluates the toxicology and residue studies and provides the risk assessment for CCPR.

The Codex system, unlike national registration agencies, does not register pesticides; its purpose is to set standards that apply to food commodities, but it relies on data already generated for national registration systems. From this perspective, it is most efficient for the JMPR to make the best use of all available data rather than to proceed through a tier-based system as used in some national systems.

---

[10]See also Chapter 10.

In response to concerns within CCPR that the ADI was not an appropriate toxicological standard for acutely toxic pesticides, the JMPR estimated its first acute RfDs (for monocrotophos and aldicarb) in 1995 (FAO, 1996).

A Joint FAO/WHO Consultation on *Food Consumption and Risk Assessment of Chemicals* (WHO, 1997a) recommended procedures for short-term dietary intake estimates for use at the international level for pesticide residues. This Consultation recognized the variations in residue levels occurring in individual units of fruit and vegetables within a single lot or consignment and proposed a variability factor for use in the intake estimates.

The three possible cases taken into account in JMPR short-term intake assessment have already been described under the section on the Australian approach.

The JMPR decided to use the highest residue level from the set of residue trials as the starting point for the calculations. This highest residue level was subsequently given the acronym 'HR'. This is defined in the FAO Manual (FAO, 2002a) as follows:

The HR is the highest residue level (expressed as mg/kg) in a composite sample of the edible portion of a food commodity when a pesticide has been used according to maximum GAP conditions. The HR is estimated as the highest of the residue values (one from each trial) from supervised trials conducted according to maximum GAP conditions, and includes residue components defined by the JMPR for estimation of dietary intake.

The HR was preferred over the MRL for short-term dietary intake estimation for the following reasons:

- The HR applies to the residue in the edible portion while the MRL applies to whole commodity, which makes a large difference where residues are mostly on the inedible portion, e.g. on bananas and oranges.
- The residue definition for dietary intake estimation applies to the HR, whereas the MRL residue definition is designed for analysis and testing. Sometimes, the two definitions are different.
- The MRL may be a 'rounded up' value and rounding up is undesirable at an intermediate stage of a calculation. The use of the HR is direct use of data for intake estimation.

Case 1

Composite sampling data reflect the residue level in a meal-sized portion of the food (commodity unit weight is below 25 g). A large portion will include a number of units, so the residue is likely to be well represented by the composite sample:

$$\text{IESTI} = \frac{\text{LP} \times \text{HR}}{\text{bw}} \tag{8.2}$$

Case 2

Composite residue data do not necessarily reflect the residue level in a meal-sized portion of the food (raw commodity unit weight exceeds 25 g). The meal-sized portion may be only one unit or a few units. The conservative assumption is that the first unit is the high residue unit (residue level = HR × v) and the remaining units have the same residue as the composite sample (residue level = HR).

Case 2a

Unit edible weight of raw commodity is less than a large-portion weight, i.e. more than one unit is consumed:

$$\text{IESTI} = \frac{(U \times HR \times v) + [(LP-U) \times HR]}{bw} \qquad (8.3)$$

Case 2b

Unit edible weight of raw commodity equals or exceeds large portion weight:

$$\text{IESTI} = \frac{LP \times HR \times v}{bw} \qquad (8.4)$$

Case 3

Processed commodity, where bulking or blending means that the STMR–P represents the likely highest residue. This case applies to commodities such as flour, vegetable oils and fruit juices where the primary commodities have originated from a number of farms:

$$\text{IESTI} = \frac{LP \times STMR-P}{bw} \qquad (8.5)$$

In Equations (8.2)–(8.5):

- IESTI – international estimate of short term intake (mg/kg bw/d)
- LP – highest large portion reported (97.5th percentile of eaters) (kg food/d)
- HR – highest residue in composite sample of edible portion found in the supervised trials used for estimating the maximum residue level (mg/kg)
- bw – body weight (kg) provided by the country from which the LP was reported
- U – unit weight of the edible portion (kg) provided by the country where the trials which gave the highest residue were carried out
- v – variability factor, i.e. the factor applied to the composite residue to estimate the residue level in a high-residue unit

- STMR – supervised trials median residue, i.e. the median residue in composite samples of edible portion found in the supervised trials used for estimating the maximum residue level (mg/kg)
- STMR–P – supervised trials median residue in processed commodity (mg/kg)

The large portion size was chosen as the 97.5th percentile consumption per day for eaters of that food. Dietary information was compiled for the general population (all ages) and children (six years and under) from data supplied by several national governments. For JMPR evaluation purposes, the highest national 97.5th percentile consumption was chosen for each commodity because food in international trade, irrespective of where it is produced, might be consumed in any country, including the one with the highest consumption.

The food commodity unit weight in Case 2 calculations has a strong influence on the calculated intake. Typical unit weights were provided by several national governments. The unit weight is chosen from the region where the trials and registered uses support the Codex MRL. The unit weight and large portion size are expressed as edible portion weights rather than as whole commodity in the calculations.

The variability factor was devised to deal with the situation where the residue in the composite sample, say five to ten fruits making up the 1–2 kg chopped for analysis, could be imagined to arise from only one of the units of fruit. Then, the residue in the single unit would be, on this conservative assumption, at a level of five to ten times as great as that in the composite.

The UK (PSD, 1998) has generated numerous data on individual units of crops such as apple, banana, carrot, kiwifruit, nectarine, orange, peach, pear, plum and potato. The majority of variability factors were in the 2–4 range, but occasionally higher values were found. A generic variability factor of 4 (based on a ratio of 97.5th percentile to mean) would be acceptable in most cases, but a conservative value of 7 was chosen for unit weights in the 25–250 g range. Leafy vegetables and residues arising from granular soil treatments were seen as carrying more variability and a variability factor of 10 was retained (Table 8.3). The 2002 JMPR received extensive unit data on residues in head lettuce (Kaethner, 2001) and adopted a variability factor of 3 for head lettuce and head cabbage.

The IESTI values produced by the JMPR are non-probabilistic calculations of intake. The possibility of pursuing probabilistic estimates at international level is currently precluded by insufficient relevant data.

The JMPR in 2001 was provided with copious single-unit residue trial data for aldicarb residues in potatoes (FAO, 2001c). The single-unit data were used to provide a highest single-unit value directly in the short-term intake calculation in place of the usual HR × v value.

No variability factor is applied in the IESTI calculations for animal commodities (e.g. meat) because, under Codex guidelines, primary samples of meat and poultry products are analysed for residues (FAO, 2002b). The final sample analysed for other commodities is a composite of primary samples.

Table 8.3 Variability factors currently (2002) used by the JMPR

Commodity situation	Variability factor
Unit weight < 25 g	Case 1
Unit weight (large items) > 250 g	5
Unit weight < 250 g, but > 25 g	7
Leafy vegetables, unit weight < 250 g	10
Head lettuce and head cabbage[a]	3
Granular soil treatments, unit weight < 250 g	10

[a]The specific factor for head lettuce and head cabbage is used instead of the general factor.

The JMPR publishes its detailed acute dietary risk assessments in the JMPR Reports each year for children up to six years and for a general population and summarizes its findings in each specific compound report. Attention is drawn to those situations where, based on deterministic calculations, the estimated intake exceeds the ARfD.

SUMMARY AND CONCLUSIONS

We should assess the potential short-term (acute) risks of pesticide residues in food in addition to their chronic or long-term risks. Methods are still being explored and will continue to develop.

Acute reference doses (ARfDs) have been estimated by national agencies and at the international level for many pesticides since the mid-1990s and the broad guidelines are now generally agreed.

Dietary exposure or intake methods divide generally into two procedures – deterministic and probabilistic. The deterministic methods currently rely on strongly conservative assumptions but require little extra data to be generated, thus allowing authorities to finalize risk assessments for many pesticides relatively quickly. Probabilistic methods need rather more data, e.g. monitoring data and per cent crop treated, and so cost more to finalize. Probabilistic assessments are currently (2002) possible in only a few countries and not at the international level.

A vital component of risk assessment is risk communication. The short-term risks of pesticide residues occurring in food should be explained to consumers, preferably as a comparison with other food-borne risks. Communication of tiered risk assessments needs special attention because the process gives a first impression of 'repeated calculations until the desired result is achieved'.

In the future, we might see more flexibility in exposure assessment methods in which hybrid deterministic–probabilistic methods are designed for individual cases to make best use of limited available data.

Further experiments and theory will help to refine default variability factors. Ultimately, variability factors should reflect 'natural variability' that cannot be improved even with the most careful and well-designed application techniques.

The methods developed for short-term risk assessment of pesticide residues have already been applied to the assessment of veterinary drugs at injection sites in meat. In future, the methods may be applied to contaminants or natural toxicants such as solanum alkaloids in foods, requiring the generation of data on the variability of toxicant levels between food units.

ACKNOWLEDGEMENTS

The authors would like to thank Raj Bhula (Australian Pesticides and Veterinary Medicines Authority, formerly National Registration Authority for Agricultural and Veterinary Chemicals), Janis Baines, Steve Crossley, Judy Cunningham and Tracy Hambridge (Food Standards Australia New Zealand), and Dugald MacLachlan (Agriculture, Fisheries and Forestry Australia) for their advice on the conduct of dietary intake assessments.

REFERENCES

ABS (1999). *National Nutrition Survey: Foods Eaten, Australia*, 1995, Australian Bureau of Statistics.

Ambrus, A. (2000). Within and between field variability of residue data and sampling implications, *Food Additives Contam.*, **17**(7): 519–537.

Billington, R. and Carmichael, N. (2000). Setting of acute reference doses for pesticides based on existing regulatory requirements and regulatory test guidelines, *Food Additives Contam.*, **17**, 621–626.

Bowles, P. and Hamilton, D. (2000). *Information gathered on unit weights of individual fruit and vegetable commodities*, Report A&PH.PB.2000.1, Queensland Department of Primary Industries, Brisbane, Australia (unpublished).

Callahan, B. G., Burmaster, D. E., Smith, R. L., Krewski, D. D. and Barbara, A. F. (1996). Commemoration of the 50th anniversary of Monte Carlo, *Human Ecol. Risk Assess.*, **2**, 627–1034.

Crossley, S. J. (2000). Joint FAO/WHO Geneva consultation – acute dietary intake methodology, *Food Additives Contam.*, **17**, 557–562.

Dewhurst, I. C. (2000). The use and limitations of current 'standard' toxicological data packages in the setting of acute reference doses, *Food Additives Contam.*, **17**, 611–615.

Douglass, J. S. and Tennant, D. R. (1997). Estimating dietary intakes of food chemicals, in *Food Chemical Risk Analysis*, Tennant, D. R. (Ed.), Blackie Academic and Professional, London, pp. 195–215.

FAO (1996). 2.6 Assessment of acute dietary risk, *Pesticide Residues in Food – 1995 Report of the JMPR*, FAO Plant Production and Protection Paper 133, Food and Agriculture Organization of the United Nations, Rome, pp. 12–14.

FAO (1999). 2.13 Procedures for estimating an acute reference dose, *Pesticide Residues in Food – 1998 Report of the JMPR*, FAO Plant Production and Protection Paper 148, Food and Agriculture Organization of the United Nations, Rome, pp. 14–17.

FAO (2001a). Annex 5, Proposed tests guideline for studies with single oral doses (for use in establishing acute reference doses for chemical residues in food and drinking water), *Pesticide Residues in Food – 2000 Report of the JMPR*, FAO Plant Production and Protection Paper 163, Food and Agriculture Organization of the United Nations, Rome, pp. 207–215.

FAO (2001b). 4.9 Dinocap, *Pesticide Residues in Food – 2000 Report of the JMPR*, FAO Plant Production and Protection Paper 163, Food and Agriculture Organization of the United Nations, Rome, pp. 67–68.

FAO (2001c). 4.1 Aldicarb, *Pesticide Residues in Food – 2001 Report of the JMPR*, FAO Plant Production and Protection Paper 167, Food and Agriculture Organization of the United Nations, Rome, pp. 23–26.

FAO (2002a). *Submission and Evaluation of Pesticide Residues Data for the Estimation of Maximum Residue Levels in Food and Feed*, FAO Plant Production and Protection Paper 170, Food and Agriculture Organization of the United Nations, Rome, p. 107.

FAO (2002b). *Submission and Evaluation of Pesticide Residues Data for the Estimation of Maximum Residue Levels in Food and Feed*, FAO Plant Production and Protection Paper 170, Food and Agriculture Organization of the United Nations, Rome, p. 91.

Goldman, L. R., Beller, M. and Jackson, R. J. (1990). Aldicarb food poisonings in California, 1985–1988: toxicity estimates for humans, *Arch. Environ. Health*, **45**, 141–147.

Harris, C. A. (2000). How the variability issue was uncovered: the history of the UK residue variability findings, *Food Additives Contam.*, **17**, 491–495.

Hill, A. E. (2000). Residue variability and sampling – practical problems and consequences for residue monitoring, *Food Additives Contam.*, **17**, 539–546.

Kaethner, M. (2001). *Determination of Residues Variability in Head Lettuce following a Tank-Mix Application of Anilinopyrimidine, Triazole, Pyrethroid, Organophosphate, Carbamate and Dicarboximide, Crop Protection Products, France/Germany 2000 to 2001*, Summary Report, Residues Expert Group, European Crop Protection Association, Brussels, Belgium (Unpublished).

Marrs, T. C. (2000). The health significance of pesticide variability in individual commodity items, *Food Additives Contam.*, **17**, 487–489.

PSD (1998). *Report of the International Conference on Pesticide Residues Variability and Acute Dietary Risk Assessment*, The Pesticides Safety Directorate, York, UK (1–3 December, 1998).

Roberts, G. S., Cook, C. R., McAllister, J. T. and Rose, G. (2002). Unit-to-unit variability of residues on apples post-harvest treated with diphenylamine, iprodione and carbendazim, presentation given at the *10th IUPAC International Congress on the Chemistry of Crop Protection*, 4–8 August 2002 Basel, Switzerland, Abstracts of Posters v^2 p. 256, Abstract 6b.10.

WHO (1997a). *Food consumption and Exposure Assessment of Chemicals*, WHO/FSF/FOS/97.5, Report of a FAO/WHO consultation, Geneva, Switzerland, 10–14 February, 1997, World Health Organization, Geneva.

WHO (1997b). *Guidelines for Predicting Dietary Intake of Pesticide Residues (revised)*, WHO/FSF/FOS/97.7 prepared by the Global Environment Monitoring – Food Contamination Monitoring and Assessment Program in collaboration with the Codex Committee on Pesticide Residues, World Health Organization, Geneva.

9 Natural Toxicants as Pesticides

JOHN A. EDGAR

Livestock Industries, CSIRO, Geelong, Australia

NATURAL PESTICIDES: THEIR ORIGINS

Many natural secondary chemicals function as pesticides protecting the producing organisms from predators and competitors (Fraenkel, 1959; Vining, 1990; Ames *et al.*, 1990). This chapter compares and contrasts natural pesticides with synthetic

Pesticide Residues in Food and Drinking Water: Human Exposure and Risks.   Edited by Denis Hamilton and
Stephen Crossley
© 2004 John Wiley & Sons, Ltd   ISBN: 0-471-48991-3

or manufactured pesticides. The relative mammalian toxicity and regulatory safeguards that apply to both are described.

THE RAISON D'ÊTRE OF SECONDARY METABOLITES

The substances that are present in organisms fall into two main categories. Of primary importance are those that are essential to life. The organism would cease to function without them. In addition to these primary substances, some organisms have evolved biosynthetic pathways that produce what have been referred to as secondary substances. These are not involved in vital processes and could be removed without immediately causing death. Their role must, however, be sufficiently important to survival of the organism to compensate for the energy that goes into their production.

More than 125 years ago, Sachs, a pioneer of plant science, made reference to the presence of secondary substances in plants but was unable to ascribe a role for them (Sachs, 1873 – cited in Hartmann, 1999). It has subsequently been shown that at least some secondary substances, if not the majority, have a defensive role or they in other ways enhance the competitiveness of the producing organism. These substances are well tolerated by the organisms that produce them but can be repellent or overtly poisonous to other organisms.

While this chapter focuses on secondary substances that can be considered natural pesticides, this may not be their only role, given that multifunctional substances are favoured by evolution on the grounds of efficiency nor is it necessarily the role of all secondary substances. No attempt has been made to cover the field of natural pesticides in its entirety. Rather, the evolution, relative toxicity, human exposure, risk analysis, and regulatory status of natural pesticides are discussed, using a few specific examples familiar to this author. Comparisons are made with manufactured pesticides.

Evolution of Natural Pesticides

The substances required for life are common to all living things, allowing one organism to 'eat' another in order to obtain the nutrients it needs to live. Plants are a source of nutrients for many organisms, particularly insects. It follows that natural selection of biosynthetic pathways in plants leading to the production of secondary chemicals that repel or kill herbivores is an evolutionary consequence of herbivory (Fraenkel, 1959).

Consider, for example, a plant that is subjected to overwhelming attack by herbivorous insects so that it is in danger of failing to reproduce and becoming extinct. A chance mutation results in a pathway of primary metabolism being diverted to production of a substance or substances that repel or kill insects without endangering the plant and with minimal diversion of energy. The mutated plant will survive while its undefended relatives are eaten to extinction.

Adaptation and Loss of Potency Over Time

Just as some insects evolve resistance to manufactured pesticides so do some insect populations eventually develop resistance to natural pesticides. In the latter case, however, the natural products failing to perform as pesticides cannot be as easily 'withdrawn' from the market. The organisms producing them must suffer the evolutionary consequences of their inadequacy. Some superseded natural pesticides may still confer protection against many herbivores but not those that have become resistant to their effects.

Over time, insects that have been excluded from feeding on a plant by the chance appearance and subsequent selection of hazardous secondary chemicals may develop mechanisms that allow them to tolerate previously repellent or harmful secondary chemicals (Blum, 1982). Mutations in the insect could result, for example, in the ability to digest harmful natural pesticides to harmless products or the ability to selectively excrete these substances, or even an ability to channel them into tissues or vesicles where they can do no harm.

If an insect evolves a capacity to tolerate or deal with secondary substances that are harmful to other insects, this trait will, in a world bristling with defensive secondary metabolites, provide a survival advantage over insects without this capability. Insects with such a tolerance will gain an exclusive food source not able to be exploited by less tolerant organisms. For this advantage to be fully exploited, subsequent mutations which enable tolerant insects to find the plants that utilize the particular chemical defence that they tolerate will be selected for. Evolution of sensory receptors for the secondary substances and orientation behaviour, such as following volatile chemical indicators up wind to plant sources, will be favoured as will triggering of egg laying on the plants that release these volatile indicators (Dethier, 1941; Feeny et al., 1983). This is thought to be the basis of the specificity of some insects in regard to selection of their food plants.

This then is the current status of some natural insecticides in plants. They no longer act as defensive substances against all insects. Rather, they attract some adapted insects that lay their eggs on and exploit such plants as sources of nutrients and more.

Co-evolution

The process of co-evolution, where two (or more) closely interacting organisms evolve stepwise in response to changes in each other, provides a mechanism for growing complexity of secondary substances leading to refinements and improvements in exploiting constitutive weaknesses in each other's defences (Ehrlich and Raven, 1964). The chemical structures of secondary substances in plants and microbes have apparently evolved through co-evolution to be structurally complex and, at least for a time, be amazingly effective in interfering with specific, essential processes of competing organisms. It is difficult, if not impossible, to completely unravel the evolutionary pressures that have influenced the secondary chemistry of modern plants and microbes.

A classical example that illustrates some aspects of co-evolution mediated by plant secondary chemistry is the association of a range of insects with plants producing 1,2-dehydropyrrolizidine alkaloids (pyrrolizidine alkaloids) (Edgar, 1975a, 1984; Boppre, 1986; Hartmann and Ober, 2000).

Pyrrolizidine alkaloids show a very wide but sporadic phylogenetic distribution in plants (Smith and Culvenor, 1981). They are considered to have developed independently in several plant taxa as a result of the mutation of a highly conserved enzyme, deoxyhypusine synthase, found in all eukaryotes and archaebacteria, to homospermidine synthase (Hartmann, 1999). The biosynthesis of pyrrolizidine alkaloids requires the rare polyamine homospermine which is synthesized by homospermidine synthase. The mutation of deoxyhypusine synthase to homospermidine synthase and the consequent capacity to elaborate pyrrolizidine alkaloids has apparently occurred independently in a number of plants. Pyrrolizidine alkaloids deter many insect herbivores, thus suggesting that insects may have provided an important selection pressure for the retention of the biosynthesis pathway leading to pyrrolizidine alkaloids in these plants. However, there are many present-day insects that have evolved a tolerance to pyrrolizidine alkaloids and some insects that seek out pyrrolizidine alkaloids and depend upon them.

A number of insect species, e.g. moths belonging to the family Arctiidae, feed as larvae exclusively on pyrrolizidine-alkaloid-containing plants. Some insects that feed on pyrrolizidine-alkaloid plants as larvae also store these plant secondary substances in adult tissues to deter predators such as spiders (Aplin et al., 1968; Benn et al., 1979; Dobler et al., 2000; Eisner, 1982; Brown, 1984). Other insects not only feed on pyrrolizidine alkaloid sources and store pyrrolizidine alkaloids in their tissues but they also use volatile metabolites of pyrrolizidine alka- loids as pheromones (chemicals used in communication) involved in courtship behaviour (Culvenor and Edgar, 1972; Conner et al., 1981). There are also but- terflies of the sub-families Danainae and Ithomiinae that do not feed as larvae on pyrrolizidine alkaloid plants but, as adults, they seek out and find sources of pyrrolizidine alkaloids. They ingest and store pyrrolizidine alkaloids in their tissues for defence and to elaborate courtship pheromones (Edgar and Culvenor, 1974; Pliske and Eisner, 1969; Edgar, 1975a,b; Edgar et al., 1971, 1973, 1976a, 1976b, 1979; Brown, 1984, 1987).

One particularly remarkable role of pyrrolizidine alkaloids is their morphogenic action in moths of the genus *Creatonotos*. The size of the pheromone dissem- inating organs in adult male *Creatonotos*, as well as the level of pyrrolizidine alkaloid-derived pheromone, is determined by the amount of pyrrolizidine alka- loids the insects acquire as larvae (Boppre and Schneider, 1985, 1989).

It is interesting to speculate on, and to seek evidence of how some of these pyrrolizidine alkaloid dependencies have evolved. A sequence can be envisaged from the initial role of pyrrolizidine alkaloids in plants as a defence against herbivores and the subsequent development of tolerance to pyrrolizidine alkaloids in some insects. Storage of pyrrolizidine alkaloids in insect tissues as a defence against predators and evolution of sensory mechanisms for pyrrolizidine alkaloids in adapted species to enable them to locate plants containing pyrrolizidine alkaloids follow, leading to development of volatile pyrrolizidine alkaloid metabolites released by males as pheromones to demonstrate 'fitness' to potential mates whose offspring would inherit the survival advantages conferred by pyrrolizidine alkaloids.

Evidence for this sequence in butterflies of the sub-families Danainae and Ithomiinae has come from the discovery of some of the predicted evolutionary stages among their present-day plant associations and behaviours (Edgar *et al.*, 1974; Edgar, 1975a, 1975b, 1982, 1984).

That pyrrolizidine alkaloids are still produced by many plants indicates that, despite attracting a number of pyrrolizidine-alkaloid-tolerant and pyrrolizidine alkaloid-dependent insect species, they retain some ability to repel non-adapted herbivores and provide benefits for the survival of the plants producing them.

CURRENT USE OF NATURAL PESTICIDES BY MAN

Secondary substances associated with plants and microbes have been known for thousands of years to have useful properties, including value as pesticides. There is considerable research directed at identifying natural products from plants and microbes that can be safely used as pesticides. Several commercial pesticides approved for use in a number of countries come directly from plants or microbes. These include pyrethrins, rotenone, sabadilla, ryania and neem insecticides, for example, and the avermectin and milbemycin anthelmintics, insecticides and acaricides (Hedin and Hollingworth, 1997).

Many other natural pesticides are being, or have been used, as lead compounds for development of synthetic or semi-synthetic pesticides (Hedin and Hollingworth, 1997).

Among the characteristics that must be met before natural pesticides and their synthetic analogues can be commercially exploited is the primacy of safety to humans. Nature does not ensure safety except to the producing organism so that the vast majority of pesticides in nature are hazardous to humans. If natural pesticides are being considered for use in a food production system, or they occur as intrinsic components of plants consumed as foods or they occur as contaminants in foods, they should be subjected to risk assessment. If a risk to public health is demonstrated, as with other hazardous substances in food, risk communication and risk management strategies should be developed and applied in the interest of public health and safety (FAO/WHO, 1995).

MYCOHERBICIDES AND OTHER BIO-PESTICIDES

Isolating the active principles from organisms with known pesticide activity for development as manufactured pesticides is one approach to generating new pesticides. Another approach has been to consider the microbial enemies of pests directly as biological control agents without identifying the active principles. Strategies for using such bio-pesticides generally involve application of massive doses of inoculum to create a fast and high-level epidemic, e.g. in a weed population (Charudattan, 1982, 1991). Such strategies could, however, prove to be hazardous if the active components (natural pesticides) produced by the bio-pesticide are not known and particularly if the mechanisms that are involved are not understood.

For example, the fungus *Phomopsis emicis* was being investigated as a possible mycoherbicide for the weed *Emex australis*. Work ceased when it was demonstrated that the fungus is a producer of hazardous phomopsin mycotoxins that are subject to food standards in Australia. It was considered that there was a risk that phomopsins could contaminate the human food chain if *P. emicis* was applied widely as a mycoherbicide in agricultural production systems (Shivas *et al.*, 1994a, 1994b; ANZFA, 2000). The phomopsin mycotoxins inhibit plant growth (Edgar, unpublished) and could be the active principles that contribute to the herbicidal activity of *Phomopsis* species such as *P. emicis*.

Biological control pathogens, e.g. potential mycoherbicides such as *P. emicis*, need to be shown to be toxicologically safe prior to field evaluation. They are not inherently safer than manufactured herbicides as is sometimes popularly believed.

REGULATORY CONSIDERATIONS

For regulatory purposes, poisonous natural pesticides in foods, that are not intrinsic components of the food, fall into the category of chemical 'contaminants'. In contrast, approved and registered manufactured pesticides that are deliberately used in food production form chemical 'residues'. This important difference in purpose or lack of it determines not only terminology but also how each is dealt with in a regulatory sense.

SETTING INTERNATIONAL STANDARDS FOR MANUFACTURED AND NATURAL PESTICIDES[1]

The Codex Alimentarius Commission is an international, inter-governmental body aimed at facilitating international trade while protecting the health of consumers. It establishes internationally agreed Food Standards for both manufactured and natural pesticides as well as for other hazards such as veterinary chemical residues and microbial contaminants (FAO/WHO, 1995, 1997, 1998). Codex Alimentarius Commission Standards are an important element in conducting international trade in agricultural products and foods under the Sanitary and Phytosanitary Agreement of the General Agreement on Tariffs and Trade. The Sanitary and Phytosanitary Agreement aims to reduce arbitrary trade restrictions

[1]See also Chapter 10.

based on health grounds by placing emphasis on scientific assessment of risk for specified hazards such as chemicals.

Two Codex Alimentarius Commission committees, i.e. the Codex Committee on Pesticide Residues (CCPR) and the Codex Committee on Food Additives and Contaminants (CCFAC) are, respectively, responsible for identifying manufactured and natural pesticides for risk analysis. They also provide advice to the Codex Alimentarius Commission on appropriate Standards for these potential food-borne health hazards.

The CCFAC requests and receives scientific risk assessment advice on poisonous natural pesticide contaminants of foods from the Joint FAO/WHO Expert Committee on Food Additives and Contaminants (JECFA). The latter is not a component of the Codex Alimentarius Commission but made up of independent scientific specialists serving as individuals and not as representatives of their governments or their employers. The JECFA is thus able to focus on scientific risk assessment and avoid possible conflicts of interest where trade issues are concerned.

The Joint FAO/WHO Meeting on Pesticide Residues (JMPR) provides parallel scientific risk assessment advice on manufactured pesticides to the CCPR.

As well as responding to requests from the CCFAC and CCPR, the JECFA and JMPR sometimes also consider and evaluate direct requests from governments, other interested organizations and pesticide manufacturers to undertake risk assessment of particular chemicals.

SETTING PRIORITIES

There are many substances in food that could justifiably be subject to risk analysis in the interest of public health and safety. Ranking of hazards for risk assessment and risk management priorities is a challenging task that must take into account many factors. As well as scientific considerations, the public perception of risk, political issues and willingness to tolerate risk are taken into account in setting priorities.

These latter factors, in particular, appear to have tilted food safety concerns towards manufactured pesticides and away from natural pesticides. While it is evident that the hazardous properties of substances are linked to their chemical structure and not their origin, in the public mind 'chemicals' (manufactured pesticides) are of greater concern than 'natural products' (natural pesticide contaminants). Thus, a pesticide created by nature is deemed to be 'safer' even though this view does not withstand scientific scrutiny (Ames et al., 1990).

WHO INITIATES AND PAYS FOR RISK ASSESSMENT AND MONITORING OF NATURAL PESTICIDES?

In the case of manufactured pesticides the industries that are to profit from sale of the pesticide provide the primary driving force for risk analysis to be undertaken and completed. The equivalent drivers in the case of poisonous natural pesticides are probably consumer groups, if they are aware of the potential hazard that natural pesticides represent. There are in both cases other interested parties and stakeholders. Governments and government agencies, for example, play a key role and bear a significant duty of care for the health of their citizens and others.

Pesticide manufacturers are compelled by government requirements to undertake the research that demonstrates safety and efficacy, as well as providing the analytical tools needed for government agencies to monitor pesticides in the environment to ensure they remain at safe or acceptable levels. The manufacturers of pesticides are motivated to provide the toxicological and other data required for registration of their products by the prospect of profiting from sales of those products.

On the other hand, there can be a disincentive for agricultural industries and sometimes governments to adequately address natural pesticide contamination issues. This is especially the case when profit margins for agricultural products are small and it is perceived that economic losses caused by natural pesticides, for example, livestock poisoning, can be tolerated. Potential human food safety issues related to natural pesticide contamination can easily be discounted, especially if latent, long-term chronic effects rather than short-term acute toxic effects, as is commonly the case, are the likely consequences of human exposure.

Considerable public expenditure is usually required to show scientifically that a natural pesticide does not represent a threat to health. Risk management, if shown to be required, can further erode profits for farmers. Without consumer or government awareness and action, poisonous natural pesticide contamination issues can remain without being properly investigated for many years.

The 'carrot' of profit and the 'stick' of government regulations ensure food safety in the case of manufactured pesticides. Poisonous natural pesticides are more dependent on the 'stick' alone and the motivation to use the 'stick' may not be particularly strong if there is little or no consumer concern.

DATA FOR RISK ASSESSMENT: MANUFACTURED PESTICIDES VERSUS NATURAL PESTICIDES

The JECFA and JMPR request quantitative toxicology and exposure data from interested parties to enable a risk assessment of particular natural and manufactured pesticides to be performed. Data of an acceptable international standard, generated by using approved protocols and analytical methods, are essential in undertaking an adequate scientific risk assessment (FAO/WHO, 1995).

Toxicological data provided by industry and government chemical registration agencies make the task easier for manufactured pesticides. In contrast, attention is generally drawn to hazardous natural pesticides as a result of poisoning of livestock. In this case, agricultural industries and government agricultural authorities are responsible for generating toxicity data. Willingness to provide funds for this is judged against the economic benefit for the agricultural production system. Public health and safety is often a secondary concern. The toxicity data that come from investigating natural pesticide poisoning of livestock is therefore generally inadequate for more than a qualitative risk assessment if there is a perceived risk to humans from contamination of agricultural products. Evidence of human health effects or fatalities caused by natural pesticides is usually required before adequate toxicity data are generated and even this is sometimes not sufficient motivation.

Dietary exposure data for natural pesticides can also be inadequate in comparison to the available data for manufactured pesticides. In the case of manufactured pesticides, their level in representative diets is routinely measured on a national basis to confirm that agreed maximum residue limits (MRLs) and acceptable daily intake (ADI) values are not exceeded. This is carried out for both trade and domestic public health and safety reasons. Several natural pesticide mycotoxins such as aflatoxins, for which Codex Alimentarius Commission Food Standards exist, and others, such as ochratoxin A and fumonisin B, currently being considered by the CCFAC, are also monitored in some countries. However, there are other poisonous natural pesticides, such as pyrrolizidine alkaloids, that are only occasionally measured in foods despite considerable evidence suggesting that monitoring may be justified (see below).

Indeed, sufficiently sensitive, reliable and validated analytical methods for determining the distribution of poisonous natural pesticides in foods may not be available.

REGULATORY LEVELS: COMPARISON OF NATURAL AND MANUFACTURED PESTICIDES

Maximum residue limits (MRLs) are specified for manufactured pesticides in agricultural products. Such limits are based on residue levels that are present in an agricultural product when a pesticide is used according to the principles of Good Agricultural Practice (GAP). They are therefore not food safety standards *per se* as is commonly assumed by the public. They are set after it has been established on the basis of determination of an acceptable daily intake (ADI) that the known toxicological hazard, if the product is used as directed on the label, do not constitute a risk to human health. MRL setting processes provide a very conservative margin of food safety for consumers (FAO/WHO, 1995).

Rather than having an ADI, poisonous natural pesticide contaminants are determined to have a Provisional Tolerable Intake (PTI) (FAO/WHO, 1995). The term 'tolerable' is considered to be more appropriate for contaminants because it signifies permissibility rather than acceptability. 'Provisional' indicates the tentative nature of most evaluations (FAO/WHO, 1995). This is due to the normal paucity of toxicity data for natural pesticides in comparison to the data for manufactured pesticides and the likelihood of the value being changed as new toxicological data become available (Speijers, 1995).

This latter point is clear evidence of the relative neglect of poisonous natural pesticides in food compared to manufactured pesticides where a full toxicological data set is required before a product is considered for registration and sale.

PTI values, like ADIs, are based on the determination of a No-Observable-Adverse-Effect Level (NOAEL), usually in a rodent or the most sensitive animal species (FAO/WHO, 1995). To convert an NOAEL to a PTI involves safety factors, typically a factor of ten for the work being done using experimental animals rather than humans and another factor of ten to allow for biological variation between people. Thus, the PTI values are usually 100 times less than the NOAEL value (IPCS, 1987). Conversion of PTI values to maximum levels or maximum permissible concentrations in particular foods involves consideration of food consumption patterns of groups at greatest risk (FAO/WHO, 1995).

TOXICITY THRESHOLDS: NATURAL VERSUS MANUFACTURED PESTICIDES

NOAELs are premised on there being a threshold of toxicity, i.e. a dose level below which no significant health risk exists. Examples of toxic substances without a threshold level include genotoxic carcinogens, that is, chemicals that directly cause genetic mutations leading to cancer. Theoretically, a single molecule of a genotoxic carcinogen could cause a mutation that is ultimately fatal. It is extremely unlikely that a manufactured pesticide without a toxicity threshold, e.g. a genotoxic carcinogen, would ever be registered for use.

A number of poisonous natural pesticides are known to be genotoxic carcinogens. These include aflatoxins and pyrrolizidine alkaloids. In the case of natural pesticides that are genotoxic, it is recommended that the contaminant be kept 'as low as reasonably achievable', ideally zero (FAO/WHO, 1995).

RELATIVE RISKS FROM EXPOSURE TO NATURAL AND MANUFACTURED PESTICIDES

MECHANISMS OF ACTION AND MAMMALIAN TOXICITY OF POISONOUS NATURAL PESTICIDES

The evolutionary forces at play in the biosynthesis of natural pesticides in plants and microbes are directed at protecting the plant or microbe from all possible competitors and enemies. The most effective and durable natural pesticides will be those directed at highly conserved biochemical targets, that is, at the most vital mechanisms that are common to all living things. Thus, a high proportion of natural pesticides are effective against a range of organisms. For example, some natural pesticides combine antiviral, antibacterial, insecticidal and herbicidal properties, as well as being toxic to animals, including humans.

The broad-spectrum biological activities of some natural pesticides are illustrated below using the previously mentioned pyrrolizidine alkaloids

(Figure 9.1), the corynetoxins (Figure 9.2) and the phomopsin mycotoxins (Figure 9.3) as examples. These three types of natural pesticides are produced, respectively, by certain plants, bacteria and fungi that occur in agricultural production systems. They are all known to be potential contaminants of foods. Of these, only the phomopsin mycotoxins are subject to food safety regulations and then only in Australia (FAO, 1997).

Pyrrolizidine Alkaloids

Drug-metabolizing enzymes (cytochromes P450) in the mammalian liver that normally lead to detoxication and clearance of foreign substances convert pyrrolizidine alkaloids to chemically reactive pyrrolic metabolites (Mattocks, 1968, 1986; Culvenor et al., 1969; Jago et al., 1970). The metabolites produced in the liver are biological alkylating agents. They react chemically, shortly after formation, with nucleophilic sites on DNA and proteins. Because the pyrrolizidine

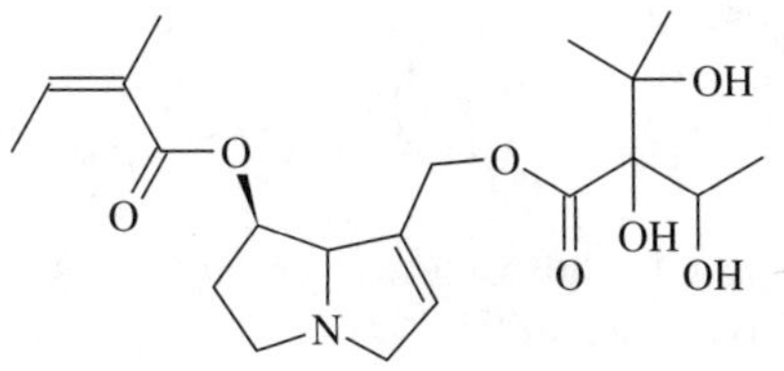

Figure 9.1 Structure of echimidine, a typical 1,2-dehydropyrrolizidine found in the plant genera *Echium* and *Symphytum*

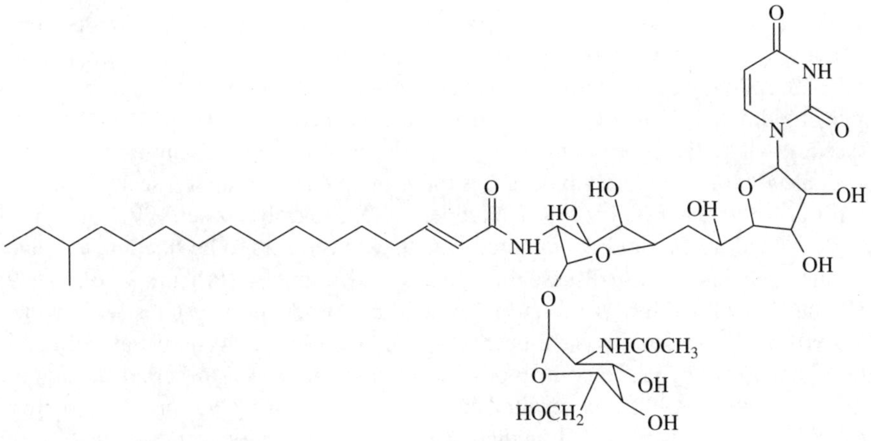

Figure 9.2 Structure of corynetoxin U17a, one of several corynetoxins produced by the bacterium *Rathayibacter toxicus*

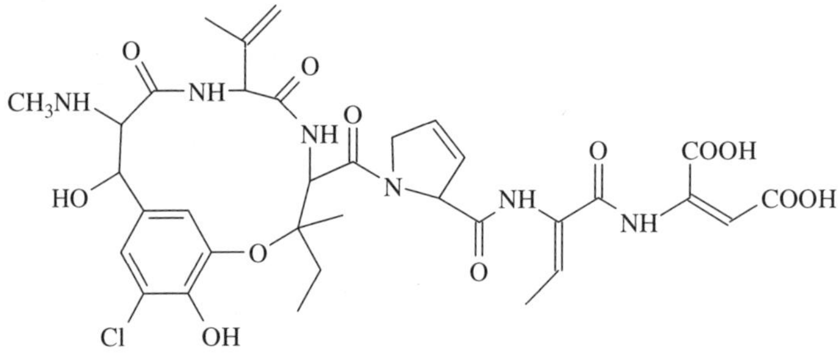

Figure 9.3 Structure of phomopsin A, a mycotoxin produced by the fungus *Diaporthe toxica*

alkaloid metabolites are bifunctional, cross-links can be formed between macro-molecules such as DNA and proteins (Curtain and Edgar, 1976; Petry *et al.*, 1984, 1986; Hincks *et al.*, 1991; Yang *et al.*, 2001a, 2001b). In mammals, the chemically reactive pyrrolizidine alkaloid metabolites primarily cause damage in the liver where they are first formed but some longer-lived pyrrolizidine alkaloid metabolites cause damage to lungs and brain (IPCS, 1988; Huxtable and Cooper, 2000).

Pyrrolizidine alkaloids can cause acute liver damage in animals that are exposed to them in their diet (Peterson and Culvenor, 1983; Gaul *et al.*, 1994). In humans, the classical acute symptoms are abdominal pain and rapidly developing accumulation of fluid in the abdominal cavity (ascites), along with lassitude, diarrhoea, oedema, emaciation, liver enlargement, spleen enlargement and mild jaundice (IPCS, 1988; Prakash *et al.*, 1999). Acute mortality in human poisoning, e.g. from grain contamination, is reported to be 15–20 % (IPCS, 1988; Mayer and Lüthy J, 1993). There are also chronic toxic effects attributable to the gene-damaging properties of pyrrolizidine alkaloids that manifest long after a single dose or following long-term, low-level exposure. The known chronic effects of pyrrolizidine alkaloids include cirrhosis and cancer (IARC, 1976). Veno-occlusive disease, where the centrilobular veins and the smaller vein tributaries of the liver are partially or completely blocked, is the principal manifestation seen in humans (Hill *et al.*, 1959; IPCS, 1988; Huxtable, 1989; Prakash *et al.*, 1999).

Pyrrolizidine-alkaloid-containing plants have been said to be 'the leading plant toxins associated with disease in humans and animals' (Prakash *et al.*, 1999; Huxtable, 1989). They are found in agricultural production systems worldwide.

Pyrrolizidine alkaloids have been subjected to a risk analysis in several countries, not for their presence in foods but for their presence in herbal medicines. In Germany, in 1992 this resulted in regulations limiting the presence of pyrrolizidine alkaloids in herbal medicines to 0.1 mg per daily oral dose and 10 mg per day for topical use (German Federal Health Bureau, 1992). The German regulations specify that such products should not be prescribed for pregnant and lactating women because of the known susceptibility of foetuses and infants. Higher levels are allowed, e.g. 1 mg orally and 100 mg topically, where use is limited to a maximum of six weeks per year. Similar regulations on pyrrolizidine alkaloids in herbal products exist in Switzerland and Austria, and some other countries (Roeder, 2000).

These herbal regulations contrast with the lack of regulations on pyrrolizidine alkaloids in foods such as grains, honeys, milk, eggs and meat where levels considerably in excess of these limits can occur (see below) (Edgar and Smith, 1999).

Corynetoxins

Corynetoxins (see, for example, Figure 9.2) are produced by a plant pathogenic bacterium, *Rathayibacter toxicus* that colonizes nematode-induced galls in the seed heads of a pasture grass *Lolium rigidum* (annual ryegrass) and some other grass species (Edgar *et al.*, 1982, 1994; Frahn *et al.*, 1984; McKay and Ophel, 1993). Corynetoxins and their closely related and toxicologically equivalent analogues, e.g. the tunicamycins, mimic the transition-state of the substrates of the first enzyme involved in the lipid-linked *N*-glycosylation of proteins (Tkacz, 1983). The toxins occupy and bind tightly, essentially irreversibly, to the active site of the enzyme, effectively preventing formation of glycoproteins (Jago *et al.*, 1983; Jago, 1985; Stewart *et al.*, 1997).

Given their mechanism of action, it is not surprising that corynetoxins, and their closely related analogues the tunicamycins, display powerful antiviral and antibacterial, as well as mammalian, toxicity (Jago *et al.*, 1983). These toxins also have an inhibitory effect on protein glycosylation in plants (Ericson *et al.*, 1977; Elbein, 1980) and influence plant growth (Edgar, unpublished).

The antibiotic properties of the corynetoxins, tunicamycins and related toxins (e.g. streptovirudins, antibiotic MM19290 and mycospocidins) (Cockrum and Edgar, 1983; Eckardt, 1983) no doubt provide an advantage to the microbes producing them against microbial competitors. In the case of *R. toxicus*, the corynetoxins may also facilitate invasion of host plants by disrupting plant glycoprotein biosynthesis, thus allowing the toxin-producing microbes to grow at the expense of plant tissues (Scheffer and Livingston, 1984).

In animals, some aspects of acute corynetoxin poisoning are consistent with reticuloendothelial (blood forming) system failure and especially vascular leakage, blood vessel constriction, pin-point haemorrhages and accumulation of injected colloid in the lungs rather than the liver. Serum levels of the multifunctional glycoprotein fibronectin are among the glycoproteins depressed by corynetoxins and a reduced concentration of fibronectin is believed to be the cause of reticuloendothelial system failure (Jago, 1985). Many animals, however, die without manifesting any pathological changes (Edgar, 1993). Given the systemic importance of glycoproteins, there could be a number of possible causes of death (Tkacz, 1983).

In livestock (cattle, sheep and pigs), acute corynetoxin poisoning generally manifests itself as a neurological disease. Animals stagger, collapse, convulse and show other nervous signs, including head nodding, fore-limb, hind-limb and digital extension, and convulsions (Berry and Wise, 1975). Neurological effects can result from capillary obstruction and localized anaemia (ischemia) of brain tissues (Berry *et al.*, 1980; Leaver *et al.*, 1988) or other, as yet unidentified, mechanisms. There may, for example, be a direct toxic effect on peripheral (sympathetic) neurones. Thus, critically important glycoprotein receptors, such as the receptor for nerve growth factor that is essential for the maintenance of neurones, are not produced in the presence of these toxins and programmed cell death (apoptosis) of neurones ensues when neurone cell cultures are exposed to them (Chang and Korolev, 1996).

Male rats suffered irreversible infertility after being given tunicamycins, toxicologically equivalent analogues of corynetoxins, at a dose that caused no clinical sign of poisoning (Peterson *et al.*, 1996). Tunicamycins, by preventing protein N-glycosylation, also interfere with embryo development. The maintenance of pregnancy is reliant on N-linked glycoproteins (Surani, 1979; Lennarz, 1993; Ioffe and Stanley, 1994) and tunicamycins, when administered on day 15 of pregnancy, caused abortions and death of pregnant rats (Stewart *et al.*, 2002).

Lehmann and Harris (1997) showed that interference with protein N-glycosylation by tunicamycins predisposed prion protein to adopt the pathogenic form associated with spongiform encephalopathy.

Being amphipathic (having both fat-soluble and water-soluble properties), tuni-camycins and corynetoxins accumulate *in vivo* in membranes (Kuo and Lampen, 1976). A lethal oral dose of corynetoxins in sheep, between 3.2 and 5.6 mg per kg body weight, can be given as a single dose or as multiple small doses over a period of months (Jago and Culvenor, 1987; Than *et al.*, 1998a). Two half-lethal oral doses given up to nine weeks apart have caused death in sheep (Jago and Culvenor, 1987). By subcutaneous injection, the approximate LD_{50} of coryne-toxins in sheep is 20 to 25 µg per kg body weight (Than *et al.*, 1998a). This amount can be administered either as a single dose, causing deaths to occur after a delay of three to four days, or at a rate of 1 µg per kg body weight per day for five days a week. Deaths begin to occur when the total dose passes 20 µg per kg body weight (Than *et al.*, 1998a).

No food safety risk assessment of corynetoxins has been published although some of the quantitative toxicity data available in the scientific literature suggest that an oral NOAEL in sheep is between 5 and 7 µg per kg body weight per day (Jago and Culvenor, 1987; Davies *et al.*, 1996). Using the normal safety factor of 100 (10 for non-human species and 10 for biological variation) this suggests a provisional tolerable daily intake (PTDI) of 0.05 or 0.07 µg per kg body weight per day, a low level if these poisonous natural pesticides contaminate foods as seems possible (see below).

Phomopsin Mycotoxins

The phomopsin mycotoxins (see, for example, Figure 9.3), like the corynetoxins, also target a fundamental mechanism *in vivo*. They are considered to be among the most potent anti-microtubule agents known (Lacy *et al.*, 1987; Luduena *et al.*, 1990; Bai *et al.*, 1990). Microtubules are intra-cellular filaments formed by the assembly of tubulin proteins. By binding to a key site on tubulin, phomopsins cause depolymerization of, and prevent the assembly of, microtubules (Tonsing *et al.*, 1984). This mechanism gives them considerable potential to adversely affect all eukaryotes that are dependent on microtubules for such critical processes as cell division and cytoplasmic organization.

In animals, the phomopsins are liver-damaging, embryotoxic and carcinogenic (Peterson, 1978, 1983, 1986, 1990) and also adversely influence plant growth at low concentrations (Edgar, unpublished), thus suggesting a possible role in colonization of the plant by the fungus (Scheffer and Livingston, 1984; Shankar *et al.*, 1999).

A maximum level of phomopsins in lupin seed for human consumption of 5 parts per billion (5 μg per kg) has been determined by the Australia New Zealand Food Authority (ANZFA, 2000).

Pyrrolizidine alkaloids, corynetoxins and phomopsin mycotoxins provide three examples demonstrating that natural pesticides that have evolved through natural selection will not necessarily be benign to mammals generally or humans in particular. Natural pesticides are not safer than manufactured pesticides. While nature has provided a rich repository of pesticides, they need to be assessed by the risk analysis processes that are currently applied to all chemicals found in the human food chain according to the international mechanisms that are in place and have been described above.

STABILITY OF POISONOUS NATURAL PESTICIDES

Biodegradability is a desirable characteristic of manufactured pesticides. This is not, however, a characteristic that nature necessarily selects for in the evolution of natural pesticides. The latter can be highly resistant to biodegradation and to temperature, pressure and fermentation. Thus, pasteurization and cooking, effective means of decontaminating foods harbouring harmful micro-organisms, do not normally reduce to any significant degree the toxicity of low-molecular-weight natural pesticides or indeed many manufactured pesticides.

A good example, in the case of natural pesticides, is the poisonous properties of bread baked using flour contaminated by pyrrolizidine alkaloids. Such bread has been responsible for large-scale poisonings of people in a number of countries (Steyn, 1933; IPCS, 1988; Chauvin *et al.*, 1993). Corynetoxins too have resisted autoclaving and years of storage at ambient temperature (Stuart *et al.*, 1994). Phomopsin mycotoxins are also stable to chemical and temperature degradation and they also have resisted food processing (Cockrum and Edgar, 1994).

HUMAN EXPOSURE TO POISONOUS NATURAL PESTICIDES VIA FOOD

Natural pesticides produced by plants and microbes are widely distributed in food production systems. They can enter the food supply chain in many ways. However, when compared to the data that exist for manufactured pesticides, their distribution and concentration in foods, with the exception of data for a small number of mycotoxins, are not well documented.

IDENTIFYING NATURAL PESTICIDES FOR FOOD SAFETY RISK ANALYSIS

Livestock are exposed to natural pesticides in pastures and feed grain. All of the mycotoxins currently subjected to national and international regulations, and those being considered by the CCFAC, cause livestock poisoning. In many cases, the attention of food safety regulators is first drawn to hazardous natural pesticides as a result of livestock poisoning episodes. Aflatoxins, for example, were first recognized and investigated when peanut meal caused poisoning of turkeys (Allcroft, 1969). The fumonisins were also identified as a cause of disease in animals and subsequently recognized as a food safety threat for humans (Marasas, 1995).

Thus, where natural pesticides cause significant poisoning of food-producing animals it is apparent that they are both poisonous and that there is opportunity for them to be transferred into animal products used as foods (meat, milk and eggs).

There are several natural pesticides likely to contaminate foods that warrant risk analysis on the basis of their poisoning of livestock even though there is currently no evidence of human illness that can be attributed to them. The corynetoxins and phomopsin mycotoxins, described above, are examples of these. There are others, for example, pyrrolizidine alkaloids, that are not only known to poison livestock (Peterson and Culvenor, 1983) but they also have a long record of causing significant outbreaks of human poisoning (IPCS, 1988; Huxtable, 1989; Chauvin *et al.*, 1993; Mayer and Lüthy, 1993).

CONTAMINATION OF GRAINS, OIL SEEDS AND ANIMAL PRODUCTS BY POISONOUS NATURAL PESTICIDES

Some poisonous natural pesticides enter the human food chain as contaminants of staple foods. Pyrrolizidine alkaloids, for example, are contaminants of cereal crops in many countries and contaminated grains have caused acute pyrrolizidine alkaloid poisoning with high mortalities in humans involving many thousands of people in cases where grain quality standards have been lax (IPCS, 1988). While a number of countries have product integrity standards for pyrrolizidine-alkaloid-containing seeds in grain, it is easy to calculate that the levels of pyrrolizidine-alkaloid-seeds allowed by these standards would amount to hundreds of micrograms of pyrrolizidine alkaloids per kilogram of grain. This would make such grain unsafe according to the regulations for pyrrolizidine alkaloids in herbal medicines in countries such as Germany where the maximum daily intake is 0.1 μg per day. For example, it was calculated by an expert committee convened by the International Programme on Chemical Safety (IPCS, 1989) that guidelines in the former USSR for a tolerance of 0.1 % for *Heliotropium lasiocarpum* seeds in stored grains could result in 1820 μg of pyrrolizidine alkaloids per kg of grain, equivalent to an intake of approximately 300 μg per day for consumption of 150–200 g cereal per day (cf. German herbal medicine regulations allowing a maximum of 0.1 μg of pyrrolizidine alkaloids per day). Similar grain standards for contamination by pyrrolizidine-alkaloid-containing seeds are found in other countries.

The presence of contaminating pyrrolizidine alkaloid seeds in grains is only an indication that pyrrolizidine alkaloid plants were present in the crop at harvest. Dust from the plants and from broken seeds adhering to grain is also a major source of poisonous natural pesticides such as pyrrolizidine alkaloids and complete removal of contaminating seeds to meet grain quality standards or prior to milling does not necessarily render grain safe (Edgar and Smith, 1999).

Honey is also a significant dietary source of pyrrolizidine alkaloids. Honey from the flowers of plants that produce pyrrolizidine alkaloids has been shown in some cases to contain more than 3900 μg of pyrrolizidine alkaloids per kg of honey (Deinzer *et al.*, 1977; Culvenor *et al.*, 1981; Roeder, 1995; Crews *et al.*, 1997). These honeys could not be eaten without the consumer exceeding the German herbal medicine standard which allows a maximum of 0.1 μg of pyrrolizidine alkaloids per day. Pregnant and lactating women in Germany would not be allowed to consume these honeys if they were herbal medicines. However, the standards for pyrrolizidine alkaloids in herbal products do not apply to foods although there is no logical reason why they should not. Ethical issues are raised if pyrrolizidine-alkaloid-containing honey is diluted by honey free of pyrrolizidine alkaloids to meet the standards extant for herbal medicines. In a practical sense too, honeys completely free of pyrrolizidine alkaloids may not always be available in the quantities required to achieve the desired dilution given that pyrrolizidine-alkaloid-producing plants have been estimated to represent 3 % of all flowering plants (Smith and Culvenor, 1981).

Studies on the plants used in honey production reported in the scientific literature demonstrate that pyrrolizidine-alkaloid-containing plant genera, for example, *Symphytum, Echium, Borago, Senecio, Crotalaria, Ageratum, Cynoglossum, Eupatorium, Heliotropium* and *Chromolaena* species, are used for honey production (Edgar *et al.*, 2002). Some of these plants are recommended for commercial honey production on the basis of the quantity and appearance, flavour and other qualities, of the honey they produce (e.g. Pel'menev *et al.*, 1983; Lozano *et al.*, 1995; Thapa and Wongsiri, 1997; Aira *et al.*, 1998). Honeys from these plants can be expected to contain pyrrolizidine alkaloids.

Eggs (Edgar and Smith, 1999) and milk (Schoental, 1959; Dickinson *et al.*, 1976; Lüthy *et al.*, 1983; Candrian *et al.*, 1984; Molyneux and James, 1990) from animals exposed to pyrrolizidine-alkaloid-containing feed can be contaminated to the extent of hundreds of μg of pyrrolizidine alkaloids per kg and toxic effects have been seen in young animals through mothers' milk (Schoental, 1959). Goats' milk is suspected of causing pyrrolizidine alkaloid poisoning in human infants (Hippchen *et al.*, 1986). Naturally contaminated human milk has also caused pyrrolizidine alkaloid poisoning of an infant (Roulet *et al.*, 1988).

Shevchenko and Fakhrutdinova (1971) reported that meat from cows fed *Trichodesma*, a pyrrolizidine-alkaloid-containing genus, when fed to dogs, caused signs of poisoning within three to four months, resulting subsequently in death or irreversible pathological changes. The milk from these cows also poisoned dogs and calves.

Corynetoxins, being cumulative, membrane-associating substances, could also enter the human food chain as contaminants of animal products such as meat and milk although no poisoning has been attributed to either of these sources (Bourke and Carrigan, 1993). Levels of biologically available corynetoxins have not been measured in milk or meat because suitable analytical methods have until recently not been available (Than *et al*, 1998b). It has been observed, for example, that livestock can accumulate three quarters of a lethal dose without showing overt signs of poisoning that would otherwise exclude them and their products from entering the human food chain. A lethal dose also takes three to four days before animals show any behavioural signs of poisoning, giving the opportunity for lethally poisoned animals to be slaughtered for human consumption (Edgar, 1997a, 1997b).

While there appears to be some opportunity for corynetoxins to enter the human food chain via animal products, of greater concern is the potential for corynetoxins entering the food chain directly in grains, such as wheat and barley, and oil seeds, such as canola (Edgar, 1997a, 1997b).

The principal host grass for *R. toxicus, Lolium rigidum*, is a significant and persistent weed of crops in Australia and corynetoxin poisoning has been reported in stock with access to cereal crops and stubble (Berry and Wise, 1975; Chapman, 1989; Anon, 1994). Harvested grain (wheat, barley and canola) can also be contaminated by poisonous annual ryegrass seeds (Chapman, 1989; Fisher *et al.*, 1979). Farmers are warned to exercise caution in feeding crop 'fines', produced by 'cleaning' grain of *Lolium rigidum* seeds, to livestock because of the danger of corynetoxin poisoning (Anon, 1994; Allen and Roberts, 1996). This raises the question of how much corynetoxin remains in the 'cleaned' grain entering the human food chain (Edgar, 1997a, 1997b).

Lupin seeds are the only food commodity known to be contaminated by phomopsin mycotoxins (Edgar, 1991). However, *Phomopsis* and *Diaporthe* species are common pathogens of many other plants used as foods (Uecker, 1988). Some of these could potentially also be contaminated by phomopsin mycotoxins.

These examples illustrate that reasonably well-studied, extremely poisonous natural pesticide contaminants of foods are less effectively managed than are lower-risk manufactured pesticides. Manufactured pesticides with the toxicological characteristics and food contamination potential of the pyrrolizidine alkaloids, corynetoxins and phomopsins can be expected to have undergone a thorough risk analysis and would be subjected to effective risk management, including extensive dietary exposure monitoring to ensure food safety.

INTRINSIC NATURAL PESTICIDES

Natural pesticides can be intrinsic to the vegetables and fruits we consume. Well known examples include glycoalkaloids in potatoes and furanocoumarins in celery. Inherent natural pesticides have in general been reduced to below toxicity thresholds in human food plants by selective plant breeding. However, while intrinsic natural pesticides in established foods may be of minor concern, it is important that all new food plant varieties, whether selected by classical plant breeding or produced by genetic modification, be assessed for natural pesticide levels to confirm that they conform to or are below the levels that have previously been tolerated (Speijers, 1995).

Examples where this was not done include programs to develop new disease-resistant varieties of celery and potatoes. While the plant breeders succeeded in their aim, the new varieties were subsequently found to have unacceptably high levels of natural pesticides associated with them.

A variety of potatoes, Lenape, resistant to certain potato beetles was released in the USA but had to be withdrawn from the market shortly after when it was discovered that it accumulated high levels of the alkaloids solanine and chaconine (Zitnak and Johnston, 1970; Fenwick *et al.*, 1990). These alkaloids are normally associated with inedible wild potatoes that are disease and insect resistant and potatoes that have turned green and are therefore considered unsafe to eat.

In the case of celery, a new variety with improved insect resistance was found to produce phototoxic effects in celery pickers and processors due to the high levels of the natural furanocoumarin pesticides (Fenwick *et al.*, 1990).

CONCLUSIONS

An attempt has been made to demonstrate in this chapter, by describing the evolutionary origins of poisonous natural pesticides and using specific examples where a food safety risk is apparent, that natural pesticides currently represent a significantly greater threat to food safety than manufactured pesticides. The relative paucity of toxicological and exposure data for natural pesticides relative to manufactured pesticides is, in part, a result of relatively lower consumer awareness and hence lack of concern with natural pesticides. The lack of effective mechanisms for generating critical quantitative toxicity and dietary exposure data, when compared to manufactured pesticides, is also a factor in their relative neglect by food safety regulatory systems.

REFERENCES

Aira, M., Horn, H. and Seijo, M. C. (1998). Palynological analysis of honeys from Portugal, *J. Apicultural Res.*, **37**, 247–254.

Allcroft, R. (1969). Aflatoxicosis in farm animals, in *Aflatoxin: Scientific Background, Controls and Implications*, Goldblatt, L. A. (Ed.), Academic Press, New York, Ch. IX, pp. 237–264.

Allen, J. and Roberts, D. (1996). Annual ryegrass toxicity – update, in *Animal Health – Animal Disease Surveillance and Preparedness Newsletter*, Industry Resource Protection Program, Agriculture, WA, (ISSN 1322–4484).

Ames, B. N., Profet, M. and Gold, L. S. (1990). Dietary pesticides (99.9 % all natural), *Proc. Natl. Acad. Sci. USA*, **87**, 7777–7781.

Anon, (1994). In *Ryegrass Matters*, Roberts, D. (Ed.), Meat Research Corporation and Department of Agriculture Western Australia, pp. 7–8 (ISSN 1323–2126).

ANZFA (2000). *Review of the Maximum Permitted Concentrations of Non-Metals in Food*, Proposal P158, Australia New Zealand Food Authority, Canberra, Australia, pp. 28–31.

Aplin, R. T., Benn, M. H. and Rothchild, M. (1968). Poisonous alkaloids in the body tissues of the cinnabar moth (*Callimorpha jacobaeae L.*), *Nature (London)*, **219**, 747–748.

Bai, R., Pettit, G. R. and Hamel, E. (1990). Dolastatin 10, a powerful cytostatic peptide derived from a marine animal. Inhibition of tubulin polymerization mediated through the *Vinca* alkaloid binding domain, *Biochem. Pharmacol.*, **39**, 1941–1949.

Benn, M., DeGrave, J., Gnanasunderam, C. and Hutchins, R. (1979). Host-plant pyrrolizidine alkaloids in *Nyctemera annulata* Boisduval: their persistence through the life-cycle and transfer to a parasite, *Experientia*, **35**, 731–732.

Berry, P. H. and Wise, J. L. (1975). Wimmera ryegrass toxicity in Western Australia, *Aust. Vet. J.*, **51**, 525–530.

Berry, P. H., Howell, J. Mc. C. and Cook, R. D. (1980). Morphological changes in the central nervous system of sheep affected with experimental annual ryegrass (*Lolium rigidum*) toxicity, *J. Comp. Path.*, **90**, 603–617.

Blum, M. S. (1982). Detoxication, deactivation, and utilization of plant compounds by insects, *Plant Resistance to Insects*, Hedin, P. A. (Ed.), ACS Symposium Series 208, American Chemical Society, Washington, DC, pp. 265–275.

Boppré, M. (1986). Insects pharmacophagously utilizing defensive plant chemicals (pyrrolizidine alkaloids), *Naturwissenschaften*, **73**, 17–26.

Boppré, M. and Schneider, D. (1985). Pyrrolizidine alkaloids quantitatively regulate both scent organ morphogenesis and pheromone biosynthesis in male *Creatonotos* moths (Lepidoptera: Arctiidae), *J. Comp. Physiol.*, **157**, 569–577.

Boppré, M. and Schneider, D. (1989). The biology of *Creatonotos* (Lepidoptera: Arctiidae) with special reference to the androconial system, *Zool. J. Linn. Soc.*, **96**, 339–356.

Bourke, C. A. and Carrigan, M. J. (1993). Experimental tunicamycin toxicity in cattle, sheep and pigs, *Aust. Vet. J.*, **70**, 188–189.

Brown, K. (1984). Adult-obtained pyrrolizidine alkaloids defend ithomiine butterflies against a spider predator, *Nature (London)*, **309**, 707–709.

Brown, K. S. (1987). Chemistry at the Solanaceae/Ithomiinae interface, *Ann. Missouri. Bot. Gard.*, **74**, 359–397.

Candrian, U., Lüthy, J., Graf, U. and Schlatter, C. (1984). Mutagenic activity of the pyrrolizidine alkaloids seneciphylline and senkirkine in *Drosophila* and their transfer into rat milk, *Food Chem. Toxic.*, **22**, 223–225.

Chang, J. Y. and Korolev, V. V. (1996). Specific toxicity of tunicamycin in induction of programmed cell death of sympathetic neurons, *Exp. Neurol.*, **137**, 201–211.

Chapman, H. M. (1989). Annual ryegrass toxicity, in *Proceedings of the Second International Congress for Sheep Veterinarians*, Massey University, New Zealand, February 12–16, 1989, pp. 268–276.

Charudattan, R. (1982). Regulation of microbial weed control agents, in *Biological Control of Weeds with Plant Pathogens*, Charudattan, R. and Walker, H. L. (Eds), Wiley, New York, pp. 175–188.

Charudattan, R. (1991). The mycoherbicide approach with plant pathogens, in *Microbial Control of Weeds*, TeBeest, D. O. (Ed.), Chapman & Hall, New York, pp. 24–57.

Chauvin, P., Dillon, J.-C., Moren, A., Tablak, S. and Barakaev, S. (1993). Heliotrope poisoning in Tadjikistan, *Lancet*, **341**, 1663.

Cockrum, P. A. and Edgar, J. A. (1983). High performance liquid chromatographic comparison of tunicaminyluracil antibiotics corynetoxin, tunicamycin, streptovirudin and MM 19290, *J. Chrom.*, **268**, 245–254.

Cockrum, P. A. and Edgar, J. A. (1994). Isolation of novel phomopsins in lupin seed extracts, in *Plant-Associated Toxins: Agricultural, Phytochemical and Ecological Aspects*, Colegate, S. M. and Dorling, P. R. (Eds), CAB International, Wallingford, Oxon., UK, pp. 232–237.

Conner, W. E., Eisner, T., Vander Meer, R. K., Guerroro, A. and Meinwald, J. (1981). Precopulatory sexual interaction in an arctiid moth (*Utetheisa ornatrix*): role of a pheromone derived from dietary alkaloids, *Behav. Ecol. Sociobiol.*, **9**, 227–235.

Crews, C., Startin, J. R. and Clarke, P. A. (1997). Determination of pyrrolizidine alkaloids in honey from selected sites by solid phase extraction and HPLC-MS, *Food Additives Contam.*, **14**, 419–428.

Culvenor, C. C. J. and Edgar, J. A. (1972). Dihydropyrrolizine secretions associated with the coremata of *Utetheisa* moths (family Arctiidae), *Experientia*, **28**, 627–628.

Culvenor, C. C. J., Downing, D. T., Edgar, J. A. and Jago, M. V. (1969). Pyrrolizidine alkaloids as alkylating and antimitotic agents, *Ann. N.Y. Acad. Sci.*, **163**, 837–847.

Culvenor, C. C. J., Edgar, J. A. and Smith, L. W. (1981). Pyrrolizidine alkaloids in honey from *Echium plantagineum* L., *J. Agric. Food Chem.*, **29**, 958–960.

Curtain, C. C. and Edgar, J. A. (1976). The binding of dehydroheliotridine to DNA and the effect of it and other compounds on repair synthesis in main and satellite band DNA, *Chem.–Biol. Interactions*, **13**, 243–256.

Davies, S. C., White, C. L., Williams, I. H., Allen, J. G. and Croker, K. P. (1996). Sublethal exposure to corynetoxins affects production of grazing sheep, *Aust. J. Exp. Agric.*, **36**, 649–655.

Deinzer, M. L., Thompson, P. A., Burgett, D. M. and Isaacson, D. L. (1977). Pyrrolizidine alkaloids: their occurrence in honey from tansy ragwort (*Senecio jacobaea* L.), *Science*, **195**, 497–499.

Dethier, V. G. (1941). Chemical factors determining the choice of food plants by *Papilio* larvae, *Am. Nat.*, **75**, 61–73.

Dickinson, J. O., Cooke, M. P., King, R. R. and Mohamed, P. A. (1976). Milk transfer of pyrrolizidine alkaloids in cattle, *J. Am. Vet. Med. Assoc.*, **169**, 1192–1196.

Dobler, S., Haberer, W., Witte, L. and Hartmann, T. (2000). Selective sequestration of pyrrolizidine alkaloids from diverse host plants by *Longitarus* flea beetles, *J. Chem. Ecol.*, **26**, 1281–1298.

Eckardt, K. (1983). Tunicamycins, streptovirudins and corynetoxins, a special subclass of nucleoside antibiotics, *J. Nat. Prod.*, **46**, 544–550.

Edgar, J. A. (1975a). Danainae (Lep.) and 1,2-dehydropyrrolizidine alkaloid-containing plants – with reference to observations made in the New Hebrides, *Philos. Trans. R. Soc. B London*, **272**, 467–476.

Edgar, J. A. (1975b). Pyrrolizidine alkaloids in *Parsonsia* species (family Apocynaceae) which attract danaid butterflies, *Experientia*, **31**, 393–394.

Edgar, J. A. (1982). Pyrrolizidine alkaloids sequestered by Solomon Island danaine butterflies. The feeding preferences of the Danainae and Ithomiinae, *J. Zool. London*, **196**, 385–399.

Edgar, J. A. (1984). Parsonsieae: ancestral larval foodplants of the Danainae and Ithomi-inae, in *The Biology of Butterflies*, Vane-Wright, R. I. and Ackery, P. R. (Eds), Academic Press, London, pp. 91–93.

Edgar, J. A. (1991). Phomopsins: antimicrotubule mycotoxins, in *Handbook of Natural Toxins*, Vol 6, *Toxicology of Plant and Fungal Compounds*, Keeler, R. F. and Tu, A. T. (Eds), Marcel Dekker, New York, pp. 371–395.

Edgar, J. A. (1993). Annual ryegrass toxicity. Aetiology, pathology and related diseases, in *Australian Standard Diagnostic Techniques*, Corner, L. A. and Bagust, T. J. (Eds), CSIRO for Standing Committee on Agricultural and Resource Management, East Melbourne, Australia.

Edgar, J. A. (1997a). Corynetoxins-are there food safety implications?, *Adv. Food Safety.*, **1**, 26–29.

Edgar, J. A. (1997b). Corynetoxins: a food safety risk?, *Microbiol. Aust.*, **18**, 11–12.

Edgar, J. A. and Culvenor, C. C. J. (1974). Pyrrolizidine ester alkaloid in danaid butter-flies, *Nature (London)*, **248**, 614–616.

Edgar, J. A. and Smith, L. W. (1999). Transfer of pyrrolizidine alkaloids into eggs: food safety implications, in *Natural and Selected Synthetic Toxins, Biological Implications*, Tu, A. T. and Gaffield, W. (Eds), ACS Symposium Series 745, American Chemical Society, Washington, DC, pp. 118–128.

Edgar, J. A., Culvenor, C. C. J. and Smith, L. W. (1971). Dihydropyrrolizine derivatives in the hairpencil secretions of danaid butterflies, *Experientia*, **27**, 761–762.

Edgar, J. A., Culvenor, C. C. J. and Robinson, G. S. (1973). Hairpencil dihydropy-rrolizines of Danainae from the New Hebrides, *J. Aust. Ent. Soc.*, **12**, 144–150.

Edgar, J. A., Culvenor, C. C. J. and Pliske, T. E. (1974). Coevolution of danaid butterflies with their host plants, *Nature (London)*, **250**, 646–648.

Edgar, J. A., Cockrum, P. A. and Frahn, J. L. (1976a). Pyrrolizidine alkaloids in *Danaus plexipus* L. and *Danaus chrysippus* L., *Experientia*, **32**, 1535–1537.

Edgar, J. A., Culvenor, C. C. J. and Pliske, T. E. (1976b). Isolation of a lactone, structurally related to the esterifying acids of pyrrolizidine alkaloids, from the costal fringes of male Ithomiinae, *J. Chem. Ecol.*, **2**, 263–270.

Edgar, J. A., Boppre, M. and Schneider, D. (1979). Pyrrolizidine alkaloid storage in African and Australian danaid butterflies, *Experientia*, **35**, 1447–1448.

Edgar, J. A., Frahn, J. L., Cockrum, P. A., Anderton, N., Jago, M. V., Culvenor, C. C. J., Jones, A. J., Murray, K. and Shaw, K. J. (1982). Corynetoxins causative agents of annual ryegrass toxicity; their identification as tunicamycin group antibiotics, *J. Chem. Soc., Chem. Commun.*, 222–224.

Edgar, J. A., Cockrum, P. A., Stewart, P. L., Anderton, N. A. and Payne, A. L. (1994). Identification of corynetoxins as the cause of poisoning associated with annual beardgrass (*Polypogon monspeliensis* (L) Desf.) and blown grass (*Agrostis avenacea* C. Gemelin), in *Plant-Associated Toxins: Agricultural, Phytochemical and Ecological Aspects*, Colegate, S. M. and Dorling, P. R. (Eds), CAB International, Wallingford, Oxon., UK, pp. 393–398.

Edgar, J. A., Roeder, E. and Molyneux, R. J. (2002). Honey from plants containing pyrrolizidine alkaloids: A potential threat to health, *J. Agric. Food Chem.*, **50**, 2719–2730.

Ehrlich, P. R. and Raven, P. H. (1964). Butterflies and plants: a study of coevolution, *Evolution*, **18**, 586–608.

Eisner, T. (1982). For love of nature: exploration and discovery at biological field stations, *Bioscience*, **32**, 321–326.

Elbein, A. D. (1980). Effect of several tunicamycin-like antibiotics on glycoprotein biosynthesis in mung beans, *Vigna radiata* and suspension cultured soybean *Glycine max* cultivar mandarin cells, *Plant Physiol.*, **65**, 460–464.

Ericson, M. C., Gafford, J. T. and Elbein, A. D. (1977). Tunicamycin inhibits GlcAc-lipid formation in plants, *J. Biol. Chem.*, **252**, 7431–7433.

FAO (1997). *Worldwide Regulations for Mycotoxins 1995. A Compendium,* FAO Food and Nutrition paper 64, Food and Agriculture Organization of the United Nations, Rome.

FAO/WHO (1995). *Application of Risk Analysis to Food Standards Issues,* WHO/FNU/FOS/95.3, Report of the Joint FAO/WHO Expert Consultation, Geneva, Switzerland, 13–17 March, 1995, World Health Organization, Geneva.

FAO/WHO (1997). *Risk Management and Food Safety,* Report of the Joint FAO/WHO Expert Consultation, Rome, Italy, 27–31, January 1997, FAO Food and Nutrition Paper 65, Food and Agriculture Organization of the United Nations, Rome.

FAO/WHO (1998). *Application of Risk Communication to Food Standards and Safety Matters,* Report of the Joint FAO/WHO Expert Consultation, Rome, Italy, 2–6 February, 1998, FAO Food and Nutrition Paper 70, Food and Agriculture Organization of the United Nations, Rome.

Feeny, P., Rosenberry, L. and Carter, M. (1983). Chemical aspects of oviposition behaviour in butterflies, *Herbivorous Insects. Host Seeking Behaviour and Mechanisms,* in Ahmad, S. (Ed.), Academic Press, New York, pp. 27–76.

Fenwick, G. R., Johnson, I. T. and Hedley, C. I. (1990). Toxicity of disease-resistant plant strains, *Trends Food Sci. Technol.*, 23–25 (July).

Fisher, J. M., Dube, A. J. and Watson, C. M. (1979). Distribution in South Australia of *Anguina funesta,* the nematode associated with annual ryegrass toxicity, *Aust. J. Exp. Agric. Anim. Husb.*, **19**, 48–52.

Fraenkel, G. S. (1959). The raison d'être of secondary plant substances, *Science,* **129**, 1466–1470.

Frahn, J. L., Edgar, J. A., Jones, A. J., Cockrum, P. A., Anderton, N. A. and Culvenor, C. J. J. (1984). Structure of the corynetoxins, metabolites of *Corynebacterium rathayi* responsible for toxicity of annual ryegrass (*Lolium rigidum*) pastures, *Aust. J. Chem.*, **37**, 165–182.

Gaul, K. L., Gallagher, P. F., Reyes, D., Stasi, S. and Edgar, J. A. (1994). Poisoning of pigs and poultry by stock feed contaminated with heliotrope seed, in *Plant-Associated Toxins: Agricultural, Phytochemical and Ecological Aspects,* Colegate, S. M. and Dorling, P. R. (Eds), CAB International, Wallingford, Oxon., UK, pp. 137–143.

German Federal Health Bureau, (1992). *Bundesanzeiger, June 17, 4805, Dt. Apoth. Ztg.,* **132**, 1406–1408.

Hartmann, T. (1999). Chemical ecology of pyrrolizidine alkaloids, *Planta,* **207**, 483–495.

Hartmann, T. and Ober, D. (2000). Biosynthesis and metabolism of pyrrolizidine alkaloids in plants and specialized insect herbivores, *Topics Curr. Chem.*, **209**, 207–243.

Hedin, P. A. and Hollingworth, R. M. (1997). New applications for phytochemical pest-control agents, *Phytochemicals for Pest Control,* in Hedin, P. A., Hollingworth, R. M., Masler, E. P., Miyamoto, J. and Thompson, D. G. (Eds), ACS Symposium Series 658, American Chemical Society, Washington, DC, pp. 1–12.

Hill, K. R., Markson, L. M. and Schoental, R. (1959). Discussion on Seneciosis in man and animals, *Proc. R. Soc. Med.*, **53**, 281–288.

Hincks, J. R., Kim, H. Y., Coulombe, R. A., Segall, H. J., Molyneux, R. J. and Stermitz, F. R. (1991). DNA cross-linking in mammalian cells by pyrrolizidine alkaloids, *Toxicol. Appl. Pharmacol.*, **111**, 90–98.

Hippchen, Von C., Entzeroth, R., Roeder, E. and Greuel, E. (1986). Experimentelle untersuchungen zur lebertoxizität von senecioalkaloiden aus *Senecio vernalis* an ziegen, *Praktische. Tierarzt.*, **67**, 322–324.

Huxtable, R. J. (1989). Human health implications of pyrrolizidine alkaloids and the herbs containing them, in *Toxicants of Plant Origin, Vol. 1, Alkaloids,* Cheeke, P. R. (Ed.), CRC Press, Boca Raton, FL, USA, pp. 41–86.

Huxtable, R. J. and Cooper, R. A. (2000). Pyrrolizidine alkaloids: physicochemical correlates of metabolism and toxicity, in *Natural and Selected Synthetic Toxins, Biological Implications*, Tu, A. T. and Gaffield, W. (Eds), ACS Symposium Series 745, American Chemical Society, Washington, DC, pp. 100–117.

IARC (1976). Pyrrolizidine alkaloids, in *IARC Monograph on the Evaluation of Carcinogenic Risk of Chemicals to Man – Some Naturally Occurring Substances*, Vol. 10 International Agency for Research on Cancer, Lyons, France (see also Vol. 31 (1983)).

Ioffe, E. and Stanley, P. (1994). Mice lacking N-acetylglucosaminyltransferase I activity die at mid-gestation, revealing an essential role for complex or hybrid N-linked carbohydrates, *Proc. Natl. Acad. Sci. USA*, **91**, 728–732.

IPCS (1987). *Principles for the Safety Assessment of Food Additives and Contaminants in Food*, Environmental Health Criteria 70, World Health Organization, Geneva.

IPCS (1988). *Pyrrolizidine Alkaloids*, Environmental Health Criteria 80, World Health Organization, Geneva.

IPCS (1989). *Pyrrolizidine Alkaloids Health and Safety Guide*, Health and Safety Criteria Guide No. 26, World Health Organization, Geneva.

Jago, M. V. (1985). Causative action of corynetoxins in annual ryegrass toxicity, in *Plant Toxicology*, Seawright, A. A., Hegarty, M. P., James, L. F. and Keeler, R. F. (Eds), The Queensland Poisonous Plants Committee, Yeerongpilly, Australia, pp. 569–577.

Jago, M. V. and Culvenor, C. C. J. (1987). Tunicamycin and corynetoxin poisoning in sheep, *Aust. Vet. J.*, **64**, 232–235.

Jago, M. V., Edgar, J. A., Smith, L. W. and Culvenor, C. C. J. (1970). Metabolic conversion of heliotridine-based pyrrolizidine alkaloids to dehydroheliotridine, *Mol. Pharmacol.*, **6**, 402–406.

Jago, M. V., Payne, A. L., Peterson, J. E. and Bagust, T. J. (1983). Inhibition of glycosylation by corynetoxin, the causative agent of annual ryegrass toxicity: comparison with tunicamycin, *Chem.–Biol. Interact.* **45**, 223–234.

Kuo, S. C. and Lampen, J. O. (1976). Tunicamycin inhibition of [^{3}H] glucosamine incorporation into yeast glycoproteins: binding of tunicamycin and interaction with phospholipids, *Arch. Biochem. Biophys.*, **172**, 574–581.

Lacy, E., Edgar, J. A. and Culvenor, C. C. J. (1987). Interaction of phomopsin A and related compounds with purified sheep brain tubulin, *Biochem. Pharmacol.*, **36**, 2133–2138.

Leaver, D. D., Schneider, K. M., Rand, M. J., Anderson, R. M., Gage, P. W. and Malbon, R. (1988). The neurotoxicity of tunicamycin, *Toxicology*, **49**, 179–187.

Lehmann, S. and Harris, D. A. (1997). Blockade of glycosylation promotes acquisition of scrapie-like properties by the prion protein in cultured cells, *J. Biol. Chem.*, **272**, 21479–21487.

Lennarz, W. J. (1993). Glycoprotein synthesis and embryonic development, *CRC Crit. Rev. Biochem.*, **14**, 257–272.

Lozano, M., Montero Espinosa, V., Osorio, E. and Sanchez, J. (1995). Honeys of Extremaduran pastureland: physico-chemical characteristics and pollens, *Vida Apicola*, **72**, 40–46.

Luduena, R. F., Roach, M. C., Prasad, V. and Lacey, E. (1990). Effect of phomopsin A on the alkylation of tubulin, *Biochem. Pharmacol.*, **39**, 1603–1608.

Lüthy, J., Heim, Th. and Schlatter, Ch. (1983). Transfer of [^{3}H] pyrrolizidine alkaloids from Senecio vulgaris L and metabolites into rat milk and tissues, *Toxicol. Lett.*, **17**, 283–288.

Marasas, W. F. O. (1995). Fumonisins: their implications for human and animal health, *Nat. Toxins*, **3**, 193–198.

Mattocks, A. R. (1968). Toxicity of pyrrolizidine alkaloids, *Nature (London)*, **217**, 723–728.

Mattocks, A. R. (1986). *Chemistry and Toxicology of Pyrrolizidine Alkaloids*, Academic Press, London, UK.

Mayer, F. and Lüthy, J. (1993). Heliotrope poisoning in Tadjikistan, *Lancet*, **342**, 246–247.

McKay, A. C. and Ophel, K. M. (1993). Toxigenic *Clavibacter/Anguina* associations infecting grass seedheads, *Annu. Rev. Phytopath.*, **31**, 153–169.

Molyneux, R. J. and James, L. F. (1990). Pyrrolizidine alkaloids in milk: thresholds of intoxication. *Vet. Hum. Toxicol.*, **32**(Suppl.), 94–103.

Pel'menev, V. K., Kharitonova, L. F. and Babarykina, A. N. (1983). Nectar plants of the borage family, *Pchelovodstvo*, **8**, 15–17.

Peterson, J. E. (1978). *Phomopsis leptostromiformis* toxicity (lupinosis) in nursling rats, *J. Comp. Path.*, **88**, 191–203.

Peterson, J. E. (1983). Embryotoxicity of phomopsin in rats, *Aust. J. Exp. Biol. Med. Sci.*, **61**, 105–115.

Peterson, J. E. (1986). The toxicity of phomopsin, in *Proceedings of the Fourth International Lupin Conference*, Western Australian Department of Agriculture for the International Lupin Association, South Perth, Western Australia, pp. 199–208.

Peterson, J. E. (1990). Biliary hyperplasia and carcinogenesis in chronic liver damage induced in rats by phomopsin, *Pathology*, **22**, 213–222.

Peterson, J. E. and Culvenor, C. C. J. (1983). Hepatotoxic pyrrolizidine alkaloids, in *Handbook of Natural Toxins*, Vol. 1, Keeler, R. F. and Tu, A. T. (Eds), Marcel Dekker, New York, pp. 638–671.

Peterson, J. E., Jago, M. V. and Stewart, P. L. (1996). Permanent testicular damage induced in rats by a single dose of tunicamycin, *Reprod. Toxicol.*, **10**, 61–69.

Petry, T. W., Bowen, G. P., Huxtable, R. J. and Sipes, I. G. (1984). Characterization of hepatic DNA damage induced in rats by the pyrrolizidine alkaloid monocrotaline, *Cancer Res.*, **44**, 1505–1509.

Petry, T. W., Bowden, G. P., Buhler, D. R. and Sipes, I. G. (1986). Genotoxicity of the pyrrolizidine alkaloid jacobine in rats, *Toxicol. Lett.*, **32**, 275–281.

Pliske, T. E. and Eisner, T. (1969). Sex pheromone of the queen butterfly: biology, *Science N.Y.*, **164**, 1170–1172.

Prakash, A. S., Pereira, T. N., Reilly, P. E. B. and Seawright, A. A. (1999). Pyrrolizidine alkaloids in human diet, *Mutat. Res.*, **443**, 53–67.

Roeder, E. (1995). Medicinal plants in Europe containing pyrrolizidine alkaloids, *Pharmazie*, **50**, 83–98.

Roeder, E. (2000). Medicinal plants in China containing pyrrolizidine alkaloids, *Pharmazie*, **55**, 711–726.

Roulet, M., Laurini, R., Rivier, L. and Calame, A. (1988). Hepatic veno-occlusive disease in newborn infant of a women drinking herbal tea, *J. Pediatrics*, **112**, 433–436.

Sachs, J. (1873). *Lehrbuch der Botanik*, W. Engelmann, Leipzig.

Scheffer, R. P. and Livingston, R. S. (1984). Host-selective toxins and their role in plant diseases, *Science*, **223**, 17–21.

Schoental, R. (1959). Liver lesions in young rats suckled by mothers treated with pyrrolizidine (*Senecio*) alkaloids, lasiocarpine and retrorsine, *J. Pathol. Bacteriol.*, **77**, 485–495.

Shankar, M., Cowling, W. A., Sweetingham, M. W., Than, K. A., Edgar, J. A. and Michalewicz, A. (1999). Screening for resistance to *Diaporthe toxica* in lupins by estimation of phomopsins and glucosamine in individual plants, *Plant Path.*, **48**, 320–324.

Shevchenko, N. Kh. and Fakhrutdinova, S. Sh. (1971). Assessment of meat and milk in cases of trichodesma poisoning, *Veterinariya, Moscow U.S.S.R.*, **8**, 104–105.

Shivas, R. G., Lewis, J. C. and Groves, R. H. (1994a). Distribution in Australia and host plant specificity of *Phomopsis emicis*, a stem blight pathogen of *Emex australis*, *Aust. J. Agric. Res.*, **45**, 1025–1034.

Shivas, R. G., Allen, J. G., Edgar, J. A., Cockrum, P. A., Gallagher, P. F., Ellis, Z. and Harvey, M. (1994b). Production of phomopsin A by a potential mycoherbicide, *Phomopsis emicis*, in *Plant-Associated Toxins: Agricultural, Phytochemical and Ecological Aspects*, Colegate, S. M. and Dorling, P. R. (Eds), CAB International, Wallingford, Oxon., UK, pp. 161–166.

Smith, L. W. and Culvenor, C. C. J. (1981). Plant sources of hepatotoxic pyrrolizidine alkaloids, *J Nat. Prod.*, **44**, 129–152.

Speijers, G. J. A. (1995). Toxicological data needed for safety evaluation and regulation on inherent plant toxins, *Nat. Tox.*, **3**, 222–226.

Stewart, P. L., May, C. and Jago, M. V. (1997). Reduction and recovery of *N*-acetylglucosamine-1-phosphate transferase activity in sheep and rat liver following a single subcutaneous dose of tunicamycin, *Aust. Vet. J.*, **76**, 20–21.

Stewart, P. L., Hooper, P. T., Colegate, S. M., Edgar, J. A. and Raisbeck, M. F. (2002). Toxic effects of a single parenteral dose of tunicamycin in late stage pregnancy in rats, *Vet. Human Toxicol.*, **44**, 211–215.

Steyn, D. G. (1933). Poisoning of humans beings by weeds contained in cereals (bread poisoning), *Onderstepoort J. Vet. Sci. Animal Ind.*, **1**, 219–266.

Stuart, S. J., Than, K. A. and Edgar, J. A. (1994). New approaches to studying the fate of corynetoxins in whole animals and their products, in *Plant-Associated Toxins: Agricultural, Phytochemical and Ecological Aspects*, Colegate, S. M. and Dorling, P. R. (Eds), CAB International, Wallingford, Oxon., UK, pp. 143–148.

Surani, A. H. (1979). Glycoprotein synthesis and inhibition of glycosylation by tunicamycin in preimplantation mouse embryos: compaction and trophoblast adhesion, *Cell*, **18**, 217–227.

Than, K. A., Cao, Y., Michalewicz, A. and Edgar, J. A. (1998a). Development of a vaccine against annual ryegrass toxicity, in *Toxic Plant and other Natural Toxicants*, Garland, T. and Barr, A. C. (Eds), CAB International, Wallingford, Oxon., UK, pp. 165–168.

Than, K. A., Cao, Y., Michalewicz, A. and Edgar, J. A. (1998b). Development of an immunoassay for corynetoxins, in *Toxic Plant and other Natural Toxicants*, Garland, T. and Barr,bnsA. C. (Eds), CAB International, Wallingford, Oxon., UK, pp. 49–54.

Thapa, R. and Wongsiri, S. (1997). *Eupatorium odoratum*: a honey plant for beekeepers in Thailand, *Bee World*, **78**, 175–178.

Tkacz, J. S. (1983). Tunicamycin and related antibiotics, in *Antibiotics*, Vol. VI, *Modes and Mechanisms of Microbial Growth Inhibitors*, Hahn, F. E. (Ed.), Springer-Verlag, Berlin, pp. 255–278.

Tonsing, E. M., Steyn, P. S., Osborn, M. and Weber, K. (1984). Phomopsin A, the causative agent of lupinosis, interacts with microtubules *in vivo* and *in vitro*, *Eur. J. Cell Biol.*, **35**, 156–164.

Uecker, F. A. (1988). *A World List of Phomopsis Names with Notes on Nomenclature, Morphology and Biology*, Contributions from the US National Fungus Collection No. 3, Mycologia Memoirs, Vol. 13, Gebr. Borntraeger, Stuttgart.

Vining, L. C. (1990). Functions of secondary metabolites. *Annu. Rev. Microbiol.*, **44**, 395–427.

Yang, Y.-C., Yan, J., Churchwell, M., Berger, R., Chan, P.-C., Doerge, D. R., Fu, P. P. and Chou, M. W. (2001a). Development of a ^{32}P-postlabeling/HPLC method for detection of dehydroretronecine-derived DNA adducts *in vivo* and *in vitro*, *Chem. Res. Toxicol.*, **14**, 91–100.

Yang, Y.-C., Yan, J., Doerge, D. R., Chan, P.-C., Fu, P. P. and Chou, M. W. (2001b). Metabolic activation of the tumorigenic pyrrolizidine alkaloid, riddelliine, leading to DNA adduct formation *in vivo*, *Chem. Res. Toxicol.*, **14**, 101–109.

Zitnak, A. and Johnston, G. R. (1970). Glycoalkaloid content of B5141-6 potatoes, *Am. Potato. J.*, **47**, 256–260.

10 International Standards: The International Harmonization of Pesticide Residue Standards for Food and Drinking Water

WIM H. VAN ECK

Ministry of Health, Welfare and Sport, The Hague, The Netherlands[1]

[1]Currently (2003) World Health Organization, Geneva, Switzerland.

Pesticide Residues in Food and Drinking Water: Human Exposure and Risks.   Edited by Denis Hamilton and Stephen Crossley
© 2004 John Wiley & Sons, Ltd   ISBN: 0-471-48991-3

INTRODUCTION

Agricultural pesticides are widely used as crop protection agents across the world in the production of food and feed. As a result, residues of pesticides end up in food for human consumption. In view of the toxicological properties of pesticides, the presence of these residues in food may pose a risk to the health of consumers.

Health protection and assuring the proper use of pesticides are the main reasons why governments establish legal limits for pesticide residues in raw agricultural commodities, usually as part of their national legislation for the regulation of pesticides. Food with actual residue levels in compliance with these limits is thought to be safe.

However, these so-called Maximum Residue Limits (MRLs) set at the national level are potential barriers to trade. Raw agricultural commodities are important items in international trade. Governments, however, are inclined to establish MRLs quite independently from their trading partners, focusing on the protection of their population, taking into account only their own national agricultural practices. As a consequence, in many cases MRLs for identical commodity–pesticide combinations vary from country to country. This can easily constitute an impediment to free trade of goods as governments sometimes prevent the import of food with residue levels exceeding the nationally established limits or take otherwise legal action claiming to protect the health of their consumers.

The potential of MRLs as trade irritants has been long recognized and the recognition triggered initiatives to harmonize these standards among countries at both the regional and international level. One of the earliest examples of regional harmonization is that of the European Community (EC) which, as early as 1976, established community legislation on residues of pesticides in fruits and vegetables. In more recent years, programmes for the harmonization of MRLs for pesticides have been implemented in other regions as well in the framework of economic cooperation. Notably Canada, Mexico and the USA, working together in the North American Free Trade Association (NAFTA), are well under way in aligning their pesticide regulatory activities, including harmonization of MRLs (synonymous with 'tolerances') among their countries. At the global level, however, the early initiative of the EC was preceded by almost ten years by the Codex Alimentarius Commission (CAC) which initiated already in 1966 a programme to establish international standards for pesticide residues in food and feed.

The CAC was founded by the Food and Agriculture Organization of the United Nations (FAO) and the Word Health Organization (WHO) to implement the Joint FAO/WHO Food Standards Programme. Since its inception in 1962, the CAC has elaborated an impressive body of standards. As one of the subsidiary bodies of the CAC, the Codex Committee on Pesticide Residues (CCPR) is responsible for the elaboration of MRLs for pesticide residues, working closely together with an independent scientific advisory body, the Joint FAO/WHO Meeting on Pesticide Residues (JMPR). The importance of Codex standards has increased considerably under the Agreement on the Application of Sanitary and Phytosanitary

Measures (SPS Agreement) of the World Trade Organization (WTO), which was established in 1995. The SPS Agreement recognizes Codex standards as references in international food trade. This enhanced status of Codex standards within the framework of the WTO prompted Codex to revisit its standard setting methodology and to further elaborate and implement risk analysis principles in its standard setting activities. As a result, the relevant advisory bodies and Codex Committees, e.g., respectively, the JMPR and the CCPR, are encouraged to properly apply these principles.

Limits for pesticide residues also exist for drinking water. Their primary purpose is the protection of consumer safety. The way these limits are derived, however, differs entirely from those established for food and feed items. As drinking water is not an important commodity in international trade, no initiative exists today to harmonize these limits between countries at the international level. Drinking water, in this case, should not be confused, for instance, with natural mineral water and other packaged waters, which are internationally traded. The World Health Organization (WHO) regularly publishes Guidelines for Drinking-water Quality, including safe limits for pesticide residues. These international Guidelines are of an advisory nature and are intended to be used by governments to establish their own national or regional limits. An example of a regional limit for pesticide residues in drinking water is that of the European Community which has had drinking water standards in force since 1980.

This present chapter will focus mainly on the elaboration of international standards for pesticide residues in food and feed within the framework of the Codex Alimentarius Commission while only briefly mentioning a few examples of regional harmonization of pesticide residue limits. The development of international Guidelines for drinking water is only briefly touched upon. For a better understanding of the policy on the elaboration of international food standards, the organizations involved and their interrelationships will be described before details of the standard setting process itself, with particular attention on the process for MRLs for pesticide residues. First, Codex Alimentarius is introduced, focusing on the standard setting principles and on the recent developments in the risk analysis philosophy and its implementation in the work of Codex. The implication of the WTO SPS Agreement to Codex is also explained. The elaboration of MRLs for pesticide residues and the scientific evaluation of toxicological and residue data by the JMPR are then presented in more detail. Finally, the consequences of MRLs being based on Good Agricultural Practices (GAP) for the risk assessment approaches to be applied, are also described.

THE CODEX ALIMENTARIUS COMMISSION

THE COMMISSION

MRLs for pesticide residues are food standards. Similar standards exist for other classes of chemicals in food as well, such as maximum residue limits for veterinary drugs or guideline levels for heavy metals and other contaminants. For food additives, maximum permitted levels are established. However, food standards are by no means limited to chemicals. Labelling requirements for foodstuffs or provisions for hygienic practices in the production and handling of food are examples of food standards as well. Another category of food standards includes the commodity standards for specific foods or classes of food. Many governments establish these types of standards under their national food legislation, primarily with the purpose of safeguarding the health of their consumers and to promote fair practices in commerce (i.e. through labelling, compositional requirements, etc.). As explained previously, these standards are potential trade irritants since they can act as non-tariff trade barriers. Food is usually traded internationally on a very large scale and food standards in importing and exporting countries are seldom identical. This prompted the need for the harmonization of food standards, not only at the regional level but above all on a global scale.

In 1962, the Codex Alimentarius Commission (CAC) was established by the FAO and WHO as an international food standard setting organization. The CAC is an intergovernmental organization and its Rules of Procedure and the descriptions of procedures are contained in the *Codex Alimentarius Commission Procedural Manual* (FAO/WHO, 2000). At the time of writing,[2] its membership stands at 165 countries. International governmental and non-governmental organizations are allowed to participate as observers in the activities of the CAC and its committees in accordance with the rules and procedures of the Codex Alimentarius Commission. The Secretariat of the CAC is located at the FAO Headquarters in Rome, Italy. To facilitate communication between the Codex Secretariat and the Member Governments, each Member Government is advised to establish a Codex Contact Point, which is responsible, among other things, for the distribution of Codex documents within the country and for the co-ordination of the views of a Member on Codex matters. The history of the Codex Alimentarius Commission, its purpose and working procedures are well described in a recently published booklet *Understanding the Codex Alimentarius* (FAO/WHO, 1999).

The CAC was founded to implement the Joint FAO/WHO Food Standards Programme, namely the elaboration of international food standards where relevant. The purpose of this programme is the protection of the health of consumers, ensuring fair practices in trade and the promotion of the co-ordination of all food standards work undertaken by international governmental and non-governmental organizations (FAO/WHO, 2000). It should be kept in mind that Codex standards are voluntary in nature and do not bind Governments legally. The *Codex Alimentarius* is a compilation of these Codex standards and other recommendations adopted by the Codex Alimentarius Commission and comprises thirteen volumes. In this chapter, reference will be made to 'Codex' where it can mean either the CAC or the published standards and "Codex standard" to mean both Codex standards and other recommendations, unless these would give rise to confusion.

---

[2] 2000.

The elaboration of standards is entrusted to the subsidiary bodies of the CAC, the Codex Committees, of which there are three types. General subject committees are assigned the task of elaborating standards of general nature applicable to all types of food, e.g. pesticide residues, food additives and contaminants, veterinary drugs, food labelling and food hygiene. A rather important general subject committee is the Codex Committee on General Principles (CCGP). This committee considers procedural and general matters, such as the General Principles, which define the purpose and the scope of the Codex Alimentarius and develops guidelines for Codex Committees.

Commodity committees are responsible for the development of standards for commodities, including milk products, fats and oils, fish and fishery products, fresh and processed fruits and vegetables, cocoa products and chocolate, which are of importance in international trade. Finally, there are six regional Codex Committees. Their responsibilities are to define the problems and needs of the region concerning food standards and food control, to exchange information within the region on proposed regulatory initiatives and on problems arising from food control actions and to strengthen the food control infrastructure. Regional Committees may wish to recommend to the CAC the development of world-wide standards for products of interest to the region. In total, there are about thirty Codex Committees. The CAC operates through a system whereby Members host a Committee. A host country accepts financial and all other responsibilities, including the organization of the meetings of the Committee, appoints a chairperson and provides for a secretariat and all other conference services. From all Sessions of the CAC and its subsidiary bodies, reports in the three official Codex languages (English, French and Spanish) are published. These Reports are available from the Codex Secretariat in Rome, can be downloaded from the Codex website and are distributed through the Codex Contact Points.

The Executive Committee of the CAC is one of its executive organs and makes certain decisions on behalf of the CAC in those years in which the latter does not meet.

The CAC requires itself and its subsidiary bodies to reach a decision on a consensus basis, which is generally achievable, although the Procedures of the Codex provide for a vote for decision making. Only in very controversial cases is voting necessary to conclude the decision making on a particular draft standard. An example of such a case was the voting at the 21st Session of the CAC (1995) on MRLs for the so-called growth hormones. At the recently concluded 23rd Session of the CAC (1999), all decisions were taken on a consensus basis. In Codex, the incidences of voting on growth hormones were one of the triggers to initiate a broader debate on the role of science in the Codex decision-making process and the extent to which other factors relevant for health protection of consumers and for the promotion of fair practices in the food trade can be taken into account. This debate was further fuelled by the WTO Agreement on the Application of Sanitary and Phytosanitary Measures (SPS Agreement) stating that sanitary and phytosanitary measures (SPS measures) should be solely based on science, thus leaving very little or no room for non-scientific considerations. The case with the growth hormones clearly indicates the political implications that international standard setting may have and the case finally resulted in an official trade dispute among several Members of the WTO.

THE ELABORATION OF STANDARDS

The elaboration of Codex standards follows an established procedure contained in the Procedural Manual (FAO/WHO, 2000). The Uniform Procedure for the Elaboration of Codex Standards and Related Texts comprises eight steps. It starts with the identification of the need to establish a standard, which is in fact a process of priority setting (step 1), and the drafting of the proposed standard (step 2). The proposed draft standard is then sent to the Members of the Commission and to interested international organizations for comments on all aspects including the possible implications for their economic interests (step 3). In Codex, a request for comments is initiated through a Circular Letter issued by the Secretariat giving the background of the proposal and a deadline for comments. The comments are presented to the Codex Committee involved to consider and to amend the proposed draft standard if necessary (step 4). The proposed draft standard is then submitted through the Secretariat to the Commission or to the Executive Committee to adopt as a draft standard while giving due consideration to any comment before taking a decision (step 5). The draft standard is then sent by the Secretariat to all Members and international organizations for comments, again on all aspects, including the eventual impact on their economic interests (step 6). The comments are next presented to the responsible Committee to consider and to amend the draft standard where appropriate (step 7). Finally, the draft standard is submitted through the Secretariat to the Commission for its final adoption as a Codex standard (step 8). After the adoption, the standard is published in one of the Volumes of the *Codex Alimentarius*. Usually, the *Codex Alimentarius* is updated every two years, resources permitting, after each Session of the Commission. In practice, any decision of a Codex Committee on a draft standard is a recommendation to the CAC to consider the draft standard for adoption as a Codex standard. The Codex 'Step Procedure' is summarized in Table 10.1, using the establishment of an MRL for a pesticide residue following recommendations by the CCPR and JMPR as an example.

The development of a Codex standard through the Codex step procedure is a thorough but relatively slow process taking several years. A proposal may stay in a certain step for some time or can even be referred to a previous step in cases of major amendments, in many cases as a result of widely diverging views among the Members of the Codex or when new information has become available. Many Codex Committees, including the Executive Committee, meet only once a year and some Committees meet even less frequently. The Commission itself meets only once every two years. To speed up the process, steps 6 and 7 can be omitted, provided that this is supported by a two-thirds majority of the relevant

Table 10.1   Summary of the Codex 'Step Procedure'

Stage	Description
Step 1	Recommendation of priority compounds by the CCPR
Step 2	First evaluation of the compound by the JMPR – estimation of an ADI and of MRLs
Step 3	Submission of the proposed Codex MRLs to governments for a first round of comments
Step 4	First discussion of the proposed MRL by the CCPR in the light of the comments received
Step 5	Submission of the proposed Codex MRL to the Codex Alimentarius Commission (CAC) in the light of the CCPR discussion, for consideration
Step 6	Submission of the proposed Codex MRLs to governments for a second round of comments[a]
Step 7	Final discussion of the proposed Codex MRLs by the CCPR in the light of comments received[a]
Step 8	Consideration by the CAC for adoption of the proposal as a Codex MRL (CXL)

[a] Steps 6 and 7 can be omitted, provided that this is supported by a two-thirds majority of the CCPR or the CAC.

Codex Committee or the Commission. The possibility of proposals at step 5 being advanced by the Executive Committee, which meets every year, to the next step is helpful in that regard. In order to respond to the need for facilitation of the elaboration process, another procedure, the Uniform Accelerated Procedure for the Elaboration of Codex Standards and Related Texts, comprising only five steps, was implemented in 1993.

The adoption of a standard is followed by an acceptance procedure whereby Members notify the Commission whether they can officially accept the standard or not. The underlying idea of the acceptance procedure is to monitor the use and impact of Codex Standards and to identify areas where problems may still occur.

As time passes, there may be a need to revise or amend existing standards in the light of new developments, changed policies or new scientific information. The Codex procedure provides for such an approach. When the need for an update of an existing standard has been identified, a proposal for revision or amendment follows the same step procedure as described above, up to its adoption by the CAC. During the deliberations of a revised or amended draft standard in Codex, the original standard stays in force to avoid international trade disruptions.

THE CODEX COMMITTEE ON PESTICIDE RESIDUES

The Codex Committee on Pesticide Residues (CCPR) is entrusted with the elaboration of MRLs for pesticide residues in food and feed. The term 'pesticide' in

Codex encompasses a wide range of products for use as crop protection agents, applied in any stage of crop production, storage or transport, and includes products administered to animals for the control of ectoparasites.

For compounds that had been widely used as pesticides in agriculture before their uses were discontinued but which persist in the environment for a relatively long period of time, Extraneous Maximum Residue Limits (EMRLs) are established. These substances are environmental contaminants that can be taken up by plants or animals and end up in foodstuffs in detectable quantities warranting further consideration. Development of an MRL or EMRL begins when the CCPR prepares for approval by the CAC or its Executive Committee a priority list of pesticides for evaluation by the Joint FAO/WHO Meeting on Pesticide Residues (JMPR). After the JMPR reviews pesticide chemistry and residue trials data and recommends maximum residue levels (or extraneous maximum residue levels) for those compounds on the priority list, these recommendations are sent to Member Governments at step 3 of the Codex Procedure for comments. A further responsibility of the CCPR is in the elaboration of methods of sampling and analysis for the determination of pesticide residues in food.

The CCPR is hosted by The Netherlands and was convened for the first time in 1966. In May 2000, the 32nd Session was held. In all of these years, the CCPR has met in The Netherlands, with the exception of 1993 when the venue was moved to Havana upon the invitation of the Government of Cuba. To date, the CCPR has considered about 200 active ingredients, including some groups of compounds and at present 2466 MRLs are in effect. In addition, six compounds have been considered for extraneous residues and 50 EMRLs are established. There are also around 500 MRLs under consideration by the Committee. A list with all MRLs and EMRLs adopted by the CAC is published in Volume 2B of the *Codex Alimentarius* (FAO/WHO, 1998). This list is also available at the website of the FAO (FAO STAT Database Collections).

THE JOINT MEETING ON PESTICIDE RESIDUES

The Joint FAO/WHO Meeting on Pesticide Residues (JMPR) is an expert body administered by the FAO and WHO and is independent and separate from Codex but with which the CCPR maintains a very close working relationship. *The Codex Alimentarius Commission Procedural Manual* includes a section on the Uniform Procedure for the Elaboration of Codex Standards and Related Texts where it is stated 'In the case of Maximum Limits for Residues of Pesticides, the Secretariat distributes the recommendations for maximum limits, when available from the Joint Meetings of the JMPR'. The WHO and FAO Joint Secretaries of the JMPR establish its agenda and co-ordinate the activities of the JMPR with those of the CCPR. The JMPR has a long history, even going back before the time that the CAC was established. The history of the JMPR is briefly summarized in the FAO Manual (FAO, 1997a).

At present, the JMPR comprises the WHO Core Assessment Group and the FAO Panel of Experts on Pesticide Residues in Food and the Environment. The WHO and FAO invite independent experts to participate in a Session of the JMPR. Members of the latter usually have a regulatory or scientific background and are invited in their personal capacities as experts and do not represent their organizations or Governments. In addition, they must not have any conflicts of interests on the subject to be considered.

The independence of the experts of the JMPR is crucial for the reliability of Codex. The latter commission therefore requires an increased transparency in the way in which experts of international committees such as the JMPR are nominated and appointed. The WHO Core Assessment Group falls under the WHO International Programme on Chemical Safety (IPCS). This Group is responsible for reviewing toxicological data of pesticides and establishing values for toxicological end-points, such as the Acceptable Daily Intake (ADI) and the acute reference doses (acute RfDs), where relevant. The ADIs and the acute RfDs are indispensable for the risk analysis of pesticide residues in food. The FAO Panel is responsible for reviewing data on plant and animal metabolism, use patterns of pesticides, results of supervised field trials data and other data pertaining to estimating pesticide residue levels in food and feed commodities. In the case of environmentally persistent compounds which were used as pesticides, but have been banned or 'de-registered', EMRLs are estimated on the basis of monitoring data.

The maximum residue levels estimated by the JMPR are recommended to the Codex, Member countries of the FAO and WHO and other interested parties. The CCPR is, in fact, the main client of the JMPR and considers these JMPR recommendations on maximum residue levels to develop them as Codex MRLs. The JMPR recommendations are also useful for individual Governments, in particular those which lack a national infrastructure for the regulation of pesticides. The JMPR meets generally every year in the autumn, alternately in Rome, Italy, and Geneva, Switzerland, at the headquarters of the parent organizations, FAO and WHO. The results of a meeting of the JMPR are published in a Report, while the detailed monographs are published as Evaluations in two separate volumes, i.e. one of toxicological monographs, the other of monographs on residues.

It is planned that Reports and Evaluations will become routinely available on the Internet. In the past, the Evaluations of the JMPR were highly summarized, but upon request of the CCPR since the early 1990s the Evaluations became more and more detailed, adding to the transparency of the elaboration process of standards. The improved quality of the documents of the JMPR has contributed considerably to the acceptance of Codex MRLs by individual Governments. The JMPR is currently considering aligning its documents with formats of other international organizations involved in reviewing pesticides, such as the Organization for Economic Co-operation and Development (OECD). Reviews and evaluations by international organizations and Governments will become more mutually acceptable, thus making best use of limited resources available worldwide.

The recommendation of maximum residue levels by the JMPR and their further elaboration by the CCPR heavily rely on the availability of data. The manufacturer of the pesticide is an important source of data for evaluation by the JMPR. Data from this source are usually confidential and not published, and are considered proprietary by the data owner, namely the manufacturer. These data, requiring a significant investment, are generated to support a registration and MRLs at the national level. The data owner therefore wishes to see the data sufficiently protected from non-authorized use by his competitors. Both the FAO and WHO have recognized the need to protect the proprietary rights of the manufacturer submitting data to the JMPR for review. IPCS, on behalf of the WHO, has adopted operating procedures to guarantee proprietary rights. The JMPR reports and monographs include a statement that cautions Governments not to grant pesticide registrations based on JMPR evaluations which rely on proprietary data unless authorization for use has been obtained from the data owner.

THE WORLD TRADE ORGANIZATION

It is worthwhile to look at a possible relationship between Codex and the World Trade Organization (WTO) because Codex, through the mechanism of establishing international standards for food, facilitates the free trade of food between countries. The WTO is an international organization responsible for the free trade of goods and services and the protection of intellectual property rights. The organization is seated in Geneva, Switzerland, and was established in 1995 after the successful conclusion of the General Agreement on Tariffs and Trade (GATT) Uruguay Round on Multilateral Trade Negotiations with the signing of the Marrakesh Agreement in 1994. The WTO can be considered as the successor of the GATT, but bringing with it a much broader mandate. In February 1999, 134 countries were Members of the WTO.

Under the Uruguay Round, many separate agreements have been concluded and, among others, agreements for agriculture and agricultural products. For many agricultural commodities, tariff barriers existed which were significantly reduced as a result of the negotiations. Non-tariff barriers, such as food standards, still exist and have become relatively more important. They can form a serious impediment to international trade, certainly when mis-used for protectionist reasons. Two agreements, i.e. the Sanitary and Phytosanitary (SPS) Agreement and the Technical Barriers to Trade (TBT) Agreement, try to address these concerns. Both agreements are of importance to the Codex, but only the relevance of the SPS Agreement will be explained here further as the provisions of this agreement also cover the establishment and use of MRLs for pesticide residues. The TBT Agreement is applicable to both industrial and agricultural products, including foodstuffs. This sets rules to unify technical regulations, for instance, concerning packaging, labelling and marketing requirements. The relevance of the TBT Agreement for Codex is obvious, because these kinds of technical regulations are also considered by this Commission.

THE WTO SPS AGREEMENT

The WTO Agreement on the Application of Sanitary and Phytosanitary (phyto = plant) Measures, in short, the SPS Agreement, applies to all sanitary and phytosanitary measures to protect the health of consumers, plants and animals related to agricultural products and foodstuffs (WTO Home Page). The SPS Agreement, therefore, is a typically health-related agreement. The sanitary and phytosanitary measures may directly or indirectly affect international trade. All countries apply measures to guarantee the safety of these products for consumers and animals and to prevent the spread of infectious diseases among plants and animals. There are many types of so called sanitary and phytosanitary measures, abbreviated as SPS measures. Examples include inspection systems, sanitary provisions for slaughterhouses, quarantine measures for plants and animals from areas with a higher risk for infections, maximum residue limits or levels of pesticides, veterinary drugs and contaminants in agricultural produce and maximum permitted levels of additives in food stuffs.

The SPS Agreement has a significant impact on Codex. For a better understanding of how this agreement influences international standard setting, it is useful to look at it in more detail.

An important notion in the preamble to the Agreement is worth mentioning up front. Literally it reads as follows:

> *Reaffirming that no Member should be prevented from adopting or enforcing measures necessary to protect human, animal or plant life or health, subject to the requirement that these measures are not applied in a manner which would constitute a means of arbitrary or unjustifiable discrimination between Members where the same conditions prevail or a disguised restriction on international trade.*

Although not explicitly stated, this sentence can be interpreted that it is an undeniable prerogative of national Governments to establish a desirable level of protection of the health of humans, animals and plants. In terms of this Agreement, this prerogative has, of course, certain restrictions and should not be inconsistent with the provisions of the Agreement, to avoid creation of new trade barriers.

Article 2 of the Agreement, further lays down the basic rights and obligations of WTO Members. The national responsibility to select an appropriate level of protection is in many cases a political choice based upon science and other factors, such as enforceability of the measure and a risk/benefit analysis. The political nature of a protection level indicates that its harmonization at the international level is not readily achievable. This should be kept in mind when later in this chapter the process of MRL setting for pesticide residues and the risk assessment approach applied by Codex are explained further.

Returning to the SPS Agreement, we note other provisions relevant to Codex. Governments should take these SPS measures only when necessary to protect the health of human, animals and plants and ensure that these measures are based on scientific principles and on an appropriate risk assessment. Measures should be non-discriminatory between domestic and foreign products and where more options exist the least trade-distorting measure should be selected. Of particular importance is the notion of 'equivalence'. Members shall accept the sanitary or phytosanitary measures of other Members as equivalent, even if these measures differ from their own or from those used by other Members trading in the same product, if the exporting Member objectively demonstrates to the importing Member that its measures achieve the importing Member's appropriate level of sanitary or phytosanitary protection.

To harmonize measures, Members of the WTO are invited to base their measures on international standards, such as those of Codex. The International Plant Protection Convention (IPPC) and the International Office of Epizootics (OIE) are other international organizations responsible for establishing international phytosanitary and sanitary standards to protect the health of plants and animals, respectively. Governments are also encouraged to actively participate in the work of these international standard-setting organizations. Conformity with these international standards means conformity with the provisions of the SPS Agreement. Members may deviate from international standards, i.e. apply a higher level of protection, provided that there is a scientific justification to do so or as a consequence of a risk assessment appropriate to the circumstances.

As a consequence of the SPS Agreement, Codex standards (and those of the OIE and the IPPC) have gained additional status as recognized points of reference in international trade disputes. Codex standards are still voluntary in nature but through the mechanism of the SPS Agreement, so to speak, they may bind Governments.

The notions and ideas underlying the SPS Agreement in their turn have reciprocally influenced the standard-setting process of Codex considerably. This is most apparent in the area of risk analysis where, in the late 1990s, the FAO and WHO have devoted a series of Expert Consultations to clarify and to elaborate risk analysis approaches in establishing Codex standards. The leading principle in the SPS Agreement, i.e. that SPS measures should be based on science, sparked an interesting debate in Codex on the question if and to what extent other factors can be taken into account. This debate has not yet been settled by Codex and certainly will be continued in the forthcoming years.

RISK ANALYSIS IN CODEX

The FAO/WHO Conference on Food Standards, Chemicals in Food and Food Trade (FAO, 1991), recommended that Joint FAO/WHO Expert Committees and all relevant Codex Committees 'continue to base their evaluations on suitable scientific principles and ensure necessary consistency in their risk assessment determinations'. The conference recognized that 'risk assessment' is centred on the Joint Expert Committees such as the JMPR rather than the Codex Committees themselves. The WTO SPS Agreement reinforced the necessity of risk analysis methodology in the elaboration of sanitary and phytosanitary measures, including food standards. Moreover, the use of risk analysis methodology facilitates consistent and orderly decision-making. In the light of these developments, the CAC recognized the need to carefully re-visit the risk analysis approaches applied by the Codex Committees and the Joint Expert Committees. At the request of the CAC, between 1995 and 1998 three consecutive FAO/WHO Expert Consultations on the application of risk analysis to food standard issues were held (WHO, 1995a; FAO, 1997b, 1999a). It should be kept in mind that these Consultations, although primarily serving the interest of Codex and of the FAO and WHO, are of equal importance to individual Governments, which also have a responsibility to apply risk analysis consistently whenever establishing SPS measures.

RISK ASSESSMENT

Risk analysis is a process consisting of three separate but nevertheless integrated components, namely risk assessment, risk management and risk communication. The Joint Expert Consultations mentioned above were by and large devoted to these three components, respectively. The first Joint FAO/WHO Expert Consultation on the Application of Risk Analysis to Food Standards Issues, held in March 1995 in Geneva, explored the risk analysis domain and focused, in particular, on risk assessment (WHO, 1995a). The Meeting was also aware of the need for uniform terminology on risk analysis in the work of Codex and considered risk analysis definitions from different sources. The Consultation drafted definitions of risk analysis terms related to food safety and recommended them to the CAC. The latter subsequently amended these definitions, adopted them on an interim base and published the definitions in the Procedural Manual (FAO/WHO, 2000).

Risk assessment in Codex is defined as a scientifically based process consisting of the following four steps: (i) hazard identification, (ii) hazard characterization, (iii) exposure assessment, and (iv) risk characterization. What these steps in practice precisely mean, can be illustrated by using pesticide residues as an example. Selecting a new active substance for consideration by the JMPR and CCPR and subsequent inclusion in the Codex system is the hazard identification step. Only active substances that may be present as residues in a particular food or group of foods will be taken up in the priority list of the CCPR. The evaluation of the toxicological data of a pesticide is the hazard characterization. It is the qualitative and quantitative evaluation of the nature of the adverse health effects associated with the pesticide of which residues may be present in food. The toxicological evaluation usually results in a number of relevant toxicological end-points, including a no-observed adverse-effect level (NOAEL), from which an ADI and, where relevant, an acute reference dose (RfD) can be derived. The quantitative evaluation of the likely intake of a pesticide residue through food is the (dietary) exposure assessment for which in the case of pesticide residues specific international guidelines exist. Finally, the comparison of the intake with the ADI in the case of chronic exposure or with the acute RfD in case of acute exposure is the risk characterization. When the exposure is less than the ADI or the acute RfD, the risk of an adverse health effect is negligible.

The 1995 Geneva Consultation came forward with general conclusions and recommendations and with more specific recommendations related to chemical and biological hazards. An important principle of risk analysis is the functional separation of risk assessment from risk management, but in practice certain interactions between the two are essential for a systematic risk assessment approach. The Consultation in this regard reconfirmed the previous observation of the FAO/WHO Conference on Food Standards (FAO, 1991) that risk assessment primarily is a scientific task which should be carried out by the Joint Expert Meetings. In the case of pesticide residues, it means that this task primarily should be performed by the JMPR as the risk assessor while the CCPR should take its responsibility as the risk manager in the final decision on establishing MRLs for pesticide residues. In recent years, this conclusion has contributed to a further defining of the responsibilities of the JMPR and CCPR.

With regard to chemical hazards, the Consultation pointed further to the need for harmonized approaches to the risk assessment of food additives, contaminants and residues of pesticides and veterinary drugs within Codex and the responsible Expert Meetings, particularly in the assessment of exposure. Different approaches for risk assessment are currently applied by the Committees involved in establishing standards for the different classes of chemicals. These approaches sometimes differ for historical reasons only. In other cases, these differences are fully justifiable. In view of the increased importance of Codex standards under the SPS Agreement, harmonization of these approaches should be pursued to the extent possible. The Consultation made interesting recommendations with regard to exposure assessment. Where necessary, exposure assessment should be expanded to take into account differences in dietary patterns, and should include estimates of intake by especially vulnerable groups. It was recognized that frequently information on food consumption by the general population and by sub-groups of interest is lacking. These data are a prerequisite for a consistent risk assessment of chemicals. Governments were therefore encouraged to generate this type of information and to make it available to the international organizations.

RISK MANAGEMENT

A second consultation, the Joint FAO/WHO Expert Consultation on Risk Management and Food Safety, was held in Rome in January, 1997 (FAO, 1997b). Risk management in Codex is defined as 'the process, distinct from risk assessment, of weighing policy alternatives, in consultation with all interested parties, considering risk assessment and other factors relevant for the health protection of consumers and for the promotion of fair trade practices, and, if needed, selecting appropriate prevention and control options' (amended definition endorsed by the 1999 Session of the CAC (FAO, 1999b), published in the 11th edition of the Procedural Manual). The primary goal of the management of risks associated with food is to protect public health by controlling such risks as effectively as possible through the selection and implementation of appropriate measures.

The Consultation further defined a number of key risk management terms, explored elements of the risk management process and recommended general principles of food safety risk management. The CAC is currently considering these definitions and general principles for adoption as Codex definitions and Codex General Principles. These general principles as defined by the Consultation read as follows:

(1) Risk management should follow a structured approach.
(2) Protection of human health should be the primary consideration in risk management decisions.
(3) Risk management decisions and practices should be transparent.
(4) Determination of risk assessment policy should be included as a specific component of risk management.
(5) Risk management should ensure the scientific integrity of the risk assessment process by maintaining the functional separation of risk management and risk assessment.
(6) Risk management decisions should take into account the uncertainty in the output of the risk assessment.
(7) Risk management should include clear, interactive communication with consumers and other interested parties in all aspects of the process.
(8) Risk management should be a continuing process that takes into account all newly generated data in evaluation and review of risk management decisions.

Most of these principles are quite straightforward and are sufficiently self-explanatory, probably with the exception of the fourth principle. The Consultation defined risk assessment policy as 'Guidelines for value judgement and policy choices, which may need to be applied at specific points in the risk assessment process'. Setting a risk assessment policy is the responsibility of the risk manager, which should be carried out in full co-operation with the risk assessors. The purpose of this policy is to protect the scientific integrity of the risk assessment. It will be interesting to see to what extent these principles are already applied by the CCPR and JMPR when establishing Codex MRLs of pesticide residues in food.

RISK COMMUNICATION

Finally, a third consultation, i.e. the Joint FAO/WHO Expert Consultation on the Application of Risk Communication to Food Standards and Safety Matters, was held in Rome in February, 1998 (FAO, 1999a). Risk communication in Codex is defined as 'the interactive exchange of information and opinions throughout the risk analysis process concerning risk, risk-related factors and risk perceptions, among risk assessors, risk managers, consumers, industry, the academic community and other interested parties'. Risk communication includes the explanation of risk assessment findings and the basis of risk management decisions (amended definition endorsed by the 1999 Session of the CAC (FAO, 1999b), published in the 11th edition of the Procedural Manual). The objectives of the Consultation were as follows:

(1) To identify the elements of, and recommend guiding principles for, effective risk communication.
(2) To examine the barriers to effective risk communication and to recommend means by which they can be overcome.
(3) To identify strategies for effective risk communication within the risk analysis framework.
(4) To provide practical recommendations to the FAO, WHO, Governments, the CAC and other (inter)national organizations and interested parties in order to improve their communication on matters related to the risk assessment and management of food safety hazards.

Risk communication is certainly the most difficult component of the risk analysis process to deal with. The fundamental goal of risk communication is to provide explicit, relevant and accurate information in clear and understandable terms targeted to a specific audience. It includes fostering public trust and confidence in the safety of the food supply. In particular, in relation to the confidence of consumers in their food, risk communication is the art of bridging the gap between the scientific realm of the risk assessment and that of emotions with which consumption of food is associated. The Consultation made a number of conclusions and recommendations. Among those relevant to Codex are the recommendations to proceed as swiftly as possible to elaborate a Codex policy on which legitimate factors other than science should be considered in risk analysis. The Consultation recommended that the FAO and WHO identify and involve experts with a wider range of scientific perspectives in the work of international advisory bodies (such as the JMPR) and expert consultations.

ELABORATION OF MRLS FOR PESTICIDES BY CODEX AND THE JMPR

INTRODUCTION

Definitions in an international context serve a very clear purpose, namely to ensure that there is a common understanding of what is precisely meant. Codex definitions are published in the Procedural Manual (FAO/WHO, 2000).

For the purposes of the Codex Alimentarius, a pesticide residue 'means any specified substance in food, agricultural commodities, or animal feed resulting from the use of a pesticide. The term includes any derivatives of a pesticide, such as conversion products, metabolites, reaction products, and impurities considered to be of toxicological significance'.

A Codex maximum limit for pesticide residues is defined as 'the maximum concentration of a pesticide residue (expressed as mg/kg), recommended by the CAC to be legally permitted in or on food commodities and animal feeds. MRLs are based on data on Good Agricultural Practice (GAP) in the Use of Pesticides and foods derived from commodities that comply with the respective MRLs are intended to be toxicologically acceptable'.

Codex MRLs, which are primarily intended to apply in international trade, are derived from estimations made by the JMPR following (i) toxicological assessments of the pesticide and its residue, and (ii) review of data from supervised trials and supervised uses including those reflecting national good agricultural practices. Data from supervised trials conducted at the highest nationally recommended, authorized or registered uses are included in the review. In order to accommodate variations in national pest control measures, Codex MRLs take into account the higher levels shown to arise in such supervised trials, which are considered to represent effective pest control practices. Consideration of the various dietary residue intake estimates and determinations both at the national and international level in comparison with the ADI, should indicate that foods complying with Codex MRLs are safe for human consumption.

Finally, the definition of GAP in Codex reads as follows: 'GAP includes the nationally authorized safe uses of pesticides under actual conditions necessary for effective and reliable pest control. It encompasses a range of levels of pesticide applications up to the highest authorized use, applied in a manner which leaves a residue which is the smallest amount practicable'. Authorized safe uses are determined at the national level and include nationally registered or recommended uses, which take into account public and occupational health and environmental safety considerations. Actual conditions include any stage in the production, storage, transport, distribution and processing of food commodities and animal feed.

In fact, the principles for establishing Codex MRLs for pesticide residues, as elaborated and refined in the three decades Codex has been involved in this process, are encapsulated in these definitions for a pesticide residue, a Codex MRL and GAP. For a better understanding of these principles and the underlying philosophy, as well as for a clear view of its consequences for the risk analysis, it is worthwhile to have a closer look at these definitions.

GOOD AGRICULTURAL PRACTICE IN THE USE OF A PESTICIDE

An MRL for a pesticide residue has a connotation with consumer health and food safety. The question is whether an MRL as currently established is a real health standard. In principle, an MRL can be derived from a relevant toxicological characteristic such as an ADI. The latter for a chemical is defined as the daily intake which, during an entire lifetime, appears to be without appreciable risk to the health of the consumer on the basis of all known facts at the time of the evaluation of the chemical by the JMPR (FAO/WHO, 1993). Using the ADI, the maximum toxicologically acceptable daily intake can be calculated for an adult or a child just by multiplying the ADI with the average body weight of the age group involved. The toxicological room for the dietary intake is then allocated to certain commodities or groups of commodities resulting in maximum permissible levels for pesticide residues in these products. However, in the early days of the CCPR it was decided to abandon this approach in favour of MRLs based on the GAP of a pesticide.

Frawley (1987) summarizes the reasoning for this while at the same time capturing quite nicely the atmosphere of the first sessions of the CCPR. European countries mainly followed the ADI approach. European national MRLs at this time applied to food commodities at the time of consumption. In contrast, USA tolerances were based on an entirely different approach, taking into account primarily agricultural parameters. These tolerances applied to commodities when first entering the market. So the debate went between 'dinner-plate' tolerances (based on a toxicological approach) versus 'farm-gate' tolerances (based on an agricultural approach). A crucial argument in favour of the latter was the enforceability. It is obvious that enforcement services have better access to market places and food retailers than to places where food is consumed, i.e. private houses or restaurants. The choice of a GAP-based MRL instead of a toxicologically based one has far reaching consequences, both for its very nature as well for the risk assessment methodology to be applied.

From a public health perspective, preferably no residues should be present in agricultural produce destined for human consumption. The presence of a residue is only acceptable when there is an agricultural need to apply a pesticide for crop protection purposes. A residue therefore should never be higher than needed to achieve the required level of crop protection. Experience shows that GAP-based MRLs are generally lower than MRLs derived from toxicological end-points. This is also a strong argument in favour of the GAP approach. It is an example of the 'as low as reasonably achievable' (ALARA) principle adding to a high level of consumer health protection as MRLs are not established up to the maximum toxicologically acceptable level.

What precisely constitutes the GAP of a pesticide? The Codex definition already gives a first clue. This refers to nationally authorized or recommended uses of pesticides under actual conditions to achieve an effective and reliable level of pest control. A first and important notion is that establishing GAP is part of a regulatory or official process, in most cases the approval of a pesticide by national authorities under their pesticide legislation. Once established, in practice GAP forms the basis for the risk assessment, not only for consumers but also for occupational protection and human health in general and for environmental safety as well. The range of dosages selected, the timing and frequency of application and, where relevant, the pre-harvest interval (PHI) applied, should be such that under the prevailing conditions of pesticide application an effective and reliable crop protection result is obtained. The properties of the pesticide and its formulation, the method of application, the epidemiological characteristics of the pest to be controlled, the crop varieties used and the climatic and environmental conditions under which the crop is grown or stored and the pesticide applied, are all key factors in defining the GAP of a given pesticide. GAP then is converted into the final product label directions. Residues resulting from a pesticide application in compliance with its GAP should not be higher then necessary to achieve the desired level of pest control.

One of the prerequisites to choose GAP as the basis for MRL setting at the international level is the need for the recognition that the national GAPs are established by individual Members of Codex under their legislation. This is a very fundamental notion. It should be realized that the mandate of Codex is the harmonization of food standards, including MRLs for pesticide residues. Harmonization of registered uses of pesticides is not within the remit of Codex. Usually, a Codex MRL is driven by the GAP of a country giving rise to the highest residue levels. Generally, GAP varies widely from one country to another, mainly due to differences of the conditions in which crops are cultivated. In addition, national authorities apply different approaches to establish GAP – not all governments, for instance, currently evaluate efficacy data of pesticides – also adding to further diversification of GAP. This is one of the main reasons why governments at CCPR sessions sometimes express their reservations for accepting an MRL based on the GAP from other countries. The CCPR and JMPR currently lack the expertise to reasonably compare GAP from different areas of the world. However, an effort in that direction would mean a considerable burden on the already limited resources available to both Committees while not contributing significantly to the health protection of consumers.

Nevertheless, as the basis for a sustainable crop protection policy national governments should be encouraged to officially assess efficacy data and critically appraise the GAP, in order to establish a better-defined GAP resulting in an optimal use of pesticides in agricultural production. It should also be realized that Codex in view of its mandate is not a suitable forum to actively promulgate biological control or integrated pest management and to implement these principles when establishing MRLs. In addition, in the framework of the SPS Agreement, any measures directly related to plant protection fall within the mandate of the International Plant Protection Convention (IPPC). What can be achieved in that regard in one country does not necessarily mean that the same practices can be implemented in any other country. The agricultural conditions may preclude a successful introduction of these practices and in addition the level of expertise of growers, farmers and advisory officers in a country may also be an impediment to success, at least for the time being. Therefore, the mutual acceptance of national GAP should continue to be the basis for MRL setting at the international level.

However, efforts could be made in the future to explain the basis of national GAP, and thus possibly reduce unnecessary differences which exist for historical reasons only. Particular attention could be paid to the pre-harvest interval (PHI) which can have a very large influence on the level of the MRL and is sometimes reported at only one, or even, zero days by several countries.

SUPERVISED TRIALS FOR THE ESTIMATION OF MRLs

To derive an MRL from GAP information, data are needed on the residue levels that might occur following the application of a pesticide according to a given GAP. For this reason, so-called supervised trials are performed together with other studies related to pesticide chemistry, residue behaviour and metabolism. The FAO Manual for the first time defined supervised trials for the estimation of MRLs in an FAO document (FAO, 1997a). Supervised trials are 'scientific studies in which pesticides are applied to crops or animals according to specified conditions intended to reflect commercial practice after which harvested crops or tissues of slaughtered animals are analysed for pesticide residues. Usually specified conditions are those which approximate existing or proposed GAP'.

Supervised trials are normally carried out in the course of the registration of a pesticide and by necessity are developed in accordance with intended instead of approved uses. As Codex establishes MRLs only for current GAP, manufacturers and governments submitting data to the JMPR should ensure that data from submitted supervised trials reflect current GAP. The residue levels found in supervised trials are used to derive an MRL.

An MRL should be such that a farmer following the label prescriptions may have confidence that his agricultural produce at the time of harvest or slaughter or at the time of marketing in the case of post-harvest uses, is in compliance with the MRL set. Codex therefore takes into account the higher levels shown to arise in supervised trials, which are considered to represent effective control practices. In most cases, figures are rounded up in order to fit Codex MRLs in the geometrical progression range used. Normally figures for Codex MRLs are 0.01, 0.02, 0.05, 0.1, 0.2, 0.5, 1, 2, 5, 10, etc. mg/kg. As a result of the concept that MRLs are based on GAP and data from supervised trials, such limits are derived from uses that are usually no higher than needed to achieve an effective pest control level.

Strictly speaking, an MRL based on GAP is not a health standard but a technical limit reflecting an approved use. Exceedance of an MRL in many cases is most likely the result of the use of a pesticide applied in a way which is not in conformity with the label prescriptions. However, in most cases it is certainly not an indication of food being unsafe. Basically, an MRL is an enforcement limit. Its character as a health standard is from its use as the input for the dietary risk assessment, as will be explained later. It will also become clear later why exceedance of an MRL in most cases does not pose a risk to consumers.

EXTRANEOUS MAXIMUM RESIDUE LIMITS

In contrast to MRLs, Extraneous Maximum Residue Limits (EMRLs) are based on monitoring data because GAP no longer exists for compounds for which EMRLs are established. Over the years, the JMPR has elaborated criteria for establishing EMRLs which are summarized in the FAO Manual (FAO, 1997a). It should be noted that the JMPR uses the term 'extraneous residue level' (ERL). An EMRL must be protective of the health of consumers in the first instance. An EMRL should be distinguishable from background levels but be sufficiently low to be able to detect continued uses of discontinued products. In recent years, the CCPR has considered the necessity of criteria for EMRL setting. The Committees agreed positions on this matter is published in Appendix VIII of the Report of the 31st Session (1999) (FAO, 1999c). An EMRL is not necessarily based upon the highest values in the populations of monitoring data. The JMPR does not consider extreme values as outliers in a statistical sense, but rather as values on a tail of a large distribution. The challenge is to decide which of these extreme values can be discarded without creating unnecessary trade barriers. As a rule of thumb, the JMPR considers violation rates of 0.5 to 1 % or greater generally as unacceptable. The CCPR by and large supported the approach of the JMPR in establishing EMRLs, with the exception of the approach on violation rates. The CCPR recommends that the JMPR, in applying any violation rate to the setting of EMRLs, documents its rationale.

THE CODEX CLASSIFICATION OF FOODS AND ANIMAL FEEDS

In principle, an MRL is established on an individual commodity basis. However, an MRL can also be established for a group of related commodities, such as, for instance, pome fruits or citrus fruits, provided that the commodities are botanically and horticulturally related, the residue potential in the individual commodities in the group is more or less identical and data from supervised trials are available for at least the major commodities constituting the group. The need for harmonization of commodity descriptions at the international level is apparent. Without agreement on the description of commodities and – also importantly – on the part of the commodity to which an MRL applies, the efforts of the Codex to harmonize MRLs of pesticides would be in vain. For example, if an exporting country feels the MRLs for strawberries apply to the fruits without stems and caps while the importing country is of the opposite opinion and therefore includes these parts in the residue analysis, the different outcomes of the analytical laboratories could easily result in a trade problem – the exporting country stating that its strawberries are fully in compliance with a Codex MRL, whereas the importing country may have a case in stating that imported produce is in violation of that standard (caps of strawberries tend to catch a significant amount of the residue).

To avoid this kind of confusion the CCPR developed a classification of foods and feeds. Such a classification is first intended to ensure the use of uniform nomenclature and secondly to classify foods into groups or sub-groups for the purpose of establishing group MRLs. The present Codex Classification of Foods and Animal Feeds originates from 1989 when it was adopted by the CAC. The Classification is published in Volume 2 of the Codex Alimentarius (FAO/WHO, 1993). The Codex Classification includes food commodities and animal feedstuffs for which Codex MRLs will not necessarily be established. The Classification is intended to be as complete a listing of food and feed commodities in trade as possible. In cases where the Codex has established MRLs for groups of food or feed commodities, the Classification is indispensable to knowing which individual commodities are included in a group.

The present Classification comprises three classes of primary food and feed commodities and two classes of processed food commodities, including 19 types of commodities. Each type is divided into groups, which are further sub-divided into the individual commodities for which a letter code, a numerical code and a description (common name and the scientific name) are given.

This can best be illustrated using a simple example. Strawberries belong to primary food commodities of plant origin (class A), to fruits (type 01) and to berries and other small fruits (group 004). The code for strawberries is FB 0275. The heading of this section (group 4) of the Classification reads: 'Berries and other small fruits are derived from a variety of perennial plants and shrubs having fruit characterized by a high surface:weight ratio. The fruits are fully exposed to pesticides applied during the growing season (blossoming until harvest). The entire fruit, often including seed, may be consumed in a succulent or processed form. Portion of commodity to which the MRL applies (and which is analysed): Whole commodity after removal of caps and stems. Currants, Black, Red, White: fruit with stem'. A selection of entries from the Classification most relevant to the work of the CCPR and JMPR is depicted in Table 10.2.

The Codex Classification is an essential background document, not only for Codex and the JMPR, but for others as well. Manufacturers or governments, wishing a Codex MRL to be established for a given commodity, should make sure that supervised trials are conducted in that commodity, the correct portion to which the MRL would apply is analysed, and that in the report of these trials reference is made to the correct Codex code of the commodity involved. Data on GAP submitted to the JMPR should also be unequivocal in that regard. Too often, the JMPR has difficulties in reviewing data as the crops involved cannot be properly identified. In addition, for the analytical chemists and for the enforcement practitioners it has to be clear which part of the product has to be analysed and whether the actual amount of residue found is to be considered as a violation of the standard.

In the early days of Codex and the JMPR, MRLs were frequently established for large groups of commodities such as fruits and vegetables. However, beginning several years ago it was decided no longer to develop MRLs for such large

Table 10.2 A selection of the Codex Classification of Foods and Feeds (FAO/WHO, 1993). In particular, those groups of commodities are shown for which generally Codex MRLs are established

Type	Number	Group	Group letter code
CLASS A	*PRIMARY FOOD COMMODITIES OF PLANT ORIGIN*		
01 Fruits			
	001	Citrus fruits	FC
	002	Pome fruits	FP
	003	Stone fruits	FS
	004	Berries and other small fruits	FB
02 Vegetables			
	010	Brassica (cole or cabbage) vegetables, head cabbage, flowerhead brassicas	VB
	011	Fruiting vegetables, cucurbits	VC
	012	Fruiting vegetables, other than cucurbits	VO
	013	Leafy vegetables (including brassica leafy vegetables)	VL
	014	Legume vegetables	VP
	015	Pulses	VD
	016	Root and tuber vegetables	VR
	017	Stalk and stem vegetables	VS
03 Grasses			
	020	Cereal grains	GC
04 Nuts and Seeds			
	022	Tree nuts	TN
	023	Oilseed	SO
CLASS B	*PRIMARY FOOD COMMODITIES OF ANIMAL ORIGIN*		
06 Mammalian products			
	030	Meat (from mammals other than marine mammals)	MM
	031	Mammalian fats	MF
	032	Edible offal	MO
	033	Milks	ML
07 Poultry products			
	036	Poultry meat (including pigeon meat)	PM
	039	Eggs	PE
CLASS C	*PRIMARY ANIMAL FEED COMMODITIES*		
11 Primary feed commodities of plant origin			
	050	Legume animal feeds	AL
	051	Straw, fodder and forage of cereal grains and grasses (including buckwheat fodder) [forage]	AF
	052	idem [straws and fodder dry]	AS

(continued overleaf)

Table 10.2 (*continued*)

Type	Number	Group	Group letter code
CLASS D PROCESSED FOODS OF PLANT ORIGIN			
12 Secondary food commodities of plant origin			
	055	Dried fruits	DF
13 Derived products of plant origin			
	067	Vegetable oils, crude	OC
	068	Vegetable oils, edible (or refined)	OR
CLASS E PROCESSED FOODS OF ANIMAL ORIGIN			

groups of commodities. MRLs for large groups of commodities necessarily are relatively high as they are driven by the MRL of an individual commodity in that group with the highest residue level. Large-group MRLs have other disadvantages as well. As shown earlier, an MRL is based on GAP. In the case of a large-group MRL, this relationship with GAP is diminished. In addition, from the risk assessment perspective a large-group MRL is of little relevance as it will result in most cases in a considerable overestimate of the actual dietary intake. On the other hand, strictly applying an MRL setting on a single commodity base means a serious burden for manufacturers or grower organizations to generate the necessary field trial data to support MRLs for individual commodities, in particular for minor crops, which is not always commercially justifiable. In such cases, extrapolation from one commodity to another is a possible solution to establish MRLs for (small) groups of commodities and to alleviate the minor crops problem. The JMPR and CCPR are currently considering rules for extrapolation in a more systematic way.

THE RESIDUE DEFINITION

The Codex definition of a pesticide residue refers not only to the active ingredient but also to 'any derivatives of a pesticide, such as conversion products, metabolites, reaction products, and impurities considered to be of toxicological significance'. As was the case with the uniform description of commodities, there should also be an agreed approach for defining the residue before reasonable MRLs can be established at the international level. An MRL can only execute its function as a facilitator of international trade when there is agreement on the level of the MRL, on the commodity description and the portion of the commodity to which the MRL applies and on the question of which substances are included in the residue definition.

Basically, two elements of a residue definition can be distinguished, i.e. an analytical one and a toxicological one. The basic requirements for the definition of residues are that it should be most suitable for monitoring compliance with the MRL and therefore with GAP, and it should include compounds of toxicological interest for the risk assessment. These two functions are not always compatible, sometimes resulting in fairly arbitrary definitions. Moreover, Governments have different policies towards the choice of a residue definition. From the analytical perspective first of all it is important to know whether the residue is predominantly present as an active ingredient or in the form of one or more metabolites. In the latter case, the metabolites are more relevant to define the residue. The method of analysis in this regard also plays an important role. When the residue has to be converted into a common analyte, that analyte, of course, will in most cases drive the residue definition. From the point of view of cost-effectiveness of enforcement laboratories, there is a preference for simple residue definitions for routine monitoring and enforcement of an MRL. The current practice of the JMPR and CCPR is to follow the indicator compound approach, which means that the substance which is predominantly present in the residue or in a constant ratio to the total residue forms the basis for the residue definition. Some governments deviate from this approach and prefer to also include metabolites, when present, on a regular basis in the residue definition.

The inclusion of metabolites in the definition is of particular relevance when they are of toxicological concern. This allows for considering toxicologically relevant metabolites in the dietary risk assessment although it complicates the work of the analysts. As will be explained later when describing the risk analysis policy of the CCPR and JMPR, a recently agreed approach at the international level is to establish separate definitions for enforcement and dietary intake purposes, when appropriate. For enforcement purposes, the more straightforward indicator compound approach is followed. Toxicologically relevant substances may form part of the definition for dietary risk assessment. For the sake of harmonization, Members of Codex are encouraged to align their approaches with those of Codex. The complexity of establishing a residue definition is further elucidated in the FAO Manual (FAO, 1997a).

METHODS OF SAMPLING AND ANALYSIS

The 23rd Session of the CAC (1999) adopted the revised recommended methods of sampling for the determination of pesticide residues for compliance with MRLs which replaced the previous sampling methods (FAO/WHO, 1993) and will be published in Volume 2A of the Codex Alimentarius. Sampling for the enforcement of an MRL should be consistent with the principles applied in the setting of that MRL and must be practical for the examination of lots in trade.

In the case of meat and poultry products, generally the JMPR reviews data from individual animals and their separately analysed tissues, except when combining of tissues is required to obtain an adequate sample size for analysis. Combining tissues from more than one bird is the case, for instance, for poultry organs. Hence, in the case of meat and poultry products (excluding eggs) generally so-called primary samples are analysed. For all other commodities, including eggs and milks, the recommended sampling for field trials involves collection of a bulk sample made up of a number of primary samples which are combined as the final composite sample. This sample, or a relevant part of it, is then analysed.

Thus, in the enforcement practice, the principle of applying an MRL for meat and poultry products to the residue concentration found in primary samples, and applying the MRL for most other commodities to the residue concentration found in a final sample is consistent with the way that the JMPR evaluates data to recommend MRLs to the Codex.

Volume 2 of the Codex Alimentarius contains a section with recommended methods of analysis of pesticide residues (FAO/WHO, 1993). These recommended methods for analysis can, from practical experience of the CCPR, be applied to the determination of pesticide residues for regulatory purposes. The policy of the CCPR in this regard is not to restrict analysts to use these recommended methods of analysis, but to allow alternative methods to be used, thereby allowing for quick implementation of new developments in the area of analytical chemistry and technology.

Table 10.3 gives a list of MRLs for pesticides in food and animal feed, at various steps of the Codex Procedure. The example (quintozene) is taken from a working document of the 1999 CCPR. This summarizes the recommendations of the JMPR on the ADI, the MRLs, the residue definition(s) and reflects the status of the decisions of the CCPR and the CAC (steps).

RISK ANALYSIS BY THE JMPR AND CCPR

As explained earlier, MRLs of pesticides are based on GAP and on data from supervised field trials. MRLs are, in fact, enforcement levels, only indicating whether a pesticide is used in accordance with its GAP. As an MRL is not related to any toxicological characteristic of the pesticide, the question arises if food with actual residue levels up to the MRL is safe for consumers. Time and again this problem has been discussed by the CCPR when establishing MRLs for a given pesticide: will these MRLs result in a situation in which the relevant toxicological parameter, the ADI or the acute RfD be exceeded or not? A definitive answer to this question can only be obtained by means of dietary intake studies, e.g. total diet studies. Unfortunately, these types of studies are generally not available and certainly not at the international level. Therefore, this question can only be solved by predicting the dietary intake of pesticide residues on the basis of available data. Once predicted, the intake of a pesticide is then compared with the relevant toxicological parameter. There are two important steps in the risk assessment of pesticide residues, i.e. the dietary exposure assessment and the final risk characterization.

The first guidelines for predicting the dietary intake of pesticide residues were elaborated by a FAO/WHO Consultation held in Geneva in 1987 and were published in 1989 (WHO, 1989). These Guidelines focused on long-term or chronic

Table 10.3 List of MRLs for pesticides in food and animal feeds, at various steps of the Codex Procedure (example taken from a working document of the 1999 CCPR)

---

64   QUINTOZENE
Main uses[a] 5 FUNGICIDE
JMPR[b]   69, 73, 74R, 75, 95', 98R
ADI[c]   0.01 mg/kg body weight for quintozene containing less than 0.1 % hexachlorobenzene (1995)
RESIDUE[d] • Plant commodities, quintozene; animal commodities, sum of quintozene, pentachloroaniline and methyl pentachlorophenyl sulfide, expressed as quintozene (fat-soluble)
            • These residue definitions are for enforcement purposes. For consumer risk assessment purposes both plant and animal commodities are defined as the sum of quintozene, pentachloroaniline and methyl pentachlorophenyl sulfide, expressed as quintozene (fat-soluble)

Code[e]	Commodity name	MRL (mg/kg)	Step[f]	JMPR[g]
GC 0640	Barley	0.01(*)[h]	3	98
AS 0640	Barley straw and fodder, Dry	0.01(*)	3	98
VB 0400	Broccoli	0.02	CXL	
VB 0400	Broccoli	0.05	3(a)	98
PM 0840	Chicken meat	0.1(*)(fat)[i]	3	98
PO 0840	Chicken, Edible offal of	0.1(*)	3	98
VL 0482	Lettuce, Head	3 1[j]	CXL	95, 98
SO 0703	Peanut, Whole	5	CXL	

[a] The code number 5 refers to the classification of uses.

[b] The years the JMPR reviewed toxicology and residue data: a year followed by an R or a T indicates, respectively, that only residue data or data on toxicology were reviewed in that year; a year followed by' refers to the year of the periodic review: future schedules can be included.

[c] The 1995 JMPR reviewed the toxicology and changed the ADI from 0.007 to 0.01 mg/kg body weight.

[d] In the case of quintozene, the 1995 JMPR established separate residue definitions for plant and animal commodities for enforcement purposes and established a separate definition for the risk assessment.

[e] The codes refer to the Codex Classification of Foods and Feeds.

[f] The 'Step' refers to the Codex 'Step Procedure' and indicates the step that the proposal has reached: A proposal in a step with the suffix (a) refers to the periodic review procedure – such a proposal is an amendment of the existing CXL (Codex Standard adopted by the CAC) and will replace the existing CXL when it reaches step 8. Both the CXL and the new proposal are included in the table – when only the CXL is included, the JMPR confirmed the existing standard.

[g] Quintozene was reviewed by the 1995 JMPR as a compound for periodic review. As critical supporting studies were not available, the Meeting recommended withdrawal of all existing CXLs and did not recommend new or amended MRLs. As the manufacturer made a commitment to submit the lacking data, the CCPR decided to maintain the CXLs for a period of four years (in accordance with the periodic review procedure). The 1998 JMPR reviewed the critical data and recommended MRLs for consideration by Codex. Therefore, the year of reference is 1998 instead of 1995.

[h] (*) following MRLs: at or about the limit of determination.

[i] The MRL applies to the fat of meat. Other notes on the MRLs are: Po, the MRL accommodates post-harvest treatment of the commodity; T, the MRL is temporary, until required information has been provided and evaluated.

[j] Explanatory footnotes briefly describing, respectively, the recommendations and decisions of the JMPR and the CCPR (in this case, the 1995 JMPR requested additional information but recommended withdrawal of the CXL in 1998).

dietary intake of pesticide residues. For the dietary exposure assessment, two parameters are relevant, namely the residue level in food and the amount of food consumed.

Several indices of the residue level can be used to predict the intake of residues. The MRL is one such index and represents the maximum residue level that is expected to occur in a commodity following the application of a pesticide according to GAP. Using MRLs in the prediction of the dietary intake of a pesticide generally will lead to an overestimate of the actual intake since residues are mostly well below the MRLs even when a pesticide is applied at the maximum GAP. In addition, in practice a pesticide is only rarely used up to its maximum GAP, which is clearly demonstrated by the data of monitoring programmes of national governments. Residues found in food commodities at the retail level are generally well below the MRL or are even not detectable. A further decrease of residue levels may occur as a result of food processing, such as washing, peeling or cooking, and through storage and transport of food.

Although an MRL is a reasonable measure of the residue level, at least for a 'worst-case' dietary intake estimate, the selection of an appropriate measure for the food consumption is more problematic. For a comparison of the dietary intake of residues with the ADI, which is based on the acceptable intake over a lifetime, information on long-term food consumption habits is needed. Up-to-date information on average daily food consumption is scarce, even at the country level, as only a few governments regularly carry out food consumption surveys. Moreover, in view of the variability of consumption patterns between countries, this information is not very suitable for use at the international level.

For the purpose of predicting pesticide residue intake at the international level, therefore, it has been decided to rely on the five regional diets developed by the Global Environment Monitoring System/Food Contamination Monitoring and Assessment Programme (GEMS/Food) of the WHO based on average food consumption data given in the FAO Food Balance Sheets (FBS). These are generally based on a country's annual food production – adding the amount of imported food, subtracting the amount of exported food and dividing by the population. Current information on these regional diets is published in an informal document of the WHO (WHO, 1998a). The regional diets are described as Middle Eastern, Far Eastern, African, Latin American and European. Food consumption values are given for the whole raw agricultural commodity. In some cases, values are given for processed commodities, in particular, when these commodities are eaten as such. Average food consumption estimates derived from FBS are usually higher than the actual average food consumption based on more precise national surveys. However, the FBS data do not take any account of differences in consumption within the population, nor any population sub-groups. Despite these inaccuracies, the regional diets derived from FBS are at present the most reliable index for food consumption for use at the international level. The 1989 Guidelines still included a hypothetical global diet consisting of the highest average value of food consumption from each of the regional diets.

The total dietary intake of residues from any pesticide is obtained by multiplying the residue level in a food commodity by the amount of that commodity consumed and subsequently by summing the intakes from all commodities containing residues of the pesticide concerned. The 1989 Guidelines follow a tiered approach for predicting the dietary intake. The first tier is the Theoretical Maximum Daily Intake (TMDI), calculated for the hypothetical global diet or a national diet and using the MRL as the input for the residue level. The TMDI is a gross overestimate of the actual intake because (i) the proportion of a crop treated with a pesticide is usually far less than 100 %, (ii) very few of the crops treated with a pesticide contain the maximum residue level, (iii) residues normally dissipate during storage, transport and food processing, and (iv) the MRL applies to the whole raw commodity, frequently including inedible parts. Upon removal of this inedible part, a significant proportion of the residue may be discarded. Through a second and third tier, the intake estimate could be refined as outlined below.

According to the 1989 Guidelines, the second tier was called the Estimated Maximum Daily Intake (EMDI) and was a more realistic prediction of the intake based on the regional diets. This is calculated by using data on the edible portion of the commodity and takes into account the effects of processing and preparation of food. Although the EMDI is a more realistic estimate, it is still a large overestimate of the intake. It is based on the assumption that the whole crop is treated with the pesticide under consideration and the residue in the edible portion is still derived from the MRL, which is, as indicated before, not a very realistic input for an intake assessment. The third tier was the Estimated Daily Intake (EDI) which takes into account factors such as (i) data on food consumption, including that of sub-groups of the population, (ii) known uses of the pesticide concerned, (iii) known residue levels, (iv) the proportion of the crop treated, (v) the ratio between the amount of home-grown produce to imported food, and (vi) the reduction in the level of the residue during storage, processing and food preparation. It is obvious that EDI calculations can only be performed at the national level, as the type of information needed is usually only available at this level.

Immediately after publication, the 1989 Guidelines were routinely used by the JMPR and the CCPR as well as by individual governments. Since the early 1990s, the Reports of the JMPR summarized the intake estimates indicating for which pesticides the TMDI or EMDI had been calculated and whether the ADI was exceeded. After each Meeting of the JMPR, more detailed calculations were also made available by the WHO to the CCPR. The 1989 Guidelines proved to be very useful as a screening tool in answering the question as to whether MRLs under consideration by the CCPR are safe or pose a possible risk to consumers, i.e. result in a situation in which the ADI is exceeded. Paradoxically enough, the Guidelines became at the same time a serious handicap in establishing MRLs by the CCPR for those pesticides for which the intake estimates exceeded the ADI. In fact, the CCPR struggled with the question of how to implement risk analysis consistently in establishing MRLs. The Guidelines clearly state that when the

TMDI exceeds the ADI it should not be concluded that the proposed Codex MRLs are unacceptable. The TMDI is a gross overestimate of the intake and is to be considered only as a screening tool indicating the cases where a pesticide is safe for consumers. In the CCPR disputes arose, however, in cases where a further refinement of the intake estimate was warranted. Some governments wished to rely on their national monitoring data in considering the acceptability of MRLs for which the TMDI (or the EMDI) exceeded the ADI. Other governments, however, strictly followed the Guidelines, stating that further refinement of the intake estimate in many cases was not possible due to lack of data, such as residue levels in the edible portion or effect of food processing or food preparation on residue levels. A number of draft Codex MRLs, therefore, stayed in a certain step of the Codex Procedure for a couple of years pending a solution to bridge the diverging opinions expressed by governments at the CCPR.

Upon the request of the CCPR, an FAO/WHO Expert Consultation was convened in 1995 in York in the United Kingdom with the objective of reviewing the existing guidelines and to recommend feasible approaches for improving the reliability and accuracy of methods for predicting dietary intake of pesticide residues in relation to chronic risk (WHO, 1995b). The Consultation noted that the stepwise process of estimating the TMDI, EMDI and EDI of the existing guidelines is a very prescriptive one, often precluding the early use of the most realistic residue data for the prediction of the dietary intake. The Consultation, therefore, identified as a central issue that emphasis should be placed on making the best use of all available data at a given time. The Consultation agreed that the TMDI should be retained, mainly as a quick and resources-saving screening tool to identify cases of no intake concern. The TMDI should be based on the five regional diets. The EMDI as a separate step should be discontinued while the EDI concept should be retained in a modified form. To that effect, the Consultation introduced the IEDI and the NEDI, the International Estimated Daily Intake and the National Estimated Daily Intake, respectively. The IEDI incorporates those factors that can be applied at the international level and which comprise a sub-set of the factors that might be considered at the national level. The NEDI represents a refinement of the international assessment based on more realistic estimates of pesticide residue levels in food and consumption data. A second important result of the York Consultation was to do away with the MRL as the input for the intake assessment, i.e. for the estimation of the IEDI and the NEDI. It was felt that the median pesticide residue level from supervised trials – calculated from the same data base as used to derive an MRL – is a more realistic value than the MRL for the long-term intake assessment and the subsequent comparison of the intake with the ADI. Thus, the STMR – the Supervised Trials Median Residue concept – was born.

As explained previously, the residue definition of the MRL established for enforcement purposes is not necessarily suitable for the risk assessment. For dietary intake purposes it is desirable to include also metabolites and degradation products of toxicological concern. Therefore, the Consultation agreed that, during evaluations, the composition of residues should be considered for both dietary intake estimates and regulatory purposes (compliance with MRLs) and, if necessary, separate definitions of residues should be developed. Finally, the Consultation noted that if the ADI is still exceeded by the estimated intake after all factors are applied, the risk characterization should be refined by taking into consideration additional relevant data provided by manufacturers and other sources. If the ADI is exceeded by the estimated intake after all relevant factors and data are applied, the dietary intake concern becomes a risk management issue. Based on the recommendations of the successful York Consultation, revised Guidelines were prepared in collaboration with the CCPR and were published in 1997 (WHO, 1997a).

Already the 1996 JMPR began to implement the recommendations of the York Consultation and estimated in that year for the first time STMR values. The JMPR continues to recommend MRLs but routinely establishes STMRs for dietary intake purposes. Where relevant, also separate residue definitions for the risk assessment are established. The STMR concept proved to be a significant refinement of the dietary intake assessment as was illustrated in a worked example with parathion-methyl comparing the intake estimate following the 1989 Guidelines and the draft revised Guidelines, prepared at the request of the Codex Secretariat (FAO, 1997c).

The recommendations of the York Consultation and the revised Guidelines formed the basis to reach consensus in the CCPR on long-term risk assessment in establishing MRLs. In the course of the deliberations it became clear that a distinction should be made between what can be achieved at the international level and at the national level. As explained earlier in relation to the SPS Agreement of the WTO, it is a prerogative of national governments to establish a desirable level of protection for their population. This already precludes full harmonization at the international level of risk analysis approaches as different views may occur on the acceptability of a certain amount of risk.

Taking into account that Codex is an international organization which should rely on international evaluations and guidelines, the 29th CCPR (1997) agreed that their decisions should be exclusively based on MRLs, STMRs and ADIs respectively recommended and established by the JMPR (also an international organization) and following the internationally agreed guidelines on predicting the dietary intake of pesticide residues which make use of the five regional diets. The Committee concluded that, when the best estimate of the intake for each of the five regional diets does not exceed the ADI, the MRLs can be recommended to the CAC for adoption as Codex standards, irrespective of the outcome of a national risk assessment. In this situation, a government may wish to inform the CAC that draft Codex MRLs may pose a risk to its consumers, preventing the government from subsequently accepting these Codex MRLs, but their national assessment will not block the establishment of Codex MRLs.

However, when the best estimate of the dietary intake in any of the five regional diets exceeds the ADI, the draft MRLs will not advance through the Codex step procedure. Then, a further refinement of the intake assessment at the international level is needed, making use of additional data. If the ADI is still exceeded in one or more of the regional diets, after exhaustion of options for refining the dietary intake estimates, it becomes a risk management issue. The CCPR will, for example, recommend withdrawal of some of the draft or adopted MRLs to the extent the intake no longer exceeds the ADI, unless national refined intake calculations indicate that these MRLs do not pose a risk to consumers. For the chronic risk assessment, the CCPR is of the opinion that this policy reasonably strikes the balance between consumer protection and economic interests, being neither overly protective nor giving priority to trade implications.

In the case where the ADI is still exceeded, the way to select which draft and Codex MRLs should be withdrawn is still a matter of future consideration by the CCPR. The manufacturer, of course, may identify which draft or adopted MRLs are no longer of interest to the company. From an agricultural perspective, however, Governments may arrive at entirely different conclusions. A further complication is that these interests of Governments in many cases will not point in the same direction.

When the CCPR began to use the 1989 Guidelines, matters frequently had to be referred to the JMPR requesting a refinement of the evaluations, for instance, considering the impact of food processing. The JMPR, rather than the CCPR, was in a position to request this information from manufacturers when it was missing. These frequent referrals to the JMPR were inefficient and caused a further delay in elaboration of Codex MRLs. Moreover, it was felt that the risk assessment should be further integrated in the activities of the JMPR. In view of the recommendations of the 1995 FAO/WHO Consultation (WHO, 1995a) on strengthening the consideration of risk assessment in the elaboration and use of Codex MRLs, in particular, in making a clear distinction between risk assessment and risk management responsibilities, the 1997 JMPR proposed that MRLs for which the available information is insufficient for the JMPR to conclude that the ADI would not likely be exceeded, should be clearly identified (FAO, 1997d).

There is clearly a need for risk managers in Codex and at the national level to be able to make a distinction between these two types of MRLs recommended by the JMPR. The 1997 JMPR recommended that the two types of MRLs should be distinguished for new or periodic review compounds, clearly stating the information needed for the JMPR to refine the dietary intake estimates at the international level. The 1998 JMPR for the first time flagged MRLs exceeding the ADI explicitly. The CCPR generally appreciates this new approach. The 1999 JMPR furthered this approach, extending it to MRLs exceeding the acute RfD.

A further consideration in chronic intake estimates conducted at the international level is that although the consumption data of the five regional diets are related to long-term intake patterns of average populations, they do not give any indication on, for instance, intake values of individual consumers or particular population sub-groups. However, some governments can estimate dietary intake for individual consumers or population sub-groups at the national level, although in interpreting these estimates it should be remembered that the ADI is by definition the toxicologically acceptable level for exposure to a chemical over an entire lifetime.

Although the 1989 Guidelines and the 1997 revised Guidelines proved to be very useful in establishing MRLs for pesticides in general, the CCPR gradually became aware of the necessity to also consider short-term exposure in relation to acutely toxic pesticides.

Upon request of the CCPR, the 1994 JMPR considered situations in which the ADI was probably not an appropriate toxicological benchmark for assessing risks posed by short-term exposure to acutely toxic residues. The acute reference dose (acute RfD) was developed to assess acute hazards using the same basic principles and methods that are used to derive the ADI. As part of this assessment, the JMPR specifies the sub-groups of the population that are at risk on the basis of the acute toxicity end-points used to establish the acute RfD. This allows for the use of the appropriate food consumption and body weight in the short-term risk assessment. The 1998 JMPR, upon request of the CCPR, considered the principles of establishing acute reference doses and drafted a general guidance document on this subject.

The 1995 York Consultation (WHO, 1995b) briefly dealt with a number of issues related to acute exposure. The Consultation agreed that the WHO should develop a database of large-portion weights for fruits, vegetables and other selected commodities in order to conduct initial acute intake assessments. The Consultation further recommended that the JMPR should routinely assess the potential for various acute toxic effects when evaluating all pesticides and, if necessary, consider establishing acute reference doses together with ADIs. Further progress on acute intake assessment was made at a 1997 FAO/WHO Consultation in Geneva (WHO, 1997b). Being aware of the variability in residue levels in individual units of a composite sample on which MRLs are based, the Consultation distinguished two cases for acute dietary exposure assessment. These two cases reflect whether or not the composite sample data represent the residue level in the food as consumed which mainly depends on the size of the individual items in the composite sample.

The Consultation coined two new abbreviations, i.e. the IESTI and the NESTI, the international and national estimates of short-term intake, respectively. In the formulae as proposed to calculate the IESTI and the NESTI, a few factors are of utmost importance, for which at least at the international level reliable data are so far lacking. In addition, at the national level usually such data are not readily available in most countries. These crucial factors are the large portion consumption data for the commodity, the median weight of the commodity unit and a variability factor reflecting the variability of the residue present in the individual units in a composite sample.

To further implement the recommendations on acute intake assessment of the 1997 Consultation, the Government of the United Kingdom convened in 1998 in York an international conference on pesticide residues variability and acute dietary risk assessment (PSD, 1998). The Conference explored the criteria for derivation of acute reference doses, developed further the underlying science of the acute intake methodology as proposed at the 1997 Geneva Consultation and focused on the causes of residue variability and the parameters underpinning the variability factor. This Report was brought to the attention of the 1999 CCPR and the 1999 JMPR. The latter refined the methodology to calculate the IESTI and estimated for the first time IESTI values for the compounds under consideration by the Meeting. It based its estimations on databases for large-portion consumption and commodity unit weights as recently established by the WHO. It should be realized that these databases were relatively limited at the time as only data from a few countries have been made available to the WHO.

There is now at the international level an urgent need to establish an agreed methodology for acute dietary exposure assessment of pesticide residues. As described above, decision making in the CCPR on the elaboration of MRLs so far has been based solely upon chronic intake considerations. However, the Committee has already identified the acute risk assessment as an area of high priority and subsequently has asked guidance from the JMPR. The issue of acute risks will affect, in particular, the MRL setting for organophosphorus compounds and carbamates, at least for certain commodities like pome and stone fruits for which single units are eaten.

The awareness in the CCPR for acute risks began to rise in the early 1990s and was fuelled by a publication of the US National Research Council on pesticides in the diets of infants and children (NRC, 1993) which in general focused on the risks of pesticide residues for these age groups. The concerns of the Council were based on the possible differences in susceptibility toward chemicals and to differences in exposure (= intake) between children and infants and adults. The report drew the attention of governments and consumer organizations around the world. In the USA, its recommendations formed the basis for the Food Quality Protection Act (FQPA) which was passed in 1996 and which is currently being implemented by the US Environmental Protection Agency. The FQPA is considered as a cornerstone legislation to protect the health and safety of all Americans, in particular children. In the European Union, the report of the National Research Council resulted, among others, in a Directive for infant formulae and follow-up formulae, setting a generic MRL at a level for pesticide residues for these types of foods which do not allow the use of pesticides on the ingredients. In view of these and other developments at the national and regional levels, the CCPR wishes to implement expeditiously an internationally agreed methodology on acute risk assessment in elaborating MRLs for pesticide residues.

Apart from the acute intake assessment, the CCPR has in recent years also given due consideration to a number of related issues, such as aggregate exposure and the cumulative exposure to pesticides with a common mechanism of action. With regard to aggregate exposure (exposure through different routes, both dietary and non-dietary), the Committee was of the opinion that this can best be handled at the national level, as suitable data on different routes are lacking at the international level. Questions on cumulative exposure, in particular for compounds with a common mechanism of toxicity, were referred to the JMPR for further consideration and guidance.

To summarize, during the 1990s considerable progress was made by the CCPR and JMPR in improving and implementing principles of risk assessment and risk management in the elaboration of MRLs for pesticides. At the end of the second decade in the history of the CCPR, the first guidelines for predicting the dietary intake of pesticide residues were developed and published. Their use by the JMPR and CCPR provoked an intense debate in the CCPR on chronic risk assessment which was more or less settled when the revised guidelines were published, introducing the concept of the STMR as the basis for the chronic exposure assessment and emphasizing the use of all available factors, such as levels in edible portions, processing factors, etc. In the years to come, the acute risk assessment has to be addressed expeditiously.

In conclusion, while much has been achieved, much has still to be resolved. It should be realized in that regard that Codex (and the JMPR) is not a supranational regulatory body. Codex has to rely on data either primarily requested by Governments from manufacturers as part of their national regulation schemes or generated by the Governments themselves (for instance, food consumption data). The methodology for exposure assessment available at the international level, by definition, is less sophisticated in comparison with national approaches. For example, well-designed food consumption surveys at the national level will always be superior to whatever can be achieved internationally. Consequently, at the international level a larger degree of uncertainty has to be accepted in the risk assessment. Ultimately the aim should be to strike the balance between an overestimate and an underestimate of the risk for consumers. A mix-up of international methodology with national approaches should be avoided. The CCPR could only reach consensus on the chronic intake assessment when it realized that, as an international organization considering recommendations of another international organization (JMPR), it should rely only on internationally agreed methodology and estimates that can be performed internationally, based upon internationally available data and information.

PRIORITY SETTING

One element of the terms of reference of the CCPR has not yet been discussed. This is the obligation to prepare priority lists of pesticides and pesticide–commodity combinations for evaluation by the JMPR for eventual elaboration of Codex MRLs. Over the last thirty years, the CCPR has established, or is considering, MRLs (and EMRLs) for about 200 active ingredients. Almost from the very beginning of the CCPR a priority setting system for evaluating compounds has been in place. Although it is a responsibility of the CCPR to elaborate these lists based on proposals of Member Governments, which need to be adopted officially by the CAC as new work, in practice these lists are prepared in close collaboration with both Joint Secretaries of the JMPR. The final scheduling of evaluations is the responsibility of the JMPR. The Joint Secretaries are better informed on the timing of data submissions by manufacturers and Governments and have a better insight into the workload of a provisional agenda for a given year.

Over the years, the CCPR has elaborated criteria for inclusion of substances and commodities in the priority list, taking into account the general provisions of the CAC. Before a pesticide can be considered for inclusion in the priority list, the compound must be available for use as a commercial product. Moreover, the pesticide must give rise to residues in or on food or feed commodities moving in international trade. The presence of such residues must be a matter of health concern and thus create or have the potential to create problems in international trade. The commodities for which establishment of Codex MRLs or EMRLs are requested should form a component in international trade, represent a significant proportion of the diet and contain pesticide residues, as evidenced, for instance, in monitoring programmes.

Governments proposing new compounds for inclusion on the priority list should consult the manufacturers about the existence of sufficient toxicology and residue data and on the willingness of the manufacturers to submit the information in due course to the JMPR. In the late 1980s, the CCPR encouraged Governments and manufacturers to actively propose and support inclusion of new compounds in the priority list. This worked well for several years. However, in the late 1990s the number of new compounds for evaluation by the JMPR dropped again. This is partly due to the success of the periodic review of compounds already included in the system, adding considerably to the workload of the JMPR. On the other hand, manufacturers seem to be more reluctant to sponsor compounds for Codex purposes, as they are at the same time also deeply involved in the re-evaluation of their products by national or regional authorities. The pesticide re-evaluation programmes of the USA and the European Union can in particular be mentioned in this regard.

PERIODIC REVIEW

The periodic review of compounds which have been evaluated by the JMPR in the past and for which Codex MRLs have been established since 1966, is currently one of the most successful parts of the programmes of the JMPR and the CCPR. The need for a periodic review of the toxicology and the pertinent residue data of pesticides was recognized in the late 1980s. In the years thereafter a procedure was developed by the CCPR in close co-operation with the JMPR, which was ultimately endorsed by the CAC in 1997 (FAO, 1997e). The procedure is appended to the FAO Manual (FAO, 1997a). In fact, the JMPR performed its first full periodic review of a few chemicals in 1991, improving its approach in subsequent meetings.

At the national level, pesticides are commonly registered for a fixed period of time after which the manufacturer has to apply for a new registration or for the continuation of the existing registration. This gives national authorities not only the opportunity to evaluate newly generated data on toxicology, occupational safety and environmental risks in accordance with the current state of science and knowledge, but also to match a renewed approval with the up-to-date criteria for health and environmental protection. This may result in certain restrictions in the use of the pesticide in comparison with the previous registration, in a withdrawal of one or more uses or ultimately even in a total ban. In the late 1980s, some Governments, such as the USA, already had in place a national re-registration programme, or were considering such a programme. For instance, at that time the European Community was negotiating a registration directive for agricultural pesticides which also covered the review of pesticides already on the market in the Community. This directive was published in 1991.

In practice, the review of pesticides requires the generation of a substantial amount of new data which have to be provided by the manufacturer. A manufacturer therefore has to decide whether it is commercially or otherwise of interest to continue to support a chemical or at least to maintain certain uses. As a consequence of national review programmes, many pesticides have been withdrawn from the market and uses of other products have been restricted, so leaving the farmer less choice for crop protection with chemicals. In particular, in the case of minor commodities this has become a problem. Likewise, the continued support or the establishment of a national or regional MRL is under the pressure of additional data requirements. In view of these national initiatives and their consequences, it is fully understandable that Governments asked in Codex for a periodic review of pesticides for which Codex MRLs had been established in the past. A full review and a renewed risk assessment requires an update of the toxicology and residue evaluation and a review of current GAP. In the course of the life of a pesticide, its GAP may frequently change significantly. Basically, for a periodic review a full data package is needed as is the case for the review of a new compound.

The first step in the procedure is the identification of candidate chemicals for re-evaluation. In fact, this corresponds with the first step of the Codex procedure for elaborating or revising standards. Pesticides for which MRLs were first established more than ten years ago are tentatively listed for review. The candidate list then is notified to Member Governments and interested Observer organizations, representing data owners, by way of the reports of the CCPR sessions and Codex circular letters requesting information on the commitment to support the review. Recently, the provisional agendas of the JMPR have been uploaded onto the FAO web site to which data owners have direct access. Where there is a commitment, the data owner (almost always a manufacturer, or in an exceptional case, a grower organization) should submit to the JMPR and CCPR a list of all commodities being supported, a brief summary of all current GAP related to supported commodities and a list of all chemistry (residue, metabolism, animal transfer, processing, analytical sample storage stability, analytical methods, etc.), toxicology studies and other data the data owner is willing to provide to make a complete data package submission to the JMPR. This positive information on data commitment is used to schedule final JMPR reviews. In contrast, a lack of commitment to support a chemical is also brought to the attention of the CCPR. The Committee then will recommend, for endorsement by the CAC, the revocation of all existing Codex MRLs.

Once the JMPR has reviewed a chemical, three scenarios may occur. First, there were sufficient data to confirm the existing Codex MRL that remains in place, or secondly, there were sufficient data to recommend a new MRL or to amend an existing Codex MRL. The new or amended proposal then enters step 3 of the Codex procedure (see Table 10.3). The existing Codex MRL remains in place for no more than four years when it is automatically deleted. Immediate withdrawal would result in trade disruptions, which is avoided by maintaining the Codex MRL for a few years. In the third case, insufficient data have been submitted to confirm or to amend an existing Codex MRL. On being advised of this data inadequacy, the manufacturer may provide a commitment to the JMPR and CCPR to provide the necessary data for review within four years. The existing Codex MRL is maintained for a period of no more than four years pending the review of the additional data. The four-year period may be extended by the CCPR only to the extent necessary for the JMPR to schedule and complete a review of the data. After the review of data in a case where the four-year period clause applies, one of the three scenarios described above may occur again. In case of insufficient data, the existing Codex MRL is then recommended for withdrawal. A second period of four years is not granted.

The consequences of the periodic review procedure are considerable. For the JMPR it has meant a significant increase in their workload. With almost 200 active substances for pesticides in the Codex system, it readily can be understood that there is a need for further specification of the criteria to identify chemicals for periodic review to reasonably manage the review process. Over the last few years, the JMPR has reviewed more old compounds than new ones. This trend has been further strengthened by a tendency of Governments and manufacturers to sponsor fewer new compounds for inclusion in the priority list, which partly may be due to their involvement in national or regional review programmes. Limited resources prevent them from furthering both the evaluation of old and new compounds at the same time. As stated above, re-registration procedures and establishment of MRLs require a substantial investment that manufacturers are not always willing to make. Since a periodic review programme was introduced, the CCPR has recommended the withdrawal of hundreds of existing Codex MRLs for tens of pesticides due to the lack of support of these pesticides or pesticide–commodity combinations by the manufacturers. Those that remain or are newly established, however, are based on an adequate data set and have gone through a full consumer risk assessment.

The increased data requirements apply equally to new compounds and chemicals for periodic review. This increase is inevitable and logically follows the policy of national Governments with regard to the regulation of pesticides and other hazardous chemicals. For this reason, the JMPR has not established fixed data requirements as manufacturers certainly are not willing to generate data for CCPR and JMPR purposes only, if these data are not already required by Governments in the framework of their registration schemes. Nevertheless, over the years the JMPR has frequently considered the issue of data requirements. Those in relation to residue evaluation are described in the FAO Manual (FAO, 1997a). Recent views on the toxicological database are summarized in the Reports of the JMPR, most recently in 1995 (FAO, 1995).

On several occasions the CCPR has questioned the database for establishing MRLs, e.g. the number of supervised trials, and has considered the need to develop a minimum database. The 1994 JMPR gave due attention to this question (FAO, 1994). The view of the Meeting comprised basically two important elements. The first one was that the JMPR is not a regulatory body but a scientific committee reviewing all available data. Secondly, better evaluations would result from a proper understanding of the residue behaviour rather than from empirical treatment of data only. The JMPR was aware of the co-operation of other international organizations with Governments to develop internationally agreed data bases for registration of pesticides. The European Union has taken the initiative to develop such minimum data requirements for MRL setting in the framework of the OECD Pesticide Forum. Once these have been established, most likely the JMPR will consider these sympathetically, which may ultimately result in an increased acceptance of Codex MRLs by Governments. The 1999 JMPR took note of progress made on this subject since 1994 in the framework of the OECD.

The enhanced data requirements, including the number of supervised field trials, makes it increasingly more difficult to elaborate Codex MRLs, in particular, for minor commodities. As these data requirements generally are defined by developed countries with a sophisticated infrastructure for regulation of chemicals, it presents at the same time a serious problem for developing countries since sufficient data are commonly not available to establish Codex MRLs for pesticide–commodity combinations which are primarily of interest to their economies. Efforts of the CCPR to mitigate this problem have remained without success so far. As it is unthinkable to alleviate the data requirements in general and the extrapolation of residue trial data of related commodities will most likely only have a limited impact, a possible solution might be to strengthen the co-operation of countries in a given region to generate the necessary database for elaboration of Codex MRLs.

REGIONAL INITIATIVES

Brief reference will be made to a few regional regulatory activities on pesticides, including the harmonization of MRLs. As mentioned in the introduction, in 1976 the European Community agreed upon legislation to harmonize MRLs for pesticide residues on products of plant and animal origin. Since then, a considerable number of community MRLs have been elaborated and are in force.

In more recent years, other regional activities have occurred, either as technical co-operation or more firmly based on treaties between countries. Far from being exhaustive, the following examples can be mentioned. Under the North American Free Trade Association (NAFTA), the USA, Canada and Mexico have aimed to attune their regulatory activities in the area of pesticides, including the harmonization of MRLs. Similar activities are undertaken now in Latin America with Mercosur, the common market of Brazil, Argentine, Uruguay, Paraguay and Chile as one of the better known examples.

The purpose of all of these initiatives, ranging from bilateral and regional negotiations to global harmonization, is to reduce potential trade irritations while still safeguarding the health of consumers. The question arises as to whether it is still worthwhile to harmonize MRLs at the national level or to elaborate regional MRLs when, more easily, reference can be made to global MRLs, i.e. Codex MRLs. This question has become particularly relevant since Codex standards are recognized under the WTO–SPS Agreement as references in international trade.

There is no simple answer to this question. First of all, it should be realized that Codex, in view of its mandate and the criteria set to establish MRLs for pesticide residues, covers only part of the pesticides which are currently used in agriculture in any part of the world. Codex MRLs therefore do not necessarily accommodate intra-regional trade. This is, in particular, the case for commodities that do not enter international trade but which may be of importance in local trade. In addition, the lack of data for minor species may prevent Codex from establishing MRLs for these commodities. Another technical argument is that establishing MRLs through the Codex system is a fairly slow process that countries may not wish to await, so encouraging them to set MRLs among themselves. In addition, in those cases where Codex had adopted MRLs a long time ago, countries are often reluctant to adhere to them, stating that the GAP on which the MRLs are based has most likely changed in the meantime and that new toxicological data have become available since the first evaluation, thus warranting a full review of all relevant data.

Finally, political and legislative considerations may also be reasons for Governments not to accept Codex MRLs as the basis for their regulations or measures. Nevertheless, it is strongly recommended that Governments refer to Codex MRLs first at any time they consider the establishment of national or regional MRLs to see whether these limits can be used. It will save resources and by the end of the day it is an efficient way to fulfil the obligations of the WTO–SPS Agreement as compliance with Codex means compliance with the WTO.

STANDARDS FOR DRINKING WATER

Access to safe and wholesome drinking water is of paramount importance to mankind. An adequate supply of safe water is an essential component of primary health care. Generally, drinking water is obtained from ground water sources, from surface waters or from rain water. These water supplies can easily become contaminated by biological agents and chemical substances. In many countries, the principal risk to human health associated with drinking contaminated water is related to microbial organisms. Therefore, for centuries sanitation measures have been applied to reduce the risk of water-borne infection diseases. However, chemical contamination should not be ignored. A classical example is the residues of lead originating from the piping system. It should be realized, however, that certain chemicals such as minerals and heavy metals naturally occur in water supplies and that other substances are intentionally applied to drinking water for different purposes, e.g. as disinfectants or as processing aids to prepare drinking water. Pesticides can also be used intentionally, for instance, to control Aedes mosquito larvae in drinking water containers, as one example among several others. On the other hand, the presence of residues in drinking water may be due to diffuse emissions of residues into water following their agricultural or non-agricultural use.

As stated in the introduction, no international standards for pesticide residues in drinking water exist today in a sense which is comparable with MRLs for pesticide residues in food as established by Codex. First of all, drinking water, not to be confused with mineral water or bottled waters, is not an important item in international trade. Besides, the view by Governments on the presence of residues in drinking water varies considerably throughout the world so that agreement on any proposal for an international standard would be highly unlikely. Increasingly Governments consider drinking water as a product for which there is no apparent need to accept residues of pesticides. Indeed, in 1980 the European Community agreed upon a drinking water directive with a generic MRL for pesticide residues at what was effectively a zero concentration level. This was a political choice and not one based upon toxicological considerations but nonetheless is an example of a regional standard.

Despite the lack of international standards, not only for pesticide residues but also for other constituents in drinking water, there was a need for international guidelines for drinking water quality in view of its importance for consumer safety. Such guidelines should provide guidance to Governments in establishing national standards and to assist them to apply the necessary sanitation or to take the appropriate risk management measures in case of an emergency. The WHO published in 1984–1985 the first edition of the Guidelines for drinking water quality. The Guidelines comprise three volumes, namely Recommendations (Volume 1), Health Criteria and other supporting Information (Volume 2) and Surveillance and Control of Community Supplies (Volume 3). These Guidelines are regularly updated, most recently in 1998 (WHO, 1998b). In establishing the actual guideline values for individual substances, the WHO takes into account the evaluations already carried out by the IPCS in the framework of the Environmental Health Criteria documents, the JMPR, the JECFA and the International Agency for Research on Cancer (IARC). Generally, a guideline value is based upon the assumption of a daily consumption of two litres of drinking water by an adult person weighing 60 kg. In cases where, in particular, children will be exposed, these figures are adjusted for the daily consumption and body weight of infants or children. In calculating the guideline value, usually 10 % of the ADI is allocated to drinking water. A rationale for deviations from this approach is given in the Recommendations. The way in which a guideline value is derived makes it a real health-related figure. A daily exposure during a lifetime to a residue in drinking water up to the level of the guideline value is considered to be safe.

As pesticide residues may be present in drinking water it seems logical to include the WHO guideline values in a combined risk assessment when establishing Codex MRLs in food. It would be an example of a dietary exposure assessment including two different routes, namely food and drinking water. The current international guidelines for estimating the dietary intake of pesticide residues do not provide for such an approach. The 1997 Geneva Consultation on food consumption and exposure assessment of chemicals (WHO, 1997b) recommended that in addition to estimating dietary intake, other possible sources of exposure, such as drinking water, occupational exposure, environmental exposure, etc., should also be considered. The Consultation did not advance the subject at that time because exposure other than from residues in food was beyond the scope of their meeting.

As described before, the CCPR expressed as its opinion that aggregated risk assessment could best be performed at the national level, at least for the time being. Including residue limits in food and in drinking water in a combined risk assessment would be at present extremely difficult, not necessarily from a scientific perspective but from a political one. Where there is generally consensus at the international level on the need to apply pesticides in agriculture and consequently to accept residues in food, this is definitely not the case with residues in drinking water. The prevailing conditions in countries not only vary widely with regard to the need to apply pesticides in drinking water or drinking water sources, but also the political views range from acceptance of certain levels, through acceptance of temporary levels in the case of accidents, to non-acceptance at all. This excludes a successful combined risk assessment at the international level. Governments, therefore, when considering Codex MRLs of pesticide residues in food should take into account their policies towards residues in drinking water as well.

ACKNOWLEDGEMENTS

The author acknowledges the valuable comments of Dr Y. Yamada (National Food Research Institute, Tsukuba, Japan[3]) and Mr J. W. Dornseiffen (Ministry of Health, Welfare and Sport, The Hague, The Netherlands) on the manuscript.

REFERENCES

FAO (1991). *FAO/WHO Conference on Food Standards, Chemicals in Food and Food Trade, 1991*, Food and Agriculture Organization of the United Nations, Rome.

---

[3]Formerly Food and Agriculture Organization of the United Nations, Rome, Italy.

FAO (1994). *Pesticide Residues in Food, Report 1994*, FAO Plant Production and Protection Paper No. 127, Food and Agriculture Organization of the United Nations, Rome.

FAO (1995). *Pesticide Residues in Food, Report 1995*, FAO Plant Production and Protection Paper No. 133, Food and Agriculture Organization of the United Nations, Rome.

FAO (1997a). *FAO Manual on the Submission and Evaluation of Pesticide Residues Data for the Estimation of Maximum Residue Levels in Food and Feed*, Food and Agriculture Organization of the United Nations, Rome.

FAO (1997b). *Risk Management and Food Safety*, Report of a Joint FAO/WHO Consultation, Rome, 27–31 January, 1997, FAO Food and Nutrition Paper 65, Food and Agriculture Organization of the United Nations, Rome.

FAO (1997c). *Worked Example of Intake Estimate According to the Revised Guidelines*, document prepared by D. J. Hamilton at the request of the Codex Secretariat, CX/PR 97/6, Joint FAO/WHO Food Standards Programme, *Report of the Twenty-Ninth Session of the Codex Committee on Pesticide Residues*, Alinorm 97/24A, Food and Agriculture Organization of the United Nations, Rome.

FAO (1997d). *Pesticide Residues in Food, Report 1997*, FAO Plant Production and Protection Paper No. 145, Food and Agriculture Organization of the United Nations, Rome.

FAO (1997e). *Joint FAO/WHO Food Standards Programme, Report of the Twenty-Second Session of the Codex Alimentarius Commission*, Alinorm 97/37, Food and Agriculture Organization of the United Nations, Rome.

FAO (1999a). *The Application of Risk Communication to Food Standards and Safety Matters*, Report of a Joint FAO/WHO Consultation, Rome, 2–6 February, 1998, FAO Food and Nutrition Paper 70, Food and Agriculture Organization of the United Nations, Rome.

FAO (1999b). *Joint FAO/WHO Food Standards Programme, Report of the Twenty-Third Session of the Codex Alimentarius Commission*, Alinorm 99/37, Food and Agriculture Organization of the United Nations, Rome.

FAO (1999c). *Joint FAO/WHO Food Standards Programme, Report of the Thirty-First Session of the Codex Committee on Pesticide Residues*, Alinorm 99/24A, Food and Agriculture Organization of the United Nations, Rome.

FAO/WHO (1993). *Joint FAO/WHO Food Standards Programme, Codex Alimentarius Commission, Codex Alimentarius*, Volume 2, *Pesticide Residues in Food*, Food and Agriculture Organization of the United Nations/World Health Organization, Rome.

FAO/WHO (1998). *Joint FAO/WHO Food Standards Programme, Codex Alimentarius Commission, Codex Alimentarius*, Volume 2B, *Pesticide Residues in Food*, Food and Agriculture Organization of the United Nations/World Health Organization, Rome.

FAO/WHO (1999). *Understanding the Codex Alimentarius*, Food and Agriculture Organization of the United Nations/World Health Organization, Rome.

FAO/WHO (2000). *Joint FAO/WHO Food Standards Programme, Codex Alimentarius Commission, Procedural Manual*, 11th Edn, Food and Agriculture Organization of the United Nations/World Health Organization, Rome.

FAO STAT Database Collections. [http://apps.fao.org/CodexSystem/pestdes/pest_q-e.htm].

Frawley, J. P. (1987). Codex Alimentarius – food safety – pesticides, *Food Drug Cosmetic Law J.*, **42**, 168–173.

NRC (1993). *Pesticides in the Diets of Infants and Children*, National Research Council, National Academy Press, Washington, DC, USA.

PSD (1998). *Report of the International Conference on Pesticide Residues Variability and Acute Dietary Risk Assessment*, The Pesticide Safety Directorate, York, UK.

WHO (1989). *Guidelines for Predicting Dietary Intake of Pesticide Residues*, prepared by the Joint UNEP/FAO/WHO Food Contamination Monitoring programme in collaboration with the Codex Committee on Pesticide Residues, World Health Organization, Geneva.

WHO (1995a). *Application of Risk Analysis to Food Standards Issues*, Report of the Joint FAO/WHO Expert Consultation, Geneva, 13–17 March, 1995, World Health Organization, Geneva.

WHO (1995b). *Recommendations for the Revision of the Guidelines for Predicting Dietary Intake of Pesticide Residues*, Report of a FAO/WHO Consultation, 2–6 May, 1995, York, United Kingdom, World Health Organization, Geneva.

WHO (1997a). *Guidelines for Predicting Dietary Intake of Pesticide Residues* (revised), prepared by the Global Environment Monitoring System–Food Contamination Monitoring and Assessment Programme (GEMS/FOOD) in collaboration with the Codex Committee on Pesticide Residues, World Health Organization, Geneva.

WHO (1997b). *Food Consumption and Exposure Assessment of Chemicals*, Report of a FAO/WHO Consultation, 10–14 February, 1997, Geneva, Switzerland, World Health Organization, Geneva.

WHO (1998a). *GEMS/Food Regional Diets. Regional Per Capita Consumption of Raw and Semi-Processed Agricultural Commodities*, prepared by the Global Environment Monitoring System–Food Contamination Monitoring and Assessment Programme (GEMS/FOOD), World Health Organization, Geneva.

WHO (1998b). *Guidelines for Drinking-Water Quality*, Addendum to Volume 1, *Recommendations*, 2nd Edn, World Health Organization, Geneva.

WTO Home Page. [http://www.wto.org].

11 Explaining the Risks

SIR COLIN BERRY

The Royal London Hospital, London

INTRODUCTION: THE CONCEPT OF SAFETY

The evaluation of a pesticide for widespread use is essentially a risk/benefit study with a large number of targets at hazard; each of these will need a risk assessment. These targets include operators (the group with the heaviest and usually the best documented exposure), consumers, non-consumer exposed public (e.g. walkers), other mammals, birds, beneficial insects and the aquatic environment (fish, invertebrates and water plants), with the health of the soil an ill-defined but important additional factor. Having made these assessments, there is a subsequent informal summation with a large number of what may be independent variables to be integrated.

It is therefore not possible to define a risk in the global sense – **is this compound dangerous**? Risk can only be categorized in terms of a number of closely specified events (as examples, does it cause a number of incidents or specified illnesses per year, per 100 000 people, per growing season or per number of applications?).

It is not reasonable to expect members of the public to take all of these steps; even isolated and directly relevant risk assessments such as those used in evaluation of dietary intakes (Figure 11.1) produce numbers which explain very little to the uninformed. Most consumers will think in terms of single foods rather than diets and not consider their drinking water; paradoxically, most will be worried about occasional intakes where the problems that concern them have usually been identified by experiments involving lifetime daily exposures to animals.

Pesticide Residues in Food and Drinking Water: Human Exposure and Risks.   Edited by Denis Hamilton and Stephen Crossley
© 2004 John Wiley & Sons, Ltd   ISBN: 0-471-48991-3

Risk Assessment

Toxic effect   +   Exposure

Estimate initial risk

Refine risk with new data

Restrict use, make engineering or application changes

Establish new pattern of use

MONITOR

Figure 11.1   A scheme which could be used to evaluate the significance of pesticide residues in food and to modify outcomes

These uncertainties are taxing, but there is also the problem that there are selective views in particular groups with regard to the nature of hazards – in some instances and in some groups, natural toxicity is preferred to synthetic safety. This is a matter of attitude; it is clear that the perceptions of some individuals are altered when compared to the rest of their community and their perception affects their reactions – those who describe themselves as 'very worried' about local environmental conditions were ten times more likely to complain of headaches than those not so concerned (Shusterman *et al.*, 1991; see also Hall *et al.*, 1998).

Society also makes judgements – as Wessely has pointed out, attributions of the sources of harm are not neutral. Those affected by a pollutant are often described as innocent victims but if the same symptoms are attributed to a psychological process substantial blame is attached by society and guilt is experienced by the individual (Wessely, 1994). This tendency is amplified by changing perceptions in society – what Barsky and Borus (1995) have called *somatization*. This term is used to describe the reporting by patients of somatic complaints that have no pathophysiological explanation and the phenomenon has been characterized as a distinct behavioural entity. These authors have described how changes in social attitudes have reduced the tolerance of the public to mild symptoms and benign infirmities. The threshold for seeking medical attention has been lowered and discomforts and isolated symptoms are identified as diseases by patients – new 'syndromes' appear in this way. The syndrome then needs an explanation and the current concern is the focus of attention.

SAFETY

What do we mean by safe? It is clear that 'Safety' is a different concept for the public as a collective, as opposed to the individual. As defined by Sidall (1980), safety:

is the degree to which temporary ill health or injury, or chronic or permanent ill
health or injury, or death, are controlled avoided, prevented, made less frequent
or less probable in a group of people.

The above definition thus includes the vital point that one can only comment on
the effects on one set of those at risk at a time – what is believed to be safer
for consumers may be more dangerous for operators (e.g. the use of machinery
versus herbicides on farms).

This definition is clear and can be used in risk management, but does not take
into account the loading (often irrational) attached to particular perceived risks
by particular groups. Loadings are applied in this way and affect the general
perception of the problem. As a simple statement, 'Natural is good, Synthetic is
bad' would probably obtain considerable public support.

SYNTHETIC VERSUS NATURAL

The caution exhibited by the public in the last two decades in accepting the value
of synthetic chemicals is remarkable. Natural foods contain more than half a million chemicals, responsible for appearance, form, colour, taste, flavour, perfume
and so on. There are a few hundred chemicals, mainly synthetic, which are used
as pesticides and perhaps a total of two to three thousand used in food preparation
or preservation. All of the latter have been evaluated in an internationally agreed
programme of chemical testing and have reliable assay methodologies associated
with them. Only a tiny proportion of the greater number of natural chemicals
have been assessed in any way, despite the clear biological activity of many of
them, but they are generally regarded as being 'safe'.

What accounts for the difference in perception? Much of it is a simple matter
of prejudice – assessments would lead us to ban many foods if we regulated
them in the same way as pesticides. As an example, let us look at potatoes.
Because so many are eaten (around 350 million tons annually) and glycoalkaloid production by sprouting potatoes is a well-known health hazard, there are
good data and monitoring programmes for their levels. Table 11.1 shows that
the levels of glycoalkaloid approach intervention levels in popular cultivars; a
pesticide residue with this safety margin would not be tolerated in a staple food.
It is worth noting that many who grow their own food appear not to realize
that injury to potatoes (mechanical injury, fungal infection or insect damage)
may produce the same effect. The risk is not remote; an outbreak of potato
poisoning from glycoalkaloids occurred in Lewisham (in South London) only
twenty-five years ago where severe anticholinesterase effects were found in 78
school children (McMillan and Thompson, 1979). Some people eat large amounts
of potato peel in the belief that vitamins and fibre are usefully obtained from
this source.

Table 11.1 The glycoalkaloid content of popular potato cultivars (note that potatoes may not be marketed with an excess of 200 mg% of glycoalkaloid)

Cultivar	Content (mg/100 g dry wt)[a]	
	1994	1995
Epicure	110–140	110–140
Golden Wonder	120–190	130–190
Kerrs Pink	80–140	100–120
King Edward	70–110	80–120
Pentland Hawk	60–140	90–130

[a] Limit for safety: < 200 mg/100 g dry weight.

SUSPENSION OF BELIEF

It is difficult to understand why natural toxicity or overdosing with vitamins or minerals does not generate anxiety. There are many recent examples of natural products causing severe illness and deaths. A large number of cases of renal failure and a very significant number of transitional cell carcinomas have been induced in Belgium from a Chinese herbal dieting remedy (Cosyns *et al.*, 1998). A 'new' vegetable, asinasin, has caused well over 100 cases of bronchiolitis obliterans with rapidly progressive lung disease in many of those affected; lung transplants have been used in an attempt to treat the disease (Lai *et al.*, 1998). Within the last few years, the dietary supplement L-tryptophan was withdrawn following extensive use in the USA because of the sometimes fatal eosinophilia-myalgia syndrome (EMS) (Simat *et al.*, 1996). As a 'natural' product, L-tryptophan was taken for insomnia, the pre-menstrual syndrome and stress reduction; some EMS victims were taking 17 g per day. This is one example of the absurd but widely held idea that if a little of something is good for you, a lot must be better. Many compounds normally present in the diet are both essential in small amounts and dangerous in large quantities – selenium being a good example. In view of current concerns about environmental oestrogens, it is surprising that noone has started a scare about soy proteins in the diet (Setchell *et al.*, 1997) (Table 11.2).

Such problems from natural toxins have been readily identified in clinical practice (these are recent examples, although there are many others). It is important

Table 11.2 Data from Setchell *et al.* (1997) showing the phyto-oestrogen levels (ng/ml) of infant formulae and breast milk

Diet	Genistein	Diadzein
Soy formula	684	295
Cows-milk-based formula	3.2	2.1
Breast milk	2.8	1.4

to note that no comparable data exist for disease produced by consumer exposure to pesticides, despite extensive searches; however, examples of diseases caused by natural toxicants which might be prevented by pesticides are readily found. Chasseur *et al.* (1997) have described how a number of fungi, acting at different parts of the food cycle, may cause Kashin–Beck disease.

THE NATURE OF PUBLIC CONCERN

Most public concern appears to be related to the long-term effects of compounds ingested at low dose and to be associated with cancer or reproductive effects (in day-to-day regulation, action tends to focus on the relatively high exposure to contractors or to those using a particular compound repeatedly). The context in which public anxiety has arisen is often important (workplace, school, survey, etc.) and it should be remembered that the data giving rise to the concern may be supplied by pressure groups who are selective in their presentation of data. Practically, it is important to note that the problem may relate to non-pesticidal uses of compounds (such as the organophosphates in sheep dips or in the Gulf war, for example) with a 'spill-over' effect on agricultural use.

Clearly, public perception of problems is of major significance. The development of a 'tool kit' by the School of Environmental Sciences at the University of East Anglia (Norwich, UK) to help identify the concerns of consumers towards food-related risks has proved evidence that the public view risk as a multi-dimensional concept. This helpful study also indicates the nature of the task confronting us; the problems presented by the respondents are a depressing list from the scientific viewpoint. 'Harm to Future Generations' is a major concern of many groups of consumers, despite the absence of any evidence for chemical injury of this kind in eighty years of searching. Natural and traditional foods were seen as less worrying despite a lack of data on their safety. It is clear that this type of study is valuable in identifying concerns but less evident that it will contribute useful information on how to dispel them.

In dealing with the public it is probably best to seek to put human risks into a perspective of a number of health outcomes (both morbidity and mortality) and show that hazard identification contributes little to the process. It is important to get across the fact that epidemiology is necessarily a blunt tool and that reported associations of effects and exposures at less than three times background tell us little about risks requiring interventions of some kind. In any epidemiological study, if six factors are analysed and the confidence intervals are set at 95% (they usually are), then the probability of finding a false positive is one in four. Hill (1965) highlighted a number of important points about identifying the differences between association and causation which are often forgotten. For example, in order to be confident that an association is telling us something, that association should be strong (there was a very large increase in scrotal cancer in chimney sweeps in his example), it should be consistent (observable in different places by different people at different times) and specific (the disease produced is limited to particular groups of workers), there should be a biologically credible temporal relationship between exposure and effect (putative teratogens acting after the palate has closed are not likely to produce cleft palate), there should be a dose–response relationship, the effect should be biologically plausible (this cannot be mandatory – what is plausible will depend on the state of knowledge at the time), and the cause and effect proposed should not conflict with what is known about the natural history and biology of the disease concerned (e.g. transmission of non-genotoxic effects to subsequent generations). He noted that experimental evidence can be powerfully supportive and pointed out that it is fair to judge by analogy in some circumstances; the information that the rubella virus and thalidomide produce malformations increases the suspicion that other viruses and drugs may act in the same way. Attention to these points will inform any debate about risk and can be used to reassure the group who feel themselves to be at risk.

Risk assessments of the type used in evaluations of food-related intakes can be shown to give values for the vast majority of pesticides that will affect the incidence of none of the outcomes which trouble the public. This rational approach seems to be correct intuitively but the effects of investigations do not suggest that this is a sufficient approach.

THE REAL PROBLEM

Almost all health risks in the diet are microbiological in origin, with water as a major source of viral and bacterial disease (Anon, 1992). Reviews of diseases in recreational water users confirm the great range of pathogens normally found in rivers, reservoirs and the sea (Fewtrell et al., 1993). Fungal disease is widespread and in a number of non-processed food crops, as an example, mycotoxins may be found – it is clear that failure to treat with fungicides in some instances will expose the population to a significant risk (Frank, 1980) and cases of ergotism have occurred in Germany in recent years. There is no doubt that those working with stored grain are exposed to significant and potentially dangerous levels of aflatoxin B_1 (Selim et al., 1997). Patulin in apple juice caused anxiety in the United Kingdom only three years ago; it is noteworthy that a number of the naturally occurring mycotoxins are powerful carcinogens.

However, most bacterial and viral contamination is introduced during food preparation (some of the organisms involved may require more intensive preventative treatment with pesticides than has been given in the past).

The chemical problems that exist are mainly naturally occurring. Some European populations bear a severe disease burden because of natural chemical pollution of water sources – natural nitrate and arsenic levels in drinking water in Hungary have produced deaths from methaemoglobinaemia, and skin changes including hyperpigmentation and keratoses, respectively, in recent years (Csanady and Straub, 1995). Although there are many synthetic organophosphates, they also occur naturally (there is a cyanobacterial organophosphate toxin, anatoxin A, which kills fish, in algal blooms).

It is important to emphasise that the documented risk to the public from pesticides is negligible in terms of acute effects, despite years of study. It is also theoretical in terms of long-term hazards; time after time, the putative risk from any group of compounds, perhaps identified as a weak epidemiological linkage, has disappeared on further analysis (Berry, 1996). Even if highly exposed groups in manufacturing are studied for many years (Jukes, 1995; Littorin et al., 1994), adverse effects are not documented. A good example is herbicide use and soft tissue sarcoma risks, where a gradual reduction of apparent 'risk' identified in population studies has been seen, as studies are refined and confounding factors – notably exposures – are considered in detail. In a number of Scandinavian studies, questionnaires and telephone interviews were used in early (large) studies and were considered to show a strong association, but as various other methodological flaws were identified by critical reviews of the data they resulted in more and more publications with a steadily diminishing risk. As Gough (1993) has said: 'The Swedish studies have been widely cited and probably will be cited for years to come. People who cite these studies in arguing for a link between dioxin and soft tissue sarcomas will have to ignore the evidence that doesn't support the link, but that can be done'.

THE EFFECTS OF MISINTERPRETATION

If the documented risks are small, why is there so much comment in the various media? It is clear that pressure groups play a part; just as they may be valuable in identifying problems which have not been considered, so they may be destructive in maintaining untenable positions when new data fail to support their earlier assertions, as with the example cited above. This factor may be important in the development of the selectivity of our concerns.

Our scale of risk is 'set' wrongly. The perception of the size of the risks to which all individuals are exposed differs wildly from reality. If we take a simple and easily understood example, namely the chances of dying in a given year, it will be evident after questioning a very few of your colleges that there is a widespread misunderstanding of what is probable. Table 11.3 shows the number of deaths per year at particular ages and results in an often uncomfortable realization of the value of life insurance. Very few are prepared to accept that the real risk of good food appears to be overnutrition; Lutz and Schlatter (1992) and Hart et al. (1995) have argued convincingly that neither natural nor additive carcinogens explain food-related carcinogenic risks – overnutrition is the problem. A great deal of animal data support this conclusion (Shimokawa et al., 1996).

In considering these data, it is generally true to say that cognitive factors greatly affect our reactions to things we perceive as hazards. There are clear instances

Table 11.3 Age-dependent death rates for the UK in 1992. There will be one death for a number of living individuals shown in the column for each year in the age groups illustrated

Age (years)	Unicohort for 1 death/year	
	Male	Female
0–1	266	359
5–9	1883	3141
10–14	641	1415
20–24	478	1261
25–29	472	1150
30–34	398	790
35–39	282	505
40–44	198	302
60–64	24	38
80–84	5	4

where those who are most at risk consider themselves to be invulnerable; young males do not rate the chances of being killed or injured on a motor bike (the most important cause of deaths and injuries in the group) as being high on a list of comparative risks – interestingly, they think that nuclear power and pesticides afford them a greater probability of harm.

If mortality data are not well understood it is clear that data on morbidity are harder to collect and even less readily interpreted by the non-professional. The number of visits to a general practitioner or hospital accident and emergency (A&E) department can be used as surrogate markers for illness; studies of the latter type are good examples and give a proper perspective on the risk to children from pesticides (Casey and Vale, 1994; Casey *et al.*, 1994). In the two decades these authors studied, there were no deaths from accidental pesticide exposure in the UK. Unhappily, the same was not true for household cleaners or medicines.

THE COSTS OF A FALLACIOUS CONCEPT OF SAFETY

The pursuit of safety may be destructive. In strict terms, it is not possible to prove that anything is safe; at any time, new data may appear to confound previous and reasonably held positions. Low risks are particularly difficult to deal with – if we take as an example the problem of electromagnetic fields, public anxiety has been maintained by a series of indifferent studies and the public belief that radiation is a 'bad thing'; thus, all shades of it must do harm. Weak electromagnetic fields (EMFs) are as remote from therapeutic irradiation (or old nuclear weapons exposures – our current reference point) as penicillin is from lysergic acid diethylamide (LSD).

The US National Academy of Science was commissioned by Congress to examine the possible health effects of EMFs. They reviewed 500 studies from 1979 onwards and found 'results had been inconsistent and contradictory and do not constitute reliable evidence of an association' (National Research Council, 1997). The data which seem to have started this are reported in the 17-year-old study from Denver (Wertheimer and Leeper, 1979) which was flawed; it showed a weak association with leukaemia and a more 'wired' home as a surrogate for field measurement – but there was no examination of traffic density, air quality or other construction features of the homes and so on – think of radon, benzene and other hydrocarbons. Using invalidated surrogate measurements is the curse of epidemiology. Even in scientifically good studies in this area, such as that on 380 000 Finns (Verkasalo *et al.*, 1996), the authors reported an odds ratio of 1.22 for myeloma in men and 1.16 for colon cancer in women. These signals are just 'noise', and not a justification for further work and add nothing to our considerable body of knowledge about the pathogenesis of these two very different tumours. The total cost of all of this work is prodigious, while the scientific value is negligible!

Tables 11.4 and 11.5 show different calculations about the costs of ill-considered regulatory steps; remember that some of the events associated with exposure are projections and extrapolations – there are no data which justify the statement that 'Amitraz on pears has caused deaths'.

Responses to unjustified concerns are counterproductive to good and effective regulation; the use of the scientific method cannot be the mode of response to perceived hazard on some occasions and not others. Failure to use it can be damaging in many ways. At the time of the controversy about daminozide ('Alar') in 1989, a '60 Minutes' programme on American television described

Table 11.4 Cost-effectiveness of some selected US regulations

Regulation averted	Cost per premature death (US$M)[a]
Car seat belt standards	0.1
Car fuel system standards	0.4
Car side impact standards	0.8
Car rear seat belt standards	3.2
Ethylene dibromide drinking water standard	5.7
Glassplant arsenic emission standards	13.5
Asbestos ban	110.7
1,2-DCP drinking water standard	653.0
Hazardous waste, land disposal ban	4190.4
Municipal solid waste landfill standards	41 250
Atrazine/alachlor drinking water standard	92 070
Wood preserving chemicals – hazardous waste	5 700 000

[a] For the year 1990.

Table 11.5   Costs of prolonging life by one year

Intervention	Cost (US$)[a]
Car	
Driver automatic seat belts (versus manual)	0
Collapsible steering column	67 000
Airbags versus belts	120 000
Hazards	
Chlorination of drinking water	3100
Ban on asbestos pads	29 000
Amitraz 'out-of'-pears	350 000
Medicine	
Heart transplants for < 55 year olds	3600
Annual mammography for 55–64 year old women	95 000
Intensive care (14 days post-operative)	820 000

[a]For the year 1993.

it as the 'most potent cancer causing agent in our food supply'. There followed a most extraordinary collective hysteria about the supposed risk that this compound presented to children in particular, supported in the main by ill-informed and often unqualified opinions from many who were clearly unaware of the fact that there was more hydrazine compound (the putative carcinogen) in a helping of mushrooms than anyone was likely to get from apples or apple juice. There was no effective response from the US Environmental Protection Agency (EPA); actresses became the preferred communicators on the issue! My own calculations of 28 000 apples/day as the danger level, if one assumed that the mouse splenic haemangiosarcomas were relevant to Man, was matched by a calculation by the *Washington Post* that 19 000 l of apple juice would be the critical intake (Marshall, 1991). The scientific issues were never discussed seriously in any forum before supermarket chains responded by withdrawing 'Alar'-treated apples from sale – a later World Health Organization/Food and Agriculture Organization of the United Nations (WHO/FAO) view that higher levels of 'Alar' residues than those previously adopted would be safe did not save those apple growers who had been left with an unsaleable crop – one grower in the USA lost part of a farm his family had worked since 1912 due to consequent financial pressure (Whelan, 1993). Thus, we see the generation of anxiety in many parents, the destruction of some livelihoods, the decision of retailers that this kind of data presentation must produce a response rather than an analysis and a ready involvement of individuals, who one might hope to be opinion formers, in an irrational response. The subsequent measured response from *Science* (Marshall, 1991) was, necessarily, too late.

ALTERNATIVE APPROACHES

It is evident that simple misjudgment of probabilities is not the major factor in the perception of risk. Characterization of the hazard involved (dread, lack of controllability, lack of observability, new risk, etc.) affects matters – nuclear power and nerve gas rate high in 'dread' while sunbathing and bicycle riding had low scores and are perceived as 'non-risky', despite considerable knowledge. Baird (1986) found that knowledge tended to increase risk tolerance, although there are studies which suggest the opposite for radon (Kennedy *et al.*, 1991) and there are also data showing a lack of relationship between factual knowledge and risk perception in female college students with regard to the toxic shock syndrome (Colbry and Boutsen, 1993).

To paraphrase Remmington (1993): 'A regulator is a risk taker. He declares an allegiance to science and reason in an area which is profoundly human and political, and picks and chooses a pathway through the minefield of other people's prejudices and ethical systems. He will, therefore be subject to political attack'.

How does Science help? The nature of science is to demand reproducibility of results by separate investigators over time. Single observations are regarded with caution until confirmed by others, and so-called 'breakthroughs' usually take years to validate, and mostly, are not. It follows that nothing can be proved to be safe for ever, and new information can confound any opinion. This rational position may not be comforting but it is possible to respond quickly to concerns about hazards, to point to the strength of the methodology used and, above all, to illustrate how we are getting fitter and fitter, living longer and longer in better health, and that most of our problems are due to various forms of self-indulgence.

CONCLUSIONS

It is probable that we shall make progress only by persuading people to think of risk in an appropriate context and to begin to evaluate what they do day-by-day in terms of what are their aspirations about their health. I do not believe we should always evaluate our lives in these terms; that, for me, would define a psychological illness, but when confronted by what we believe to be a problem, we should think, and look for good quality data, before we act.

REFERENCES

Anon (1992). Communicable diseases control and epidemiology. Introduction, *World Health Stat. Q.*, **45**, 166.

Baird, B. R. N. (1986). Tolerance for environmental health risks: the influence of knowledge, benefits, voluntariness and environmental attitudes, *Risk Anal.*, **6**, 425–436.

Barsky, A. J. and Borus, J. F. (1995). Somatization and medicalization in the era of managed care, *JAMA*, **274**, 1931–1934.

Berry, C. L. (1996). Risks, costs, choice and rationality, in *Proceedings of the Royal Institution*, Vol. 67, Day, P. (Ed.), Oxford Science Publications, Oxford, UK, pp. 125–143.

Casey, P. B. and Vale, J. A. (1994). Deaths from pesticide poisoning in England and Wales 1945–1989, *Human Exp. Toxicol.*, **13**, 95–101.

Casey, P. B., Thompson, J. P. and Vale, J. A. (1994). Suspected paediatric pesticide poisoning in the UK. 1 – Home accident surveillance system, *Human Exp. Toxicol.*, **13**, 529–523.

Chasseur, C., Seutens, C., Nolard, N., Begaux, F. and Haubruge, E. (1997). Fungal contamination of barley and Kashin–Beck disease in Tibet, *Lancet*, **350**, 1074.

Colbry, S. L. and Boutsen, F. R. (1993). Knowledge and self-perceived risk for toxic shock syndrome, *Health Values*, **17**, 26–30.

Cosyns, J.-P., Goebbels, R.-M., Liberton, V., Schmeiser, H. H., Bieler, C. A. and Bieler, C. A. (1998). Chinese herb nephropathy-associated slimming regimen induces tumours in the fore-stomach but no interstitial nephropathy in rats, *Arch. Toxicol.*, **72**, 738–743.

Csanady, M. and Straub, I. (1995). Health damages due to water pollution in Hungary, in *Assessing and Managing Health Risks from Drinking Water Contamination*, Reichard, E. G. and Zapponi, G. A. (Eds), CAB International, Wallingford, Oxon, UK, pp. 142–152.

Fewtrell, L., Kay, D., Newman, G., Salmon, R. L. and Wyer, M. D. (1993). Results of epidemiological pilot studies, in *Recreational Water Quality Management*, Kay, D. and Hanbury, R. (Eds), Ellis Horwood, Chichester, UK, pp. 75–108.

Frank, H. K. (1980) Mycotoxinbildende Schimmelpilze als Verderber von Fruchten, *Erwerbsobstbau*, **22**, 196–199.

Gough, M. (1993). Dioxin: perceptions, estimates and measures, in *Phantom Risk: Scientific Inference and the Law*, Foster, K. R., Bernstein, D. E. and Huber, P. W. (Eds), The MIT Press, Cambridge, MT, USA, pp. 249–277.

Hall, G. H., Hamilton, W. T. and Round, A. P. (1998). Increased illness experience preceding chronic fatigue syndrome; a case control study, *J. R. College Phys. London*, **32**, 44–45.

Hart, D. A., Neuwmann, D. A. and Robertson, R. T. (1995). *Dietary Restriction: Implications for the Design and Interpretation of Toxicity and Carcinogenicity Studies*, ILSI Press, Washington, DC, USA.

Hill, A. B. (1965). The environment and disease: association or causation, *Proc. R. Soc. Med.*, **58**, 295–300.

Jukes, T. H. (1995). Effects of DDE, *Nature (London)*, **376**, 545.

Kennedy, C. J., Probart, C. K. and Dorman, S. M. (1991). The relationship between radon knowledge, concern and behavior, and health values, health locus of control and preventive health behaviours, *Health Edu. Q.*, **18**, 319–329.

Lai, R. S., Wang, J. S., Wu, M. T. and Hsu, H. K. (1998). Lung transplantation in bronchiolitis obliterans associated with vegetable consumption, *Lancet*, **352**, 117–118.

Littorin, M., Hansson, M., Rappe, C. and Kogevinas, M. (1994). Dioxins in blood from Swedish phenoxy herbicide workers, *Lancet*, **344**, 611–612.

Lutz, W. K. and Schlatter, J. (1992). Chemical carcinogens and overnutrition in diet-related cancer, *Carcinogenesis*, **13**, 2211–2216.

Marshall, E. (1991). A is for Apple, Alar and Alarmist?, *Science*, **54**, 20–22.

McMillan, M. and Thompson, J. C. (1979). An outbreak of suspected solanine poisoning in schoolboys: Examination of the criteria of solanine poisoning, *Q. J. Med.*, **48**, 227–243.

National Research Council (1997). *Possible Health Effects of Exposure to Residential, Electrical and Magnetic Fields*, National Academy Press, Washington, DC, USA.

Remmington, J. D. (1993). *Coping with Technological Risk: A 21st Century Problem*, The Royal Academy of Engineering, London.

Selim, M. I., Juchems, A. M. and Popendorf, W. (1997). Potential predictors of airborne concentrations of aflatoxin B1, *J. Agromed.*, **4**, 91–98.

Setchell, K. D. R., Zimmer-Nechemias, L., Cai, J. and Heubi, J. E. (1997). Exposure of infants to phyto-oestrogens from soy-based infant formula, *Lancet*, **350**, 23–27.

Shimokawa, I., Higami, Y., Yu, B. P., Masoro, E. J. and Ikeda, T. (1996). The influence of dietary components on occurrence and mortality due to neoplasms in male F344 rats, *Aging*, **8**, 254–262.

Shusterman, D., Lipscomb, J., Neutra, R. and Satin, K. (1991). Symptom prevalence and odor-worry interaction near, hazardous waste sites, *Environ. Health Perspect.*, **94**, 25–30.

Sidall, E. (1980). *Risk, Fear and Public Safety*, Atomic Energy of Canada Ltd.

Simat, T., van Wickern, B., Eulitz, K. and Steinhart, E. H. (1996). Contaminants in biotechnologically manufactured L-tryptophan, *J. Chromatogr. B, Biomed. Appl.*, **685**, 41–51.

Verkasalo, P. K., Pukkala, E., Kaprio, J., Keikkila, K. V. and Koskenvuo, M. (1996). Magnetic fields of high voltage power lines and risk of cancer in Finnish adults: nationwide cohort study, *Br. Medi. J.*, **313**, 1047–1051 (see comments).

Wertheimer, N. and Leeper, E. (1979). Electrical wiring configurations and childhood cancer, *Am. J. Epidem.*, **109**, 237–284.

Wessely, S. (1994). Neurasthenia and chronic fatigue: theory and practice in Britain and America, *Trans. Cult. Psych. Res. Rev.*, **31**, 173–209.

Whelan, E. N. (1993). *Toxic Terror*, Prometheus Books, Buffalo, NY, USA.

Index

Pesticide Residues in Food and Drinking Water: Human Exposure and Risks. Edited by Denis Hamilton and
Stephen Crossley
© 2004 John Wiley & Sons, Ltd   ISBN: 0-471-48991-3

Becoming Tom Thumb